# PHYSICAL GEOGRAPHY

## ENVIRONMENT AND MAN

**McGRAW-HILL SERIES IN GEOGRAPHY**

EDWARD J. TAAFFE AND JOHN W. WEBB,
*Consulting Editors*

*Broek and Webb*   A Geography of Mankind
*Carlson*   Africa's Lands and Nations
*Conkling and Yeates*   Man's Economic Environment
*Cressey*   Asia's Lands and Peoples
*Cressey*   Land of the 500 Million: A Geography of China
*Demko, Rose, and Schnell*   Population Geography: A Reader
*Detwyler*   Man's Impact on Environment
*Eliot Hurst*   Transportation Geography: Comments and Readings
*Fryer*   Emerging Southeast Asia: A Study in Growth and Stagnation
*Fryer*   World Economic Development
*Kolars and Nystuen*   Geography: The Study of Location, Culture, and Environment
*Kolars and Nystuen*   Human Geography: Spatial Design in World Society
*Kolars and Nystuen*   Physical Geography: Environment and Man
*Lanegran and Palm*   An Invitation to Geography
*Mather*   Climatology: Fundamentals and Applications
*Murphy*   The American City: An Urban Geography
*Pounds*   Europe and the Soviet Union
*Pounds*   Political Geography
*Raisz*   General Cartography
*Raisz*   Principles of Cartography
*Starkey, Robinson, and Miller*   The Anglo-American Realm
*Thoman and Corbin*   The Geography of Economic Activity
*Trewartha*   An Introduction to Climate
*Trewartha, Robinson, and Hammond*   Fundamentals of Physical Geography
*Trewartha, Robinson, and Hammond*   Elements of Geography: Physical and Cultural
*Trewartha, Robinson, and Hammond*   Physical Elements of Geography (A republication of Part I of
   the above)
*Van Riper*   Man's Physical World
*Watts*   Principles of Biogeography: An Introduction to the Functional Mechanisms of Ecosystems
*Yeates*   An Introduction to Quantitative Analysis in Human Geography

# PHYSICAL GEOGRAPHY
## ENVIRONMENT AND MAN

John F. Kolars
John D. Nystuen
*Professors of Geography*
*University of Michigan*

*Drawings by*
**Derwin Bell**
*Department of Geology*
*University of Michigan*

McGRAW-HILL BOOK COMPANY

New York    St. Louis    San Francisco    Auckland    Düsseldorf    Johannesburg
Kuala Lumpur    London    Mexico    Montreal    New Delhi    Panama
Paris    São Paulo    Singapore    Sydney    Tokyo    Toronto

*To
Jeffrey,
Leslie,
and Christine*

*This book was set in Palatino by Black Dot, Inc.
The editors were Janis M. Yates and Helen Greenberg;
the designer was J. E. O'Connor;
the production supervisor was Joe Campanella.
The printer was Federated Lithographers-Printers, Inc.;
the binder, The Book Press, Inc.*

*Cover Illustration:
From Metamorphose, M. C. Escher, Escher Foundation
Collection Haags, Gemeentemuseum, The Hague.*

*Front Matter Illustrations:
From Double Planetoid, M. C. Escher, Escher Foundation Collection
Haags, Gemeentemuseum, The Hague.*

*From Day and Night, M. C. Escher, Escher Foundation Collection
Haags, Gemeentemuseum, The Hague.*

# PHYSICAL GEOGRAPHY
## ENVIRONMENT AND MAN

1 2 3 4 5 6 7 8 9 0 F L B P 7 9 8 7 6 5

**Library of Congress Cataloging in Publication Data**

Kolars, John F
    Physical geography.

    Together with the companion vol., Human geography,
comprises a divided and expanded version of the original
work, Geography.
    Bibliography: p.
    Includes index.
    1.  Physical geography—Text-books—1945–
I. Nystuen, John D.,   date   joint author.
II. Title.
GB55.K63     910'.02     75-23033
ISBN 0-07-035290-9

# CONTENTS

*Contents*

*Contents*

**x**

*Contents*

# PREFACE

The recent visit of human beings to the moon's surface emphasized our dependence upon faultlessly working life support systems. Few of us stop to think of the earth's surface as the most vital of all space suits. But without the atmosphere, the waters, and the solid portions of the earth to protect and nurture us, we would die as quickly as a spaceman with a ruptured lifeline or punctured helmet. Moreover, the flow of energy through earth systems is just as critical to us as is the electrical circuitry aboard a space station or moon ship. We are geographers because we view the earth in this way, and we have chosen to write a book which might best be described as *environmental geography* rather than simply as *physical geography.*

*Physical Geography: Environment and Man*, like its companion publication *Human Geography: Spatial Design in World Society,* is the offspring of our basic introductory text, *Geography: The Study of Location, Culture, and Environment.* Our purpose in writing these books has not changed from that expressed in the original text:

. . . to introduce modern geography to students with no previous knowledge of the subject and to demonstrate how a geographic point of view can enhance our understanding of the world around us. To do this, we discuss social and physical systems and the interaction between them in terms of their spatial attributes, including their dimensions, densities, scale relationships, associations, and patterns. In this way, we come to our definition of geography as *the study of man-environment systems from the viewpoint of spatial relationships and spatial processes.*

While we recommend the integrated longer text as being representative of the full spectrum of geography, many people have expressed interest in shorter versions emphasizing either environmental systems or human organization. To meet these requests, we have divided the original text into its two main parts and present this volume as the second of the two shorter books. In this text we cover the same materials as the physical geography portions of our original work.

This volume is actually much more than part of the original text between new covers. Approximately 30 percent more textual material has been added, along with a proportionate amount of new figures and tables. In response to our readers, we have included a complete new chapter on landforms as well as considerable new material on food chains, webs, and pyramids, and on the flow of energy through the biotic environment. We feel that these additions not only enhance the book but that the discussions of energy flow, in particular, are given a special spatial and geographical context not found in the flood of environmental publications now inundating us.

We have been particularly influenced by the philosophy and directions provided by Ian R. Manners and Marvin W. Mikesell in their editing of *Perspectives on Environment*.[1] It has become increasingly evident that human systems of spatial organization cannot be separated from the natural systems in which they are embedded and which are so significantly altered by them. Thus, this book departs from the usual physical geography by including thorough discussions of basic spatial concepts as they apply to human ecosystems. This may seem a dramatic departure from conventional physical geography, but as geographers we attempt to understand and to explain patterns of land use and the appearance of the landscape wherever we may be. Only by keeping humankind clearly in the picture can this be done. For example, it is impossible to explain the location of agricultural land use solely by referring to climate, soil, and topography. On the other hand, the simple yet elegant land use model of von Thünen in combination with a solid grasp of physical geography makes world agriculture understandable in geographic terms. Or again, the characteristics and distribution of tornadoes are clearly physical geography's domain. But tornado death and destruction depend upon human population distribution, human awareness and perception of the threat, and the effectiveness of detection and warning systems. The latter elements are clearly within the realm of human geography. It is only when natural and human systems are considered simultaneously that geography demonstrates its insights to best advantage. That is the task this book attempts.

We have tried to include new insights and materials without destroy-

[1]Ian R. Manners and Marvin W. Mikesell (eds.), *Perspectives on Environment*, Association of American Geographers, Publication No. 13 (Washington, D.C., 1974).

ing the traditional format of physical geography. It is possible, therefore, to use this text directly in a physical geography course by beginning with Chapter 3, continuing chapter by chapter through 9, skipping Chapter 10, and selecting those portions of Chapters 11 through 13 that the instructor feels best summarize the emphasis placed upon the particular course. If the text is used for an environmental geography course or one emphasizing resource utilization, Chapters 1 and 2 make good introductory statements. Chapters 3 through 9 can be treated less intensely and greater attention paid to Chapters 10 through 13. Of course, it is our hope that the entire book will be used. For as we have said, geography to us is a combination of both natural and human systems viewed in a spatial context.

In writing a textbook, the first thing the authors learn is that the task can be accomplished only with the help of many others. We first wish to thank the students at the University of Michigan who have taken the introductory courses upon which our effort is based. The constant challenge of their critical and inquiring minds has been a spur to our ambitions. Equally important to us have been the advice and criticism of our consulting editor, Arthur Getis, who has successfully alternated as devil's advocate and staunch friend. The actual presentation of the subject matter in graphic form could not have been done without the advice and help of Waldo R. Tobler, whose new *Hyperelliptical Map Projection* was especially created to serve as the base for the world maps throughout most of the text.[2] Many of the other base maps have also been provided by Waldo Tobler, whose knowledge and ability in computerized mapping have allowed flexibility in our choice of map projections. In the same manner, we feel that the illustrations and maps prepared by Derwin Bell will long survive the words that accompany them. Finally, we wish to thank our many able assistants at the university, as well as the staff at McGraw-Hill, who have provided support, advice, and encouragement above and beyond the call of duty. If errors, omissions, or oversights vex our critics, only the authors are to blame.

*John F. Kolars*
*John D. Nystuen*

[2]W. R. Tobler, "The Hyperelliptical and Other New Pseudo Cylindrical Equal Area Map Projections," *Journal of Geophysical Research*, v. 78, No. 11, April 1973, pp. 1753–1759.

# 1 | ENVIRONMENT AND MAN:
## A GEOGRAPHIC POINT OF VIEW

The purpose of this book is twofold: to introduce physical geography to students with no previous knowledge of the subject and to demonstrate how a geographic point of view can enhance our understanding of the world around us. To do this, we discuss physical geography in the context of human activities. We also consider the spatial attributes of such interactions, including their dimensions, densities, scale relationships, associations, and patterns. In this way, we come to our definition of geography as *the study of man–environment systems from the viewpoint of spatial relationships and spatial processes.* Building on this definition, in this book we consider that physical geography should emphasize the spatial attributes of natural systems, particularly as they relate to humankind. To illustrate this, let us consider two events which show different aspects of such interactions.

On a summer afternoon, August 24, A.D. 79, the people of Pompeii, a town of about 40,000 at the foot of Mount Vesuvius, went about their lives much as city folk have done for thousands of years. Messages scrawled on those ancient walls are still preserved for us to see and give us some feeling for those vanished lives: fans praise their favorite gladiators; people urge each other to vote for their candidates; children pick out nonsense rhymes and phrases. On that fateful day, food was laid out on supper tables; goods were being bought and sold; and friends made plans for the evening's entertainment. Though the mountain above them had been grumbling, no one really anticipated what happened next. Suddenly the volcano let loose a great cloud of fiery ash that quickly buried the town. At least 2,000 people died under a 30-foot mantle of volcanic debris which preserved much of the settlement intact for the next two millennia. Now we can go as tourists and see where the life of that town so abruptly ended. Seldom are we presented with such a vivid demonstration of nature's impact on man.

Humankind, with its perceptions and values, its technology and its ability to organize space, can also be important in reshaping the natural environments we occupy. A second scene, familiar to many American readers, shows how our search for what we perceive to be most desirable can sometimes destroy the very thing we seek. An aircraft factory worker and his family move from St. Louis to Southern California in order to enjoy a high salary, the year-round sun, amd warm beaches. So have many million more recent arrivals in the "Sunshine State." Each family insists upon its right to own a private automobile and to visit the seaside and mountains whenever its members wish. But the environment cannot easily cleanse itself of the added burden of exhaust fumes and sewage that such numbers bring. The air

This photograph shows the awesome forces unleashed during an eruption of Mount Vesuvius in March 1944. Note the darker lava spilling across the roads and villages at the foot of the mountain. Lava streams often tumble and crush buildings in their paths. Pompeii perished under a mantle of ash that preserved much of the town intact. (U.S. Army Air Force photo; courtesy of the American Geographical Society)

becomes murky and unpleasant to breathe; the beaches are fouled and often unsafe for swimming. The very conditions of the natural environment which have lured a generation of Americans to California are beginning to alter under the impact of too many people. In this case, the value system under which the newcomers operate leads them to unwittingly alter and degrade the environment which they originally sought. This is Pompeii in reverse: man overwhelms nature.

## A Geographic Point of View

In order to understand the chains of linked processes which bind man and nature so effec-

tively together, we must consider both human and natural systems.[1] Many disciplines study one aspect or another of humans and their environments, but only a few have traditionally considered both man and nature in a single context. *Geography* is the oldest of such efforts to understand the total processes by means of which man will survive or perish on earth. Geography's role as a unified discipline is one thing which we wish to make clear in this text.

We attempt this in several ways. Our primary concern in this book is with spatial models of environmental systems. Unlike other physical geography texts, this book makes constant reference to mankind. We do this for a specific reason. *Man is the ecological dominant.* No other agent on earth has so modified the total environment in so brief a span of time; no other environmental agent has the potential for ending all life on earth. Only the unlikely event of a solar flare or other disaster of nearly cosmic magnitude could exceed man in his role of altering the face of the earth.

Humankind has assumed such dominance through its cultural endowment. Our ability to abstract and to make symbols, to learn from others and from other generations, provides us with both custom and history. Also, each generation learns a little more than those preceding it about how to control and manipulate its environment. Through ever-increasing technology, we humans have learned to unleash enormous reserves of energy by means of which resources are utilized and consumed, landscapes are reshaped, and the environment itself is inevitably altered for better or for worse. From a geographic point of view, our ability to spatially organize resources and human activities is extremely important. Furthermore, the basic premise in this book is that the spatial insights provided by geography apply not only to human systems but also, with appropriate modifications, to those in nature.

Therefore, our discussion begins with some basic considerations of space and time, both as the framework in which nature operates and in

[1]An expression we must accept without definition until Chapter 3.

terms of the human organization of the earth. This is followed by a presentation of some underlying facts about natural systems as they operate upon, within, and above the surface of the earth. We specifically emphasize the flow of energy, primarily solar energy in many forms, through the natural systems that shape earth environments. We then introduce the role of plants and animals in the creation of various life environments. Throughout the book we make frequent references to the impact that man has upon natural systems and to how human and natural systems in combination create the *human ecosystems* upon which we depend. In the closing chapters, we explicitly attempt to combine our view of human spatial organization with our knowledge of natural systems in order to give a more complete picture of the impact of humankind on earth environments. In this manner, we hope to show how a geographic point of view increases and enriches our understanding of the world in which we live.

## Numbers Without Dimensions

In order to understand the expression *a geographic point of view,* let us consider the following list of numbers:

49  72  34  31  31  74  69  39  60  44  56  67

This sequence might represent many things: the age of participants in a panel disucssion, the number of houses in a sample of Chinese hamlets, or the distance traveled daily by commuters to Wall Street. These numbers are, in fact, average monthly temperatures in a mid-latitude location. No discernible progression appears associated with them. Yet the list as set down is a sequence, in this case ordered by alphabetizing the months of the year. How much more logical to arrange the numbers as they occur chronologically (Figure 1-1), for the temporal arrangement carries with it additional information about the seasonal values of the figures.

In the same way, many lists are ordered alphabetically which might better have some other arrangement. This is particularly true of things relating to the earth-space in which we

In this oblique air photograph of Los Angeles, in 1967, the San Gabiel Mountains are clearly seen in the distance. However, at lower elevations a temperature inversion exists which has trapped a flat-topped smog layer visible over the city center in the middle distance of the picture. This is a typical condition in the basin, which is open to the ocean to the west but surrounded by mountains in the other directions. Automobiles are the primary source of air pollutants from the vast urban development on the basin floor. (Photo by J. Nystuen)

*Environment and Man: A Geographic Point of View*

3

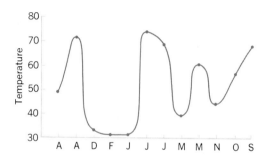

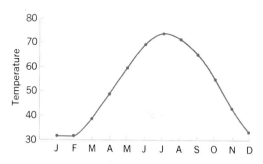

**Figure 1-1  Average monthly temperature** (*A*) Monthly temperature arranged alphabetically by month. (*B*) Monthly temperature arranged chronologically by month.

live. Census tables showing population and other characteristics of various political units often list Alaska next to Alabama and Ethiopia after England. Geographers find this practice no more logical than alphabetizing the months of the year. Examples showing the importance of the location of things in space, just as in time, are found everywhere. These would include the designation of *hurricane coasts* as *national disaster areas,* the sending of aid to *drought zones* such as the African Sahel, and the political definition of the *high seas* beyond the limits of national sovereignty.

Moreover, the fact that we recognize areas by such regional phrases as "the Corn Belt," "the Great Plains," and "the boreal forest region" immediately suggests that the human mind imposes spatial ordering upon the real world. But we must be careful about our use of space in this way. The problems of Alabama and the

problems of Alaska are very different, although their juxtaposition in alphabetized lists falsely suggests some type of continuity nonexistent in reality. Alabama and Georgia, Alaska and British Columbia, England and Scotland, and Ethiopia and the Sudan are all more logically paired than any alphabetically associated countries.

Spatial patterns, like those revealed to us on the accompanying maps, attract the geographer's attention. Almost all phenomena, from soil fertility, types of vegetation, and the level of nutrition, to population densities, cultures, and political ideologies, can be better understood when considered in a spatial context. Most of the problems facing mankind have a spatial character, an analysis of which will help with their solution. The *analytical techniques* referred to in this book are useful in understanding complicated spatial relationships, but mastery of techniques is only one part of problem solving. It is also necessary to have available the *facts* which relate to the problems. Equally important are *theories* to suggest how the facts should be arranged and which analytical techniques are most appropriate. The combination of geographic facts, techniques, and theories presented in this book establishes a *geographic point of view,* particularly useful for understanding the kaleidoscopic world in which we live.

*Nominal versus geographic locations*

The first insight associated with the geographic point of view is a simple but important one. We must distinguish between words implying places as nominal classifications and words referring to specific locations. Place terms used in nominal or nonspatial classifications cannot be located on maps. Phrases like

*the rural and urban sectors of the economy*
*the developed and developing worlds*
*the world of nature*
*vacation land*
*suburbia*

use place terms nominally. Obviously, these words indicate different sorts of environments, but beyond this distinction what do such words tell us? Where are the *rural and urban sectors* of the economy located? Are the country-dwelling commuters of Connecticut and Michigan who travel each day to offices and factories in the city better assigned to rural or urban America? How many factories must a nation have to be located in the *developed* rather than the *developing* world? Is pollution limited to the *developed nations*? And is the *world of nature* separate from *suburbia* if not *vacation land*?

The use of place words in this manner presents an abstract division of reality. There is nothing wrong in using words this way, but it is important to remember that abstract words often have as many connotations as there are people using them. Distinguishing place terms with real locations from place terms used nominally is one way of clarifying our thinking. Trying to define the geographic locations of place terms used nominally gives us additional insight and information—even if we find the real locations of certain nominal terms as elusive as *the end of the rainbow*. On the other hand, learning to think about real space in abstract terms is an important step in understanding spatial theories and analytical techniques, which, in turn, leads us to new insights about the real world.

## Dimensional Primitives: A Way of Looking at the World

We all use words referring to location in our everyday speech. Proper nouns like *Hudson Bay* and common nouns like *country* and *city,* as well as adjectives like *near* and *far,* occur in nearly every sentence. We also use a variety of spatial words with strong emotional overtones. The umpire shouts, "You're out," meaning the player must leave the field. He is literally sent out of the area of action. To be *on the in* is to be well-informed. In medieval Europe, to *sit below the salt* was to be placed in a menial position at a lord's table. *Right* and *left* have strong political connotations which we read about every day.

The problem in dealing with terms like those above is to make them exact. In this task, we may turn to mathematicians for help. We do not

intend to assign numbers to the world around us; rather, we shall consider mathematicians in their role as logicians and philosophers. In considering the nature of the world and their attempts to describe it, mathematicians long ago realized that for consistency and logic every word in every theorem must be defined. This led to the dilemma of defining every word of the definitions defining the words. Since this exercise could go on and on like the ever-diminishing images in two barber shop mirrors, a solution to the problem of ultimate definitions was needed.

The practical solution found for this problem was to agree that the meaning of certain terms should be accepted intuitively. These undefined terms, upon which the subsequent logical structures of mathematics are built, are called *primitives.* In like manner, certain basic statements which describe the fundamental conditions held to be constant throughout any sequence of mathematical reasoning are called *axioms* and are composed of *primitives.* In these axioms, each word is required for a complete description but does not duplicate the meaning of the other words; that is, each word is necessary and independent. Having been given this start, and carefully staying within the original limits of the axioms, we can evolve elaborate systems of logic from such terms.

In much the same way it is possible to describe the world in all its geographic complexity and yet to begin with only a limited number of dimensional *primitives* arranged to form a small set of *basic spatial concepts.* Consider Figure 1-2, "Cities with Over 2 Million Population." At the scale[2] of this map the entire world can be depicted on a single page. The cities which interest us appear as a series of dots or points when drawn at this scale. But in reality they cover many square miles and are of various shapes. On this map, city area and shape are not considered important. The number of very large cities and their location are the data the map is meant to present, so it is sufficient to designate each city as a point without size or

[2]Another expression which we must accept without definition until Chapter 2.

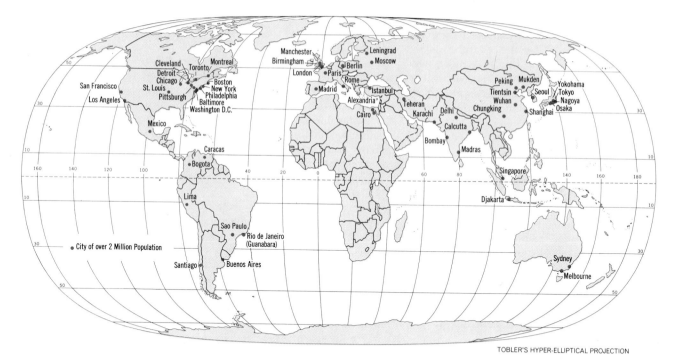

**Figure 1-2  Cities with over 2 million population**  Notice that although there is a concentration of very large cities in the Eastern United States and Western Europe, there is actually a greater total in the non-western world. Even so, these many large cities contain a much smaller proportion of the total population of the non-western world, which is still largely rural. (1970 data)

shape. Here, then, is the first of the primitives with which our geographic viewpoint can be created: A *point* is considered to be a dimensionless location.

Beginning with the concept of a point, we may then derive two additional elements so essential to geography that they can be considered primitives in their own right. A series of points arranged one after another creates a *line.* A collection of adjacent points arranged in a nonlinear fashion, or a line which closes upon itself, defines an *area.* Imagine a plain, smooth tabletop. If one grain of salt is placed on this surface and viewed from a distance, it will, for all practical purposes, appear as a dimensionless point. A row of salt grains viewed from the same distance will appear as a line. A dense sprinkling of salt, one grain in depth, when

viewed from across the room will define an area on the tabletop; when seen from only a few inches or feet away, this same area will visually decompose into individual grains or points.

Point, line, and *area* are concepts important to geographers for several reasons. Not only do they form the basic or primitive elements in various systems of geometry and topology, but they may be used to describe all manner of things in the real world. In Figure 1-2, the *points* marking the cities represent foci of human activity, concentrations of people, and sources of pollution. *Lines* on maps represent everything from transportation routes and national boundaries to rivers, continental divides, and shorelines. *Areas* stand for nations, forest regions, oceans, and continents, and every other feature extending in two dimensions.

There is a fourth dimensional primitive with which we will be concerned as our discussion unfolds. *Volume* follows logically after area. The concept of *volume* can be used to talk about spatial variation in the distribution of many things, such as populations, air temperature, and precipitation. Such distributions can be shown on two-dimensional maps just as we can depict contoured surfaces like hills and mountains. Two-dimensional, or flat, distributions are plotted along two axes designated $x$ and $y$. The third dimension is plotted on a third, or $z$, axis.

The most direct linear value that can be shown on the $z$ axis is elevation. Measures of vertical values, such as for building heights or volumetric relationships which might be needed to understand air pollution problems or the movement of air masses, are also shown on the $z$ axis. Very often the value plotted on the $z$ axis represents an abstraction—for example, population density. On the other hand, geographic locations in the form of points, lines, and areas always are shown by $x$ and $y$ coordinates.

### An orderly view of relations between dimensional primitives

The most common of everyday things can be viewed spatially. Consider the campus where you are now studying. You and your fellow students can be thought of as concentrations of biotic energy endowed with the desire to learn. At a map scale suitable for showing the entire campus, people appear as points. On good days, two things may happen: the sun will shine, and your professors will give brilliant lectures. From a geographic point of view the difference between lecturing professors and sunshine is that the professors are *punctiform,* or pointlike, while sunlight washes in upon the earth as an areal phenomenon. That is, for our purposes sunlight cannot be separated into individual points of light but covers the entire area of the campus equally. A lecture is generated at a point, the professor's head. When students wish to enjoy the sun, the geographic relationship can be described essentially as one of point to area, since they will disperse across the lawns in a fairly even distribution in order to sun themselves. If you attend a lecture, the spatial patterning is one of many points clustered around a single point.

In summary, points may be thought of as concentrations or foci. Lines represent a double function as either paths of movement or boundaries. Areas show the extent of things and represent distributions or dispersals. Another way of looking at areas is that they are used to generalize and classify the world around us. This is the campus; that is the town; beyond is the countryside. Volumes indicate the amounts, concentrations, and densities of whatever is being considered: the air we breathe, the water we drink, or the masses of waste we accumulate (Figure 1-3).

Now let us consider Figure 1-4, "Dimensional Relations," which shows all the possible combinations of points, lines, areas and volumes. One-half the table is empty; it would be redundant to fill it in completely.[3] The various relationships are illustrated by selected natural examples. The number of such examples is almost endless; each reader should develop his or her own conceptual skills by thinking of his or her own illustrations. Note that these relationships appear in the human world as well as the world of nature.

In the upper-left-hand corner is the cell representing point–point relationships. While we have already discussed one point–point example—the lecturer and his audience—natural examples are plentiful. The relationship between a tree and the seedings it produces can be thought of as a point relating to points. An example of point-line relationships would be the problems rising from the location of wells and oases along a caravan route. If they are too far apart, extra water must be carried, which in turn cuts down the payload of the animals. Point-area relationships are illustrated by the feeding territory of a bird in relation to its nest. Each bird of a particular species will need a certain optimum-sized territory. The sparser the food supply, the larger the area upon which it must depend.

[3]Any such array as this is often referred to as a *matrix.* Individual squares or boxes are called *cells.* The horizontal lines are called *rows,* and the vertical ones are called *columns.*

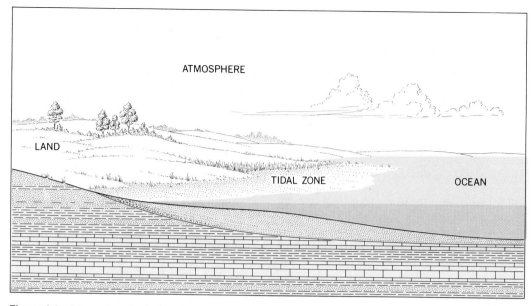

ATMOSPHERE

LAND

TIDAL ZONE

OCEAN

**Figure 1-3 Dimensional relations in nature** Life is thought to have begun in the tidal zone, which is the edge between the oceans (water volume), the atmosphere (air volume), and the land (a surface).

Conversely, too many of a species may result in crowding, smaller territories, and starvation. Point-volume relationships are also numerous. This example shows the surface vent of a volcano as the source of all the materials which make up the mountain. If the original vent is plugged, another may break out at one side, and a new volume of rock and ash accumulates to one side of the original.

At the intersection of the second row and second column, the line–line relationships found everywhere in nature could include the location of a water gap where a river penetrates across a mountain range. Such water gaps have marked critical and strategic locations throughout history. The Cillician Gates captured by Alexander the Great during his pursuit of the Persians and the Cumberland Gap, famous in American colonial history, are examples of such line–line relationships. Line–area patterns greet the eye every time we look at a satellite photograph. The branchings and subbranchings of dendritic streams illustrate this. Line–volume examples can also be important. A mountain stream depositing its burden in the valley below is an essentially linear element which creates a fan-shaped volume of alluvial debris. Such features often contain a stored volume of water useful for irrigation in arid regions.

The complexity of area–area relationships stems from the inherent difficulties of defining areas in terms of logical sets of variables as well as of defining their extent and boundaries. The interfingering of different types of vegetative environments, such as the sinuous boundary between the low, mossy plants of the arctic tundra and the needleleaf trees of the northern or boreal forests, illustrates this. Another example with direct human consequences is the area–volume relationship between the surface area of a lake or reservoir and its volume. The larger the area for a given volume, the greater will be the loss of water to the atmosphere. An extreme case would be that of playas or ephemeral desert lakes.

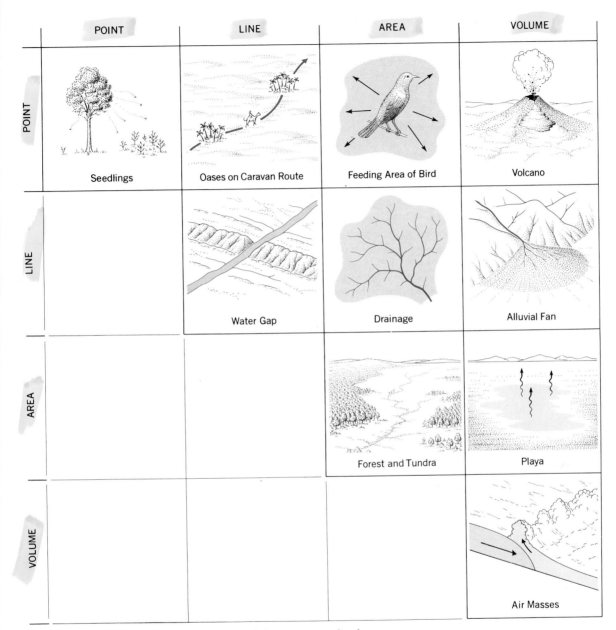

|  | POINT | LINE | AREA | VOLUME |
|---|---|---|---|---|
| **POINT** | Seedlings | Oases on Caravan Route | Feeding Area of Bird | Volcano |
| **LINE** |  | Water Gap | Drainage | Alluvial Fan |
| **AREA** |  |  | Forest and Tundra | Playa |
| **VOLUME** |  |  |  | Air Masses |

**Figure 1-4  Dimensional relations**  All spatial phenomena can be characterized by their spatial dimensions. When spatial elements interact, the dimensions involved will influence the geographic pattern that results. The table shows point–point, point–line, point–area, and other associations.

Finally, volume–volume contacts between land and sea, the atmosphere and the earth beneath, and cold and warm air masses in frontal systems with subsequent storms and precipitation affect us all.

## Some Basic Spatial Concepts

Once we have learned to think about the world in terms of its point, line, area, and volume characteristics, these dimensional primitives can be used to define basic spatial concepts of axiomatic statements. When geographers talk about space, they do not mean astronomical space. The exploration of interplanetary and interstellar space is the work of astronomers and engineers, not of geographers. Geographers also do not consider space as an object or condition which exists of itself. The space which concerns us is defined functionally by the relationship between things. It is the nature of this geometric and topological relationship that forms the basis for the study of geographic space.

Let us take a simple example that shows some geographic concepts. Imagine a stretch of open, homogeneous, rather featureless grassland in East Africa. The open plain provides few clues or directions on how animals might behave there. A lioness moves across it in search of prey. A herd of zebras grazes nearby, and though they break into flight when the great cat rushes, one less wary beast falls beneath her teeth and claws. There is a brief struggle, a cracking of bone, then silence. And with this act the homogeneity of the space is broken, and a temporary spatial order is imposed upon it. Soon the head of the pride, a great black-maned lion, joins the lioness and feeds while she and the cubs and younger cats wait, facing the king at his feast. Their turn will come quickly. Beyond the pride three or four hyenas pace nervously, while overhead, vultures and other carrion birds circle. Even farther from the carcass, a few jackals or wild dogs wait, tongues lolling, until the larger animals finish feeding. Farther still, the surviving herd animals resume their quiet grazing, avoiding the site of the kill.

An examination of this scene reveals that the different animals position themselves and treat space in different ways. What, then, are the spatial concepts which relate to this scene? First, directional orientation: there is a directional quality imposed by the site of the kill. All the animals interested in the meat not only crowd as close as they dare, but will also face inward. All vertebrate life forms have a natural orientation—a front and back—which defines a line of sight. A location or point—such as the kill—and a line of sight—or ray—are necessary and sufficient to define orientation. Thus, *direction* is of critical importance in understanding the spatial organization of the scene we are considering.

*Distance* is another important spatial quality. The lion is dangerous to all the other animals. At the same time, the smell and sight of the freshly killed zebra diminish with distance. Thus, members of the lion's pride, who might get swatted or nipped for disturbing his lordship while he dines, still press close. Other species would not be treated as gently, and so they

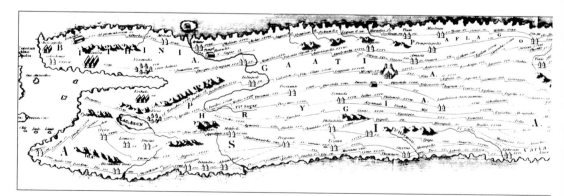

position themselves, ready for flight or ready to dash in and snatch a morsel. The weakest or the slowest carnivores will be farther off, and finally, beyond the odor of the spilled carcass graze the herds which avoid the site, and whose orientations follow different sets of rules.

Intensity of involvement falling off continuously with distance is a property shared by many, but not all, phenomena. Another way of thinking about this is to realize that transportation costs normally rise with distance. In the above example, the lion may rush at potential meat thieves, but will not go far. Or if he chooses to drag the kill to a more convenient place, even his great strength will be limited by the weight of the carcass. Other phenomena are invariant with distance—at least within some range. Legal jurisdiction is as binding at the borders of a state as at its center. Hunters cannot legally shoot an animal whether it is one meter or ten kilometers inside a game reserve.

Despite variations in behavior among the animals, all the carnivores are there because of the fresh kill. If the carcass is completely consumed, the spatial patterning of the group will dissolve, for the associations which brought them together there will no longer exist at that place. This relationship between the animals and the kill may be spoken of as a *functional association.* There is a special type of functional association specifically spatial in character

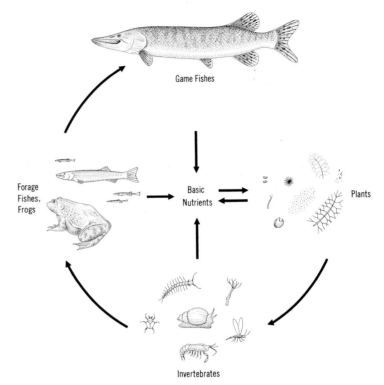

**Figure 1-5  Food web**  A single species of game fish cannot exist without a supporting web of life which includes not only the species upon which it preys but also the insects, plants, and dissolved nutrients which provide food for each life form in the entire chain.

**Figure 1-6  Peutinger map (*E. Tabula Peutingeriana*)**  The map shows routes across Asia Minor in classical times. It contains information about the route network or connections but none about true distances or directions. (*Journal of a Tour in Asia Minor*, William Martin Leake, London: John Murray, 1824)

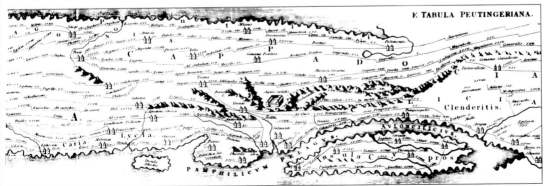

*Environment and Man: A Geographic Point of View*

which is referred to as *connectivity.* By this we mean a special relation which objects in space have to each other.

One purpose of this text is to demonstrate how solar energy is transformed by vegetation through photosynthesis and made available to all forms of animal life. Creatures which depend entirely upon plants for their sustenance are known as *herbivores* and in turn are eaten by still other animals, perhaps by *omnivores,* which eat both plants and animals, or by *carnivores,* which depend entirely upon other animals rather than plants for their sustenance. We introduce this concept in order to illustrate the importance of connectivity. It would be difficult to draw a map showing the relationships between herbivores, omnivores, and carnivores as they might exist in fields or lakes somewhere in the countryside. And yet the transfer of energy involved in these relationships sustains us all. When we eat a steak, we are the end point in a *food chain* in which the movement of energy could be diagrammed:

$$grass \rightarrow cow \rightarrow human$$

or more generally:

$$plant \rightarrow herbivore \rightarrow carnivore \ (omnivore)$$

The fact that the grass and the steer may be in Colorado, and that the steer may then be fattened on Iowa corn before being served as steak in a New York restaurant, presents some major geographic problems which will be described in the chapters to come. Meanwhile, the general functional associations involved in such food chains can be described in terms of their connectivity.

The above example is a simple one. In most of nature, several levels of carnivores may exist, and energy may pass from species to species in complex patterns of exchange. When this happens, it is more accurate to speak of *food webs* rather than of *food chains.* Part of such a food web is shown in Figure 1-5, which portrays the exchange of energy in a small stream community. Each level in the diagram represents a further step in the ascending food chain. Levels of this kind are called *trophic levels* from the Greek word *trophos,* meaning "one who feeds."

The *producer trophic level* refers to plants and algae which directly convert solar energy to forms usable by creatures which consume at higher levels. Thus, we can speak of *producers* and of primary, secondary, tertiary, and perhaps even higher trophic levels of *consumers.* We will return to these matters again and again, but the important thing to note here is that such relationships between energy users can be shown by mapping their connectivity without reference to the distances or directions separating them.

Connectivity is also important in understanding human communities and the way in which we organize our use of resources. We very often make maps or diagrams of the space in which we live, in which distance and direction are distorted, yet which are usable because the spatial connectivity is maintained. Look at Figures 1-6 and 2-3: the latter is taken from a railroad time schedule; the former is a twelfth-century copy of a third-century Roman road map. In both maps, although they were drawn seventeen centuries apart, certain spatial qualities are absent. Figure 2-3 shows the Green Bay and Western Railroad which connects Lake Michigan with the Minneapolis—St. Paul area, and Figure 1-6 shows roads in Asia Minor. In both maps, cities and roads are of major interest. In both maps, true proportional distances are lacking. In both maps, true directions or map orientation are also absent. And yet both maps contain significant amounts of information and provide keys to successful travel between the towns they show. Although true portrayals of distance and direction are absent, each shows the correct relative position or adjacency between cities. The connectivity of their space has been maintained. No matter how distorted their portrayal of the world, a traveler could find his way from city to city using these maps, for each city maintains its connectivity with its neighbors.

These, then, are the first three spatial concepts with which we are concerned: *distance, direction,* and *connectivity.* They will appear again and again in the following chapters. In combination with point, line, and area dimensional concepts, they provide a beginning for all kinds of geographic analysis.

# 2 | ENVIRONMENTAL SPACE: CONCEPTS OF SCALE

## The Importance of Scale in Viewing Earth Environments

Suppose a spaceship operated by alien creatures visits Earth, their mission to seek intelligent life on the planet. While in orbit they remain too high to observe detail on the ground. During the landing phase of the journey, heat-shield problems prevent their monitoring the surface. Once on the ground they immediately find life, but detect only creatures of low intelligence having a simple form of social organization. The alien creatures are tiny; their ship, no bigger than an acorn, has landed in a grassy meadow, and those ant-sized explorers have indeed discovered ants. Upon leaving Earth they record that the planet lacks a developed civilization. They never see humans, and so they are unable to judge what mankind has achieved.

Their scale of observation has been inappropriate. If they had been able to look from their portholes at certain elevations during the descent, they might have noted human figures and the geometric forms of man-made structures. But even then they might not have noticed us, for their own technology might produce a landscape composed of curved forms and irregular elements. In such a case, they would not recognize the crystallike and geometric structures characteristic of our settlement patterns as the products of intelligent life or life at all. Their mission might fail because of two

types of problems: one related to the scale of observation and the other to pattern recognition.

Surprisingly, all humans are like those alien creatures. Earth phenomena, both physical and cultural, often occur at scales vastly different from the range within which our unaided senses function. Thus, we frequently fail to make accurate observations of the world in which we live. Sometimes instruments allow us to change our scale of observation and to broaden the range of signals we can monitor. For example, telescopes allow us to see long distances into the hugeness of the universe, while microscopes let us probe into the world of microbes and molecules. Nevertheless, we run the risk of failing to observe the phenomena which interest us most because we may choose the wrong scales of observation. We also may not perceive a pattern when we see one if we fail to recognize that a particular form or arrangement has significance. Finally, patterns may be so obscured by unwanted information that we may need special devices to filter out all but the critical data.

### Instruments for changing the scale of observation

Telescopes and microscopes are familiar scientific instruments which change the scale of

**Figure 2-1 (A) Apollo VI space photograph of the Dallas–Fort Worth, Texas, area** Large roadways are only partially visible. The photograph is an example of a geographical data display in which distance and direction are close to being correctly represented, whereas connectivity is uncertain and fragmented. Other notable features are the light areas, which are devoted mainly to commercial and residential land uses, and the dark areas, which are agricultural fields and groves of trees. Notice how the area between the two cities is built up with urban land uses. This is a physical reflection of the interaction which exists between the two cities.

observation. At the scale at which geographers observe things, air photographs and maps are similar instruments of observation. Both maps and photos allow us to record very large areas of the earth's surface. Maps much more than air photographs, however, are selective filters which can be used to enhance important elements and to diminish or eliminate unimportant

ones. Figure 2-1*A* and *B* shows a satellite photo and a highway map of the Dallas–Fort Worth area of Texas. Notice how each presents a large, different set of information.

Space, when portrayed on maps, may also be stretched and transformed in a controlled manner. Such controlled distortions of map space are called *map transformations*. Figures 2-2 and

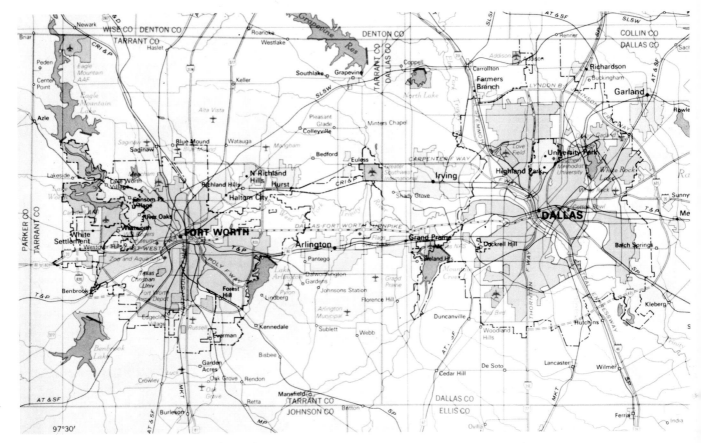

**(B) Map of the Dallas-Fort Worth area**   The map shows the same area as the space photograph above. Notice the similarities, but also notice that additional sets of data have been added to the map; to name a few: political boundaries, railroads, airfields, and place names. Connectivity of transport lines is much more complete, compared to the space shot. The photograph, on the other hand, contains information not on the map: for example, the texture, or mosaic, of land uses present and the grid pattern of the roads, which are best seen in the upper-right-hand corner. (Photo by permission of the National Aeronautics and Space Administration; map from *National Atlas of the United States of America,* U.S. Dept. of the Interior, Geological Survey, Washington, D.C., 1970)

2-3 show familiar types of map transformations: one, a Bostonian's view of the United States; and the other, the area served by a particular railroad company. In each case, the area most important to the map maker has been enlarged at the expense of other places. While these illustrate subjective map transformations, Figure 2-4 shows a map which emphasizes southern Michigan according to a mathematical formula. Conversely, inappropriate choice of transformations can obscure meaningful patterns. Thus, the Bostonian's map would not serve the needs of a traveling salesman.

Any collection of information recorded by areal units may be considered an instrument of spatial observation. A team of biologists may

*Environmental Space: Concepts of Scale*

**15**

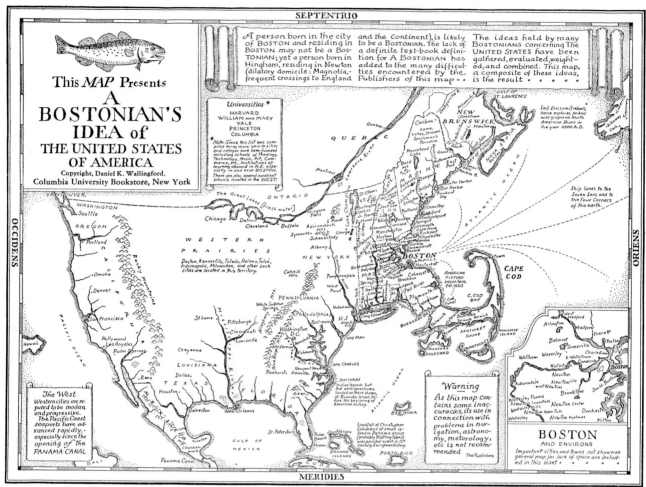

**Figure 2-2  A Bostonian's view of the United States** (Daniel K. Wallingford, Columbia University Bookstore, New York)

estimate the animal life in a field by dividing the area into units one meter square. A sample of the squares can then be taken and a careful listing made of every living creature in the selected units. Extrapolation of the figures will give a reasonable estimate of the total number of creatures involved. But the size of the animals being studied should determine the size of the unit samples. Meter squares might serve best for counts of insects and small rodents; but larger animals—lions and zebras, for example—might require survey units many miles across. When dealing with human activities we also must take the size of the units of observation into consideration. The United States Census Bureau reports information by census tracts, which are small areas within which the number of houses, the number of people by age distribution, and many other facts are listed. Information is also sometimes presented in city, county, state, or national units. For cities of more than 50,000, "block statistics," which describe the units defined by intersecting streets (i.e., city blocks), are also available.

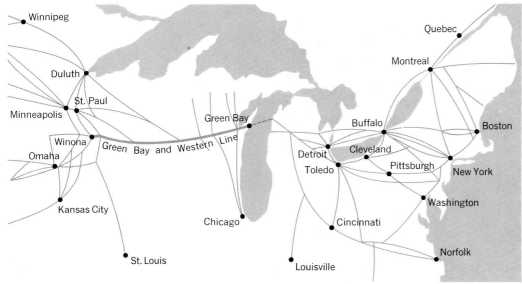

**Figure 2-3 The Green Bay and Western Railroad** Distances, direction, and area are distorted on this map, but connections are correct. The purpose of the map is to show the connectivity of the railroad company's rail network. (By permission of the Green Bay and Western Railroad Company, Green Bay, Wisconsin)

There are many similarities between mechanical instruments, such as cameras, and the charts and tables produced by census enumerators and scientists. Just as you may place a filter over the camera lens to remove unwanted wavelengths of light, census enumerators may choose to list only certain kinds of phenomena. The photographer may use other optical devices to enhance the edges of photographic images; the choice of enumeration categories may similarly enhance or delete particular census data. For example, businesses below a certain size may be omitted from an enumeration; or census tracts which have a few nonfarm families but which are still mainly used for farming may be counted as completely agricultural for some purposes.

*Limitations on scale change as an analytical device*

Just as both optical and census instruments perform transformations on data, each instrument has its own peculiar limitations which

**Figure 2-4 Map emphasizing southeastern Michigan** In this map projection centered on the state of Michigan, distance from the center is plotted as the square root of actual distance. The effect is to enlarge the center of the map at the expense of its edges.

*Environmental Space: Concepts of Scale*

distort the information transmitted. For example, data about a city are often reported only for the area within the city limits. Although city boundaries define the legal jurisdiction of the municipality, the city as a unified place of settlement may extend far beyond its limited political jurisdiction. Data reported for the political unit alone—that is, the municipality—may be incomplete in terms of the city as a regional entity with far-reaching functions. The air pollution generated by factories in a major metropolitan area can originate far outside the legal jurisdiction and limits of the central city. Conversely, a plume of smoke or smog may extend downwind far beyond the urban areas that are its source. The nature of the census unit is just as important as the speed of a camera lens, and one must be very aware of the characteristics and capabilities of the statistical instruments used in order to avoid errors of observation.

In the motion picture *Blow Up*, the story line centered on a photographer who took a picture of two people in a park. Later he discovered the real story in the picture—the image of what appeared to be a body lying in some bushes in the background. The size of the body was very small relative to the field of vision. Because of this, and to learn if there really was a body in the picture, the photographer enlarged the photograph in order to see more clearly the segment which had attracted his attention. The first enlargement he made was more enticing, but still the picture was too small for him to be certain that a body was truly shown in the photograph. Eventually, however, he enlarged the part of the photo showing the body so many times that, just as the information he sought was within his grasp, the grains of the photo emulsion and the other optical and technical characteristics of the instruments he used for enlarging and processing blurred the picture into a random pattern of gray tones. In other words, just as his choice of scale was about to reveal the truth, the ratio of unwanted to needed information became too great to retain the pattern. At the movie's end the photographer still was uncertain about what he had photographed and was unable to learn whether a crime really had been committed.

This scenario reveals the necessity for establishing upper and lower bounds on what we wish to observe. If a subject is too small a part of the total picture, it will be lost in an overwhelming flow of useless information. If it is too large, only one small portion of the whole will dominate the entire picture. The end result will be like the elephant and the blind men, each of whom could describe the elephant only in terms of the one part he could touch. In the same way, maps and other instruments used to record and display facts are heavily dependent on their inherent limitations as well as on the perceptions of the researcher using them. To have confidence in facts, one must be sure of the methods employed to gather and present them.

## The importance of scale in space-time frameworks of observation

*The frequency and timing of our observations will also affect our interpretation of whatever we analyze.* Both of these conditions can be thought of as scale differences, but one operates in time while the other operates in space. The two together constitute a space-time frame of reference of immense importance.

If we study the expansion of the United States across North America and choose to view the process only at 100-year intervals beginning in 1776, we will learn few details about the order in which statehood occurred (Figure 2-5). The time interval chosen is inappropriate for our purposes. We would do better to look at the country every decade, as in Figure 2-6. On the other hand, if we wish to study commuter patterns in metropolitan areas, we must sample the area at least every six hours. Measures made by the decade are of scant use for planning better traffic patterns and controls.

Considerations such as these apply to studies of natural systems as well as human ones. For example, the Serengeti Plains of East Africa support more than 750,000 grazing animals of many species. Such large numbers of animals must migrate from one area to another in search of grass and water as the seasons change (Figure 2-7). Even in those places where sufficient rainfall provides adequate grass growth

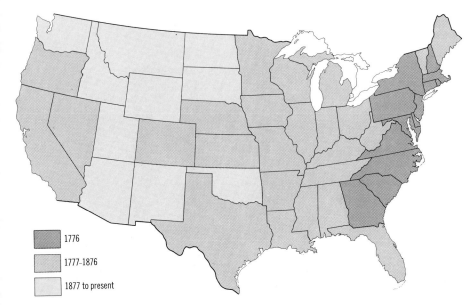

1776

1777–1876

1877 to present

Figure 2-5   Statehood at 100-year intervals.

Figure 2-6   Statehood at 10-year intervals.

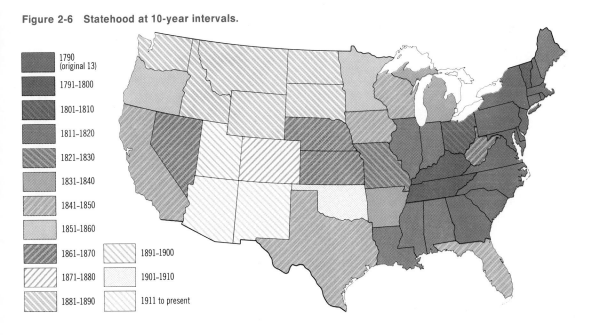

1790
(original 13)

1791–1800

1801–1810

1811–1820

1821–1830

1831–1840

1841–1850

1851–1860

1861–1870

1871–1880

1881–1890

1891–1900

1901–1910

1911 to present

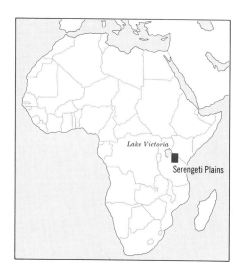

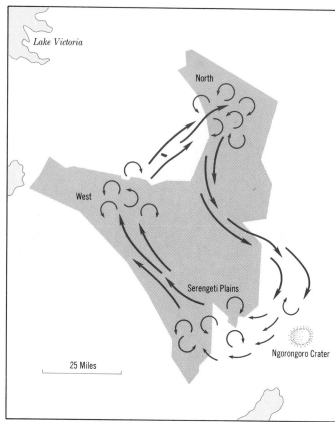

**Figure 2-7 Animal migration routes in the Serengeti Plains** The seasonal ranges of the migratory animals are related to environmental conditions. The wet season (March through May) is spent on the short grass plains region; the early dry season in the west where there is intermediate rainfall and thus intermediate grass supply; and the late dry season in the north, which is characterized by high seasonal rainfall and long grass.

throughout the year, prolonged occupance by the herds would deplete the grazing and give the grass no time to recover. In other places, grass is available only during a rainy season of brief duration. Figure 2-8 shows the amount of rainfall, the grass length, and the populations of three animal species at a given site in the western Serengeti plains. These data are based on nearly continuous observations for three years. But suppose that some sampling program at less frequent intervals were chosen. The area

**Figure 2-8 Rainfall, grass length, and animal populations in the western Serengeti Plains** Population of migrating species is shown in relation to rainfall and length of grass for nearly three seasons. The figures were obtained in the western Serengeti by a series of daily transects in a strip 3,000 yards long and half a mile wide. Successive peaks during each year mark the passage of the main migratory species in the early dry season. (Richard H. V. Bell, "A Grazing Ecosystem in the Serengeti, *Scientific American,* July 1971, copyright © 1971 by Scientific American, Inc.)

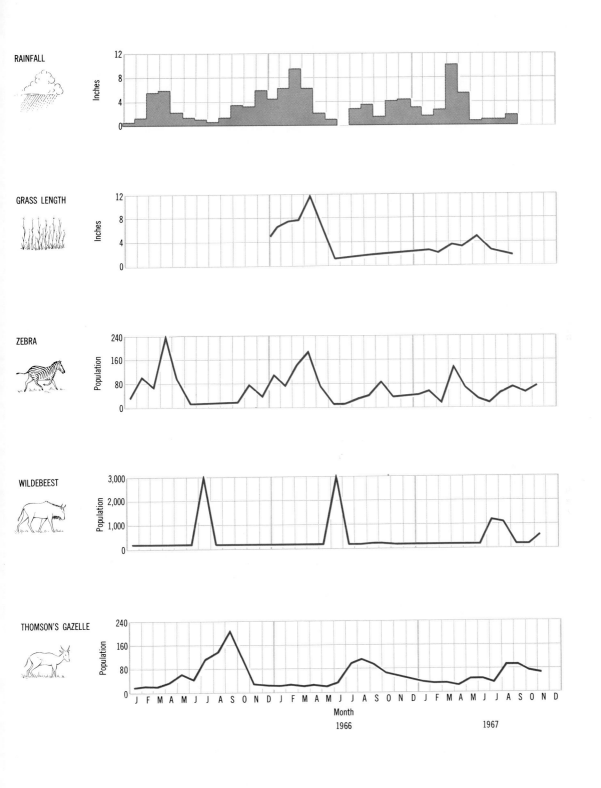

RAINFALL

Inches

GRASS LENGTH

Inches

ZEBRA

Population

WILDEBEEST

Population

THOMSON'S GAZELLE

Population

J F M A M J J A S O N D J F M A M J J A S O N D J F M A M J J A S O N D

Month

1966

1967

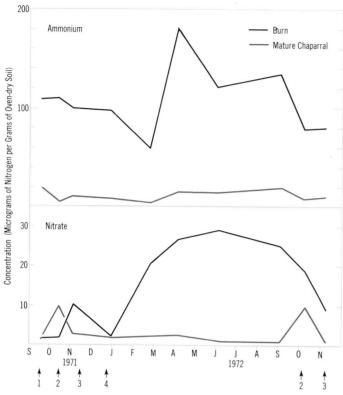

**Figure 2-9 Seasonal change in soil fertility**
Changes in ammonium and nitrate concentrations in burned and unburned chaparral soil in California reflect the occurrence of fires and the onset of the winter rainy season. Ash releases nutrients to the soil, although they may then be carried away by heavy rain, while nutrients are held in unburned vegetation. (Normal L. Christensen, "Fire and the Nitrogen Cycle in California Chaparral," *Science,* vol. 181, July 6, 1973, p. 67, © 1973, The American Association for the Advancement of Science)

might be thought of as continuously crowded or continuously deserted. Figure 2-9 illustrates the same principle for soil fertility. Fires frequently sweep across the chaparral thickets of Southern California. Ash returns ammonium and organic nitrogen to the soil, while foliage formation removes it. In the words of one researcher who has examined these soils, "The time of year and the frequency with which soil samples are taken may greatly affect the results obtained and the conclusions reached concerning soil fertility."

Returning to our spatial point of view, we must study the territorial expansion of the United States using all North America as our spatial frame and using states, whatever their size, as the statistical subdivsions. To make sense out of commuter patterns we need to look at entire metropolitan areas with reasonably small subdivisions, since most commuters leave the central cities when their work is done but at the same time do not travel far from them. As in all research the upper and lower limits of the scale of investigation should be carefully considered. We must always avoid using too many units or too few. This applies to time as well as space. Too many units—that is, too fine a filter—will hide significant relationships by isolating each observation in its own location and moment in history. Too few units—a sieve with too large a mesh—will allow clots and clusters of data to pass unobserved by hiding significant relationships within statistical averages compiled for meaningless aggregations.

## Site and Situation as Changes in Scale

Spatial observations involve both positional control and identification of local values or intensities. Positional control means keeping track of the location of each observation relative to other observations or to some fixed delineation of earth-space. Knowing the location of each observation allows preservation of spatial order, the importance of which was discussed in Chapter 1. The major reference system for knowing where things are in the world is the latitude and longitude coordinate grid, which uses the equator and the prime meridian as its reference bases. But it is not always necessary to know the latitude and longitude of a place to maintain positional control. All that is needed is information about the relative locations of elements under study. To speak of the relative location or situation of things is to comment on the properties of the space in which they are found. Spatial properties need not be constant. In one context we may be concerned only with linear distance; in another, travel time may be the best way to measure functional distances. Other attributes of the space, such as direction and connection, may also be involved. With good locational control one can speak of the shape or pattern of spatial systems. Without such control only intensity measures at given sites remain. Geographers distinguish relative location and intensities at specific places by referring to *situation* and *site* characteristics.

Site characteristics are the *in-place* attributes of a particular area. Thus, the amount of rainfall and average annual temperature of an area are considered important site characteristics. Soil type and fertility, slope, drainage, and exposure to sun and wind are also used to characterize the physical geography of each and every site. These things all relate to the amount of energy available in the physical system within which the location is incorporated. Other site characteristics could include the number of insect species, their populations, and their potential for destroying crops. The same is true for plant, animal, and human diseases. At still another level of abstraction, the human population density of an area can be considered one of the characteristics helping to determine the qualities of site. The type and intensity of pollution, the amount of built-up area, and the nature of land ownership and property fragmentation could also be included in this category. That is, any identifying characteristics can be used to describe a site. The choice of what is described depends upon the needs of the observer and the problems he or she wishes to solve. When measuring site characteristics, there is no concern for positional control. The important feature is the quality or intensity of the thing observed.

In summary, geographers speak of site and situation. Observations of site refer to the qualities or attributes of a place; observations of situation refer to the position or location of a place relative to other places. A site is best thought of as an area with particular attributes. If the same place is thought of in relation to other places, it is best considered as a point location. This involves a change of scale. Figure 2-10 illustrates such a transition from point to area.

### The representative fraction

At a scale of 1:6,600,000 the glacial landforms of the Midwest form broad zones stretching across several states. At a larger scale, 1:960,000, special types of landforms and deposits of particular materials associated with glaciation can be identified. At a still larger scale, 1:235,000, an individual gravel pit can be seen as well as the distribution of various glacial materials.

The different scales or ratios referred to above are a convenient way to keep track of the size or limits within which our observations are made. Such a ratio is called the "representative fraction." This term is used particularly by cartographers on the maps they prepare. It is a way of accurately indicating the size relationship between the real world and its graphic representation. The numerator, the number which appears before the colon, indicates one unit of *linear* measure (inches, centimeters, etc.) on the map; the second number, the denominator, represents the number of inches or centimeters (or whatever *linear* unit is specified) of the real world indicated by one map unit. Thus, if we were to draw a life-sized picture of a man, 1 inch on the paper would represent 1 inch of the man's actual measurements. If his nose were 2 inches long in reality, it would appear 2 inches long in the drawing. This would be indicated by a representative fraction 1:1, or 1/1. If we were to draw a very large map of a room in which everything appeared one-fourth actual size (e.g., a table 4 feet square would appear as a rectangle 12 by 12 inches on the paper), the representative fraction would be 1:4, or 1/4. Maps of manageable size must be quite small regardless of the areas they represent. Thus, most representative fractions are very small by the above standards. Maps which show 1 mile equal to 1 inch on paper have an R.F. (the common abbreviation for representative fraction) of 1:63,360. The range of representative fractions shown in Table 2-1 is meant to indicate the scale of the maps which could be

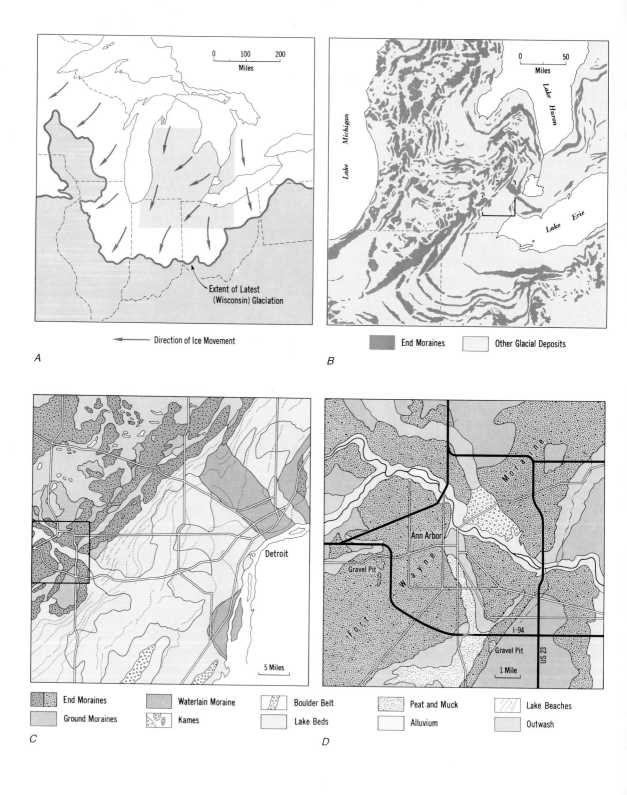

0          100          200
Miles

Extent of Latest
(Wisconsin) Glaciation

→ Direction of Ice Movement

*A*

0          50
Miles

Lake Michigan

Lake Huron

Lake Erie

End Moraines          Other Glacial Deposits

*B*

Detroit

5 Miles

End Moraines          Waterlain Moraine          Boulder Belt

Ground Moraines          Kames          Lake Beds

*C*

Moraine

Ann Arbor

Gravel Pit

Wayne

Ford

I-94

Gravel Pit

US 23

1 Mile

Peat and Muck          Lake Beaches

Alluvium          Outwash

*D*

used to plot the corresponding activities. Maps of personal space are seldom if ever made, but in order to complete the table, an R.F. of 1:1 has been included.

## The Scales at Which Things Happen

Scale considerations are important in two ways. We have seen that the scales with which we choose to observe the world influence our interpretations of it. In much the same way, the behavior of an individual can vary significantly from that of the group or population to which he or she belongs. What a particular man says as an individual may not match what he does as the elected chairman of a committee; what he does as chairman of a committee may not match how he votes as the member of a political party; how he acts as the member of a political party may be different from his response as a loyal citizen to international conditions. Such changes in behavior do not mean that this person is inconsistent or hypocritical, but rather that different types of events take place at different scales or levels of organization and his responses to those events may vary accordingly. What may be appropriate at one level can be inconsistent at the next. An appreciation of the different scales at which things happen can be useful in understanding the nature of the world.

### A hierarchy of earth space

Since we are concerned with both earth environments and the human use of the earth, it is worthwhile to consider the scales at which man and nature operate. Figure 2-11 shows the continuum of scales important to all life forms. The worldwide realm of life, which extends from

**Figure 2-10   Glacial landforms at different map scales**   *(A)* Extent of the continental glacier in the Great Lakes region, map scale 1:17,850,000 *(B)* Glacial moraines in Michigan, map scale 1:6,600,000 *(C)* Glacial features in the Detroit metropolitan area, map scale 1:960,000. *(D)* Glacial features in the vicinity of Ann Arbor, Michigan, map scale 1:235,000. Notice that the total area covered by a map gets smaller as the scale of the map enlarges. Large-scale maps are close-up views; small-scale maps are distant views.

creatures on the floor of the deepest ocean trench to spores floating miles high in the atmosphere, is called in its totality the *biosphere.* The biosphere is subdivided into smaller areas called *biochores.* These are more or less homogeneous configurations of climate, soils, and vegetation which recur in widely separated parts of the world. Thus, we can recognize grassland, desert, forest, and other biochores in several places on almost all the continents. When animals, particularly mammals, are studied in terms of the biochores they occupy, such areas may be called *biomes.* David Watts defines biomes as "major regions in which distinctive plant and animal groups usually live in harmony with each other, and are also well adapted to the external conditions of environment, so that one may make tentative, but meaningful, correlations between all three."

At still larger map scales covering smaller areas, considerable variation can be distinguished within a single type of biochore. For example, prairies, steppes, llanos, and savannas all have grassy cover and represent some but not all possible grassland environments. These variations or subdivisions are referred to as *formations* or *bioclimates.*

In order to understand the processes by means of which biochores and formations maintain themselves, we may choose to study individual *ecosystems*—that is, smaller groups of plants and animals which exchange energy and materials between themselves and with their environment. A tidal pool or a mountain meadow might each represent an ecosystem. This term comes from the Greek word *oikos,* meaning "house," and we can think of an ecosystem as a community of animals living together in the environment immediately surrounding them. However, both large and small ecosystems can occur. In fact, when we consider physical geography in a human context, the movement of food, energy, and materials into a city and the return flow of products and waste to rural areas involve both man and nature. Taken in their totality, city elements and rural elements and the relationship between them create a special kind of ecosystem.

Ecosystems, in turn, consist of several species

*Environmental Space: Concepts of Scale*

**Table 2-1 Relationship of Population Groups, Types of Activities, and Scales of Interaction**

| Type of Space | Spatial Range | Characteristics of Interaction | | | | |
|---|---|---|---|---|---|---|
| | Radius from person (representative fraction) | Primary modes of interaction | Selected controls | Type of function | Number of people involved |
| Personal space | Arm's length 1:1 | Voice, touch, taste, smell | Evasive movement | Intimate contact | 1, 2, 3, . . . |
| Living or working space (private space) | 10–50 feet 1:50 | Audio and visual (sharp focus on facial expression and slight movement or tonal changes) | Impervious walls, doors | Effective personal conversation, 1-to-1 exchanges | . . . 50–400 . . . |
| House and neighborhood space | 100–1,000 feet (line of sight) 1:500 | Audio-visual (sound amplification) | Spatial specialization | Impersonal interactions symbol recognition, many-to-one exchanges | . . . 100–1,000 . . . |
| | | Limit of Proxemic Interaction | | | |
| City–hinterland space | 4 to 40 miles (60-minute one-way commute) 1:50.000 1:63.630 = 1 inch = 1 mile | Local news media, TV, urban institutions, commuter systems | Routine channelization | Mean information field, daily contact space | . . . 50,000– 10 million |
| Regional–national space | 200–3,000 miles 1:500,000 | National network of news media, national organizations, common language desirable | Hierarchies | Legal-economic-political systems | 200+ million |
| Global space | 12,000 miles 1:50,000,000 | International communication networks, translator services, world organizations, international blocs | Travel restrictions, trade barriers and trade agreements | War, trade, cultural exchanges | 3+ billion |

of plants and animals living in relationship to each other. These subdivisions of ecosystems are called *communities*. Each community includes a number of species the members of which form *populations*. Finally, each population is composed of individuals. For completeness, Figure 2-11 includes elements below the level of the individual, but such scales are considerably less important to geographers.

Life has a variety of functions along this continuum of scale. Individuals within a species must come close together and usually touch in order to mate and care for their young. In larger areas, feeding territories are established, and the process of resource partitioning takes place whereby food supplies are divided among species. At still smaller scales covering even greater distances, migration patterns may become im-

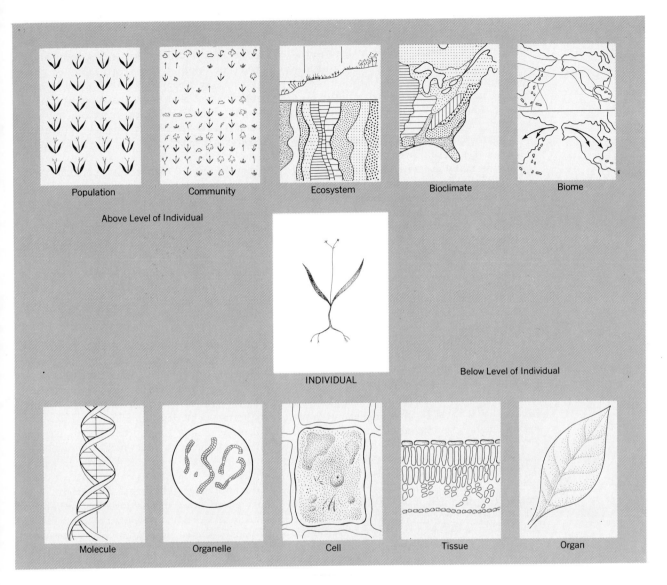

Population     Community     Ecosystem     Bioclimate     Biome

Above Level of Individual

INDIVIDUAL

Below Level of Individual

Molecule     Organelle     Cell     Tissue     Organ

**Figure 2-11  Scales at which life forms operate**  The individual serves as the central measure of this system. Below the level of the individual, microscopic observation becomes increasingly necessary. Above that level, geographic techniques are of great importance. Locational analysis can be used to study the life processes at every scale from that of populations to that of biomes.

portant, and the instincts which produce group behavior in herds and flocks, in flights of birds, and in schools of fish may help insure survival. Finally, and often at global scales, groups of animals may become separated either through the imposition of physical barriers or by sheer distance. Mutations and lack of genetic mixing can follow, and from these processes it is likely that *speciation*—the development of new species—takes place. These examples are but a few of the many events and processes tied to various levels of spatial activity. Humans, with their larger brains, their ability to think symbolically, and their inheritances of culture and history, are a special case and have particularly complex scalar relationships.

## A hierarchy of human space

The late paleontologist-philosopher Teilhard de Chardin dramatized the idea of mankind's impact on environmental processes. He suggested the concept of the *noosphere* (from the Greek *nous*, "mind") to indicate the envelope of activity which humankind has woven around the earth. His idea is an important one, and because human actions are exceedingly complex, we have included Table 2-1 as a supplement to the more general continuum of scale shown in Figure 2-11. Obvious parallels can be seen between the two schemes, but we do not mean to imply that an exact match exists.

**Personal space** The analysis of the personal space in which individuals function is of interest to anthropologists and sociologists as well as geographers, and is sometimes referred to as the study of *proxemics*. All of us surround ourselves with a small territory which we feel to be our own. We have all experienced the feeling of uneasiness when we suddenly find our elbow in competition with that of some stranger for the same armrest in a theater. This "bubble" of personal space is very important to us. We prefer that strangers "keep their distance," although friends may come closer.

**Living or work space** Beyond the more intimate limits of each person's space are the areas

of territories within which we live our daily lives. The classroom and the dormitory, as well as the office and the rooms of your own home, represent another and larger type of space which has special meaning. Within this living or work space, distances are measured from about two to three feet for conversations between casual acquaintances, to perhaps tens of feet for lectures, group activities, and parties. Conversations are carried on in normal tones, and individuals respect each others' personal "bubbles" of territory. Living spaces are controlled by the construction of soundproof walls and doors that can close off rooms for greater privacy.

These units of personal, working, and living space are the indivisible building blocks which society uses to organize the areas which it occupies. Be sure to keep them in mind as you consider how larger spaces such as urban areas are organized.

**House and neighborhood space** A group of houses upon their individual lots or several apartment buildings and other structures along a street constitute still another type of space, which might also include large lecture halls, auditoriums, gardens, and small parks. Distances here may range from 100 to perhaps 1,000 feet, the outer limit often being the limits of line-of-sight recognition. Where living units are concerned, territories of this size are often thought of as neighborhoods. Larger groups of people can meet under these circumstances. Usually an individual will not know all the other persons gathered nearby but will share some common purpose with them. Under these conditions contacts between people become more specialized and less casual. We designate certain places for meetings, others for private domiciles, still others for parks and playgrounds.

**City-hinterland space** The scale of change between neighborhood and city marks a transition where the proxemic study of face-to-face behavior gives way to more specifically geographic spatial considerations. A city consists of many neighborhoods and specialized areas which function together in a manner different from that of a cluster of households. In addition, the city occupies a central focus within a region or hinterland which it serves and which in turn is dependent upon the urban center. The extent of the city and its hinterland is fixed in large part by the distance which commuters will travel to and from work. Research has shown that the average American commuter travels about 60 minutes either going to or coming from work. Depending upon road and traffic conditions, this means that the linear commuting distances traveled in and near a city vary from 4 to 40 miles. Beyond this distance communication becomes more difficult, and it is within this range that local news media, radio and television, daily newspapers, and urban institutions—such as local and metropolitan governments—are effective.

**Regional-national space** Nations can be thought of as clusters of cities and the regions which they represent. The distances at which humans interact in terms of regional and national systems vary from a few hundred miles to several thousand. In order to tie such vast areas together, news media—such as nationwide broadcasting and television networks—must be organized. In the same way, political parties need not only "grass roots" organizations but also nationally organized superstructures to coordinate the efforts of the thousands of local groups. Nations need carefully designed chains of administrative command and clearly defined relationships between local and national governments in order to function effectively. Transportation networks physically bind the nation together. A strong constitution and legal system, which clarify and standardize numberless individual differences in human behavior, are other means of communication within the nation. Thus, the type of human activities served at this level can be thought of as nationwide transportation, political, legal, and economic systems, to name a few of the possibilities.

**Global space** Finally, individual nations and international blocs relate to each other on a global scale. The world itself is the limit of the distances involved, with 12,000 miles the farthest that one point can be from another on the surface of the globe. World affairs of all kinds are carried on at this scale. Trade agreements and cultural exchanges are the alternatives to misunderstandings and warfare. In this case, international communications systems, including satellite relay stations and trans-oceanic cables, facilitate the exchange of information. Translation from one language to another is essential, and international organizations like the United Nations and the International Red Cross help coordinate the efforts of individual countries. In the same way, trade and customs barriers, as well as visa and passport requirements, can restrict travel and the flow of goods and ideas.

*The "fit" between natural and human systems*

All the above comments apply as well to man-environment systems. Environmental problems and the political units through which society attempts to correct them must have a proper "fit." That is, the scale at which the former occur and the latter operate should be the same. Figure 2-12 indicates this in a schematic way. Five levels of government with increasing geographic areas of jurisdiction are shown as nested hexagons. At the top are listed the political units, while the lower half gives examples of the type of environmental problems best handled at the corresponding scale. For example, garbage trucks and fire engines cannot travel long distances either economically or efficiently. They must be located at frequent intervals throughout urbanized areas. Therefore, local governments are almost always accountable to householders for providing such services. At a higher level of the political hierarchy, metropolitan authorities are better able than small towns to enact land-use zoning on a rational basis. In the same way, expensive recreation facilities and parks serving widely distributed populations are also better maintained by metropolitan units. The efficient provision of pure water and the disposal of wastes involve still larger areas, and state governments

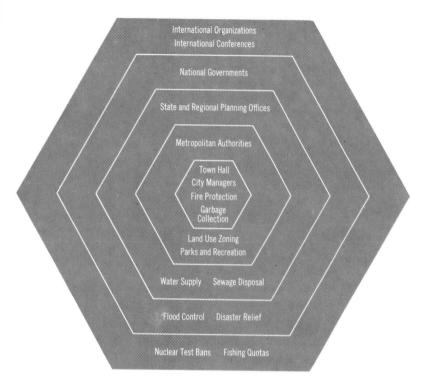

**Figure 2-12 Hierarchy of government levels and scale of environmental problems** Each type of environmental problem is best handled at a governmental level operating at the corresponding scale.

or regional planning authorities are important at this level. Finally, national and international organizations are best suited to deal with natural disasters and nuclear testing. To put this another way, peoples' interaction with nature is either expedited or hindered by the legal and political systems they use to organize space. Citizens of a single town cannot save the whales from extinction, nor can the United Nations make certain that your trash collector comes on time. The scale of the organization must fit the job on hand. We will see in the next section how the size of things can also control their shape and function.

### Size, shape, and function

*From what has already been demonstrated, you can plainly see the impossibility of increasing the size of structures to vast dimensions either in art or in nature; likewise the impossibility of building ships, palaces, or temples of enormous size in such a way that their oars, yards, beams, iron-bolts, and, in short, all their other parts will hold together; nor can nature produce trees of extraordinary size because the branches would break down under their own weight; so also it would be impossible to build up the bony structures of men, horses, or other animals so as to hold together and perform their normal functions if these animals were to be increased enormously in height; for this increase in height can be accomplished only by employing a material which is harder and stronger than usual, or by enlarging the size of the bones, thus changing their shape until the form and appearance of the animals suggest a monstrosity.*

*—Galileo*

King Kong obviously had not read Galileo before climbing the Empire State Building. The geometry of life must follow the same rules as the geometry of scale. Table 2-2 shows how area increases as the square of linear dimension, and how volume increases as the cube. If we consider the body of an animal to be roughly cylindrical in shape, then a very large gorilla two meters tall (6.56 feet) and weighing 350 kilograms (770 pounds) would be about 1/119 the weight of King Kong if he were 10 meters tall. It is unlikely that the facade or perhaps even the structural skeleton of the tower could withstand the 21 metric tons concentrated on the soles of those huge simian feet. Figure 2-13 gives some idea of these relationships.

But the above considerations are still too fanciful to be realistic. The great ape of fantasy films still would have had to walk about. His vast bulk still would have had to be supported by his legs. The bearing strength of a column or pedestal is a function of the area of its cross section. As Figure 2-13 shows, a normal gorilla can support itself on legs with a possible diameter of 15 centimeters. But what of King Kong? Compared to a real gorilla, his weight has increased 119 times but the area of his legs in cross section has grown only 28 times that of our real anthropoid. Under these conditions his legs would buckle, and he would be no better

*Physical Geography: Environment and Man*

than a stranded whale. Either that, or so much of his bulk would have to be concentrated in his legs that they would resemble those of some immense hippopotomus. In fact, all truly giant animals, such as whales and the largest dinosaurs, need or needed to be suspended in water in order to maintain their shapes.

Size and shape go together with function. We have seen how the larger the land animal, the greater must be the columns of its legs in order to support its bulk. Insects with exoskeletons on the outside of their bodies must molt those hard, confining shells in order to grow. During the molting period, their soft bodies have little or no internal support, and insects are thus limited to relatively small dimensions. Crabs and other aquatic crustaceans have similar lim-

**Table 2-2  Area and Volume Increase Relative to Linear Dimension**

| Linear Dimension $d$ | Area $d^2$ | Volume $d^3$ |
|:---:|:---:|:---:|
| 1 | 1 | 1 |
| 2 | 4 | 8 |
| 4 | 16 | 64 |
| 8 | 64 | 512 |
| 16 | 256 | 4,096 |
| 32 | 1,064 | 34,048 |

**Figure 2-13  A normal-sized gorilla versus King Kong**  Weight increases by the cube of height, whereas cross sections of leg increase only by the square of height. If King Kong were made of the same stuff as a real gorilla instead of fantasy, he would very likely not have been able to stand up under his own weight.

| Height | — 10M | Increases X5 |
| Cross Section of Leg | — 1.77M² | Increases X25 |
| Weight | — 43,750 Kg | Increases X125 |

| Height | — 2M |
| Cross Section of Leg | — .071M² |
| Weight | — 350 Kg |

NORMAL MALE GORILLA

KING KONG

*Environmental Space: Concepts of Scale*

itations, but during their molting periods they are supported by water and therefore can attain larger sizes than insects.

These rules apply as well to inanimate nature. There are no mountain peaks on earth higher than nine kilometers (about 30,000 feet). Their bases could not support the accumulated weight of such vast summits. On Mars, however, where gravity is less than on earth, recent satellite photographs have revealed a giant volcanic cone, Nix Olympica, which has an elevation of 25 kilometers. Even earth maintains its approximately spherical shape because it moves in the void of astronomical space. If some cosmic giant's child were to choose it as a marble to play with on the schoolyard of Valhala, our globe would behave as if it were a drop of water, unable to keep its shape when rested on a solid surface.

## The law of allometric growth

Not only are there absolute limits to size and function imposed by scaler relationships, there are also internal regularities relating to size in both living and inanimate nature. Thus, arms and legs are proportionate to body size. When babies are born, some parts may be more developed than others—for example, a newborn human infant's head is disproportionately large compared with that of an adult—but normal growth patterns quickly establish the correct proportions. For example, when a child is about two years old, parents can tell what its mature height will be by doubling its height at that time. Moreover, if the child is normal and healthy, all its parts will continue to share approximately the same proportion of its total stature and weight no matter how large an adult it becomes. This rule is generally true for animals of the same species and is called the *law of allometric growth.* Swedish farmers measure the chest circumference of cows in order to estimate the amount of edible meat which the animal will produce. It does not matter whether the cow is large or small: the proportion between chest girth and the final weight of butchered meat will remain constant. By the same token, paleontologists can reconstruct the general appearance of prehistoric persons from a few

bones or teeth. All that is necessary is that the organic proportions and the size limits of a particular species be known.

To summarize, we can say that, within a given species, if the size of one part is known, it is possible to predict the size of the entire individual or of its other parts. If the total size is known, it is possible to predict the size or proportion of one or more of its parts.

Stig Nordbeck has applied this law of allometric growth to many inanimate things, including volcanoes and large cities. His reasoning is that while a city grows and maintains its same general shape and function, the various parts within it will retain the same proportion to each other regardless of city size. It follows that if the size or rate of growth of one type of land use is known, it is possible to predict the total size of the city or the size of the various areas devoted to other land-using activities. Conversely, if a city planner has an estimate of the future population of the city, he or she can also estimate the total increment of streets, shopping centers, sewers, and residential areas which must be added to accommodate that future population.

## Applications of the law of allometric growth

There are many interesting applications of the idea of allometric growth. Geographers have shown that for every settlement there is a regular relationship between the size of the built-up area and its population. Although most cities are irregular in shape, this relationship can be thought of more clearly if we assume that settlements are all shaped like perfect circles and let the size of the circle be proportional to the total area of the city we wish to consider. In this case, the circle representing the city would be proportional both to the size of the built-up area and to the population living in the built-up area. The radius of such a circle can be expressed by the formula

$$r = aP^b$$

$P$ is the population, and $r$ is the radius, while $a$,

the maximum density, and $b$, the rate that density declines with distance, are coefficients based on empirical observations of the particular cities being considered. If the area of the city is known and we wish to estimate the population, the same formula can be written

$$P = \left(\frac{r}{a}\right)^{1/b}$$

Waldo Tobler has demonstrated this application of the law of allometric growth for the cities of Fort Worth and Dallas, Texas. Using an Apollo VI photograph of the two cities (Figure 2-1) he estimated their built-up areas, which appear as lighter shades of gray on the photograph. Coefficients $a$ and $b$ were supplied by previous studies of American cities. It took only 15 minutes to estimate a population for Dallas of 668,000. This number is within 20 percent of the 1970 census figure. The implications of this are important because it means that in the near future it will be possible with further refinements to take gross census measures of city populations using low-cost satellite photographs of every part of the world.

# 3 | WAYS TO WEIGHT THE WORLD: A PREFACE TO MAN-ENVIRONMENT SYSTEMS

We have complicated our trip up the ladder of scale changes by including man at every level. Most botanists, zoologists, and ecologists simplify their task by omitting man from their considerations of the world at different scales. Geographers have long recognized the interdependence of man and nature as well as the totality of the interaction between climate, vegetation and soils, and all animal life. This notion, so important that it legitimately can be described as a philosophical point of view, is called the *concept of environmental unity* and is a major theme underlying the remainder of this book.

We will discuss the world at many different scales, but communities, ecosystems, bioclimates, and biomes will be our principal concerns. At each of these scales the processes which support life can be better understood through an appreciation of the spatial relationships which exist among the actors and between actors and the lifeless, but seldom immobile, elements such as wind, water, and the earthquake-prone earth itself.

## The Continents of Man

Man utilizes the resources of the earth in order to survive and prosper. The number of people living in a given area—that is, population density (Figure 3-1)—is one measure of his ability to convert the raw stuff of nature into food, shelter, and all the other material things upon which he depends. His ability to organize space and to move materials, goods, and people from one place to another complicates the picture. For example, Manhattan Island, Hong Kong, and many other crowded but relatively resource-poor areas could not maintain their populations without importing food and exporting services and manufactured goods to pay their grocery bills. The uneven scattering of human populations across the surface of the earth is shown again in Figure 3-2A. We have omitted the outlines of the continents and other familiar reference points in order to emphasize the fascinating, lumpy distribution that results. These are, in the words of William Bunge, the *continents of man*, the places where in terms of numbers alone he is most successful. If we can understand the reasons for the settlement pattern shown on this map, we will have come much closer to understanding how man interacts with his environment and how he organizes the space he occupies.

### Environmental constraints

The presence of densely settled human populations immediately attracts our attention, but the

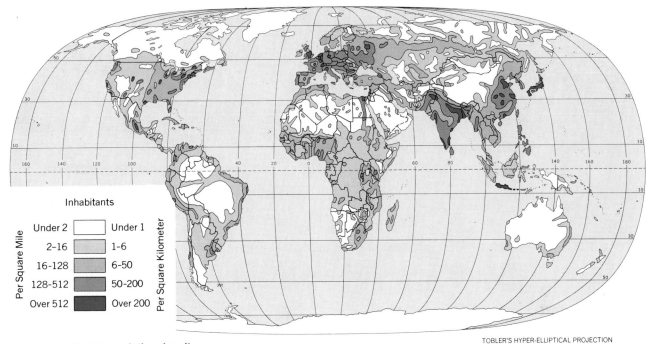

**Figure 3-1   World population density.**

TOBLER'S HYPER-ELLIPTICAL PROJECTION

empty areas on this map are as interesting as the crowded ones. Viewed at a global scale the effects of three environmental constraints which limit the extent of human settlement can be seen. Man overcomes these constraints only with great difficulty; otherwise, the result is either empty areas or sparse populations. Limitations are placed on human settlement by cold, aridity, and large bodies of water. Note first that no major concentrations of population exist poleward of lines showing the 90-day frost-free limit for agriculture (Figure 3-2*B*). Where solar energy is sparse, domesticated food crops cannot mature and ripen, and there is little incentive for human habitation. We will have much more to say on this subject in the next chapter, which discusses the relationship of the earth to its dominant energy source, the sun. In much the same way, man has difficulty supporting himself in large numbers where less than 10 inches of rain fall each year (Figure 3-2*C*). Finally, the constraint imposed by oceanic waters is obvious (Figure. 3-2*D*).

## Resources, Technology and Population Density

The environmental constraints mentioned above establish outside limits on the distribution of densely settled areas. However, within regions with tolerable natural conditions there is still great variation in population density. To understand these variations other qualities of the global ecosystem must be taken into consideration, including man's technology and his ability to organize space and to migrate in search of perceived opportunities.

Natural resources become available to man through his application of technology to raw materials within his environment. Changes in technology result in changing increments of wealth and/or population. In fact, there may be a trade-off between new wealth and new population. That is, if the output of a system increases, such an increase may sustain more people at a fixed level of living, or the level of living may increase for a fixed number of peo-

*Ways to
Weight
the World:
A Preface
to Man-
Environment
Systems*

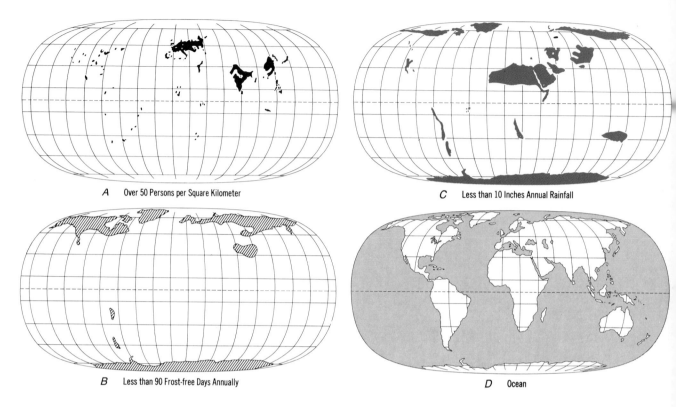

A  Over 50 Persons per Square Kilometer

C  Less than 10 Inches Annual Rainfall

B  Less than 90 Frost-free Days Annually

D  Ocean

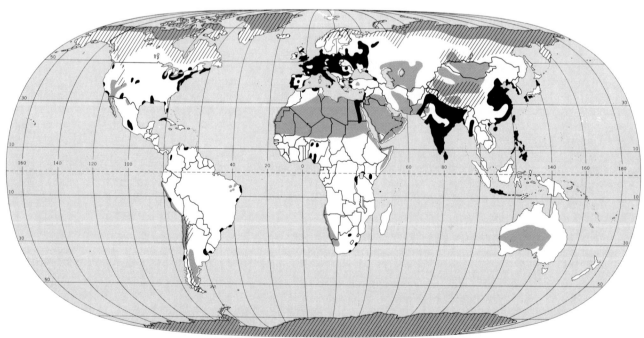

E

ple. Obviously, some middle ground exists in most cases. World population has increased at the same time that the level of living has risen in many countries. But both trends cannot continue indefinitely.

The only way societies can increase their populations *and* their prosperity is to intensify the activities which support them. Essential to such intensification is greater social and spatial organization with an accompanying emphasis upon transportation. Dense concentrations of people, in order to prosper, must have a rich resource base, effective technology, and efficient spatial organization. Otherwise, the price they pay for their numbers is poverty and squalor.

If we reconsider the world distribution of technical skills and natural resources, we can distinguish four major combinations of these elements. There will be those fortunates commanding a sophisticated technology who also control many natural resources. Americans are a good example of this group. Other people, such as the Swiss, may have well-developed technical skills which support them in environments having a limited or poor resource endowment. Next there are groups with few technical skills living in well-endowed areas. Among these would be inhabitants of the oil-rich nations of the Middle East. Finally, poorest of all are those who have few technical skills and who control a national territory, such as Somalia or Chad, with very few natural resources.

Before considering these combinations in greater detail, we must recognize two important trends in the human occupance of the earth. Homo sapiens has become enormously successful, if sheer numbers are a measure of success. At the same time, we have increasingly changed our life-styles from rural to urban forms. This rural-to-urban transition has, in turn, significant environmental implications at both local and global scales. As cities increase in number and size, they coalesce to form urban areas which may become as large as some agricultural regions. Thus, as shown in Figure 3-3, the spatial form of the dominant life-style on earth has changed from *farm areas* with scattered punctiform settlements to contemporary cities serving as major foci for human activities. In the foreseeable future, the dominant life-style will be found in vast *urban areas.*

The two processes, population increase and continuing urbanization, create overlapping sets of problems. Let us first review the nature of world population growth and then consider the shift to urban ways.

### The explosion that isn't

Perhaps the most common term used by social scientists with grim predictions for the future is the expression *the population explosion.* The cataclysmic overtones of that phrase nevertheless hold out nuances of hope. Surely after the catastrophe, some lucky survivors will pull themselves from the ruins, dust themselves off, and start over again. To a human observer far enough away to be safe, a conventional explosion displays maximum violence simultaneously with the triggering action, and thereafter the force of the blast rapidly depletes itself. In other words, if you and the structure you're in survive the initial destruction, you are among the survivors—you are alive.

This is simply not an accurate analogy for the growth of world population. The sequence of events taking place around us, to which we contribute by our own presence, is actually the reverse of any known explosion. An explosion rapidly dissipates its energy and loses its potential for further damage or further growth; world population increases daily, and every increase magnifies its potential for disaster.

The upward sweep of the world population curve, shown in Figure 3-4, is familiar to most readers. While world population growth estimates vary according to the viewpoint and

**Figure 3-2  The continents of man—the effect of environmental constraints on world population distribution.**  The seemingly haphazard distribution of people (*A*) can be explained in large part through a combination of low temperature (*B*), aridity (*C*), and oceanic waters (*D*).

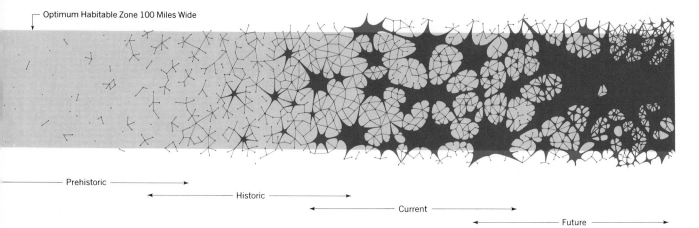

Prehistoric

Historic

Current

Future

**Figure 3-3  Evolution of urban form**  This is an abstract representation of the change in human organization of space through time. The organization proceeds from scattered punctiform settlements in prehistoric times to a predominantly rural agricultural landscape organized by a network of transport lines and urban points. Currently, the urban landscape is expanding in large areas. In the future, the dominant life-style and spatial organization may be urban. The figure is meant to show settlement in a 100-mile zone best suited for habitation. In the future, even with a greatly increased population expected, the world will have vast areas with inhospitable conditions nearly empty of people. The implication is that future populations will have to extend the urban order into the poor resource areas to some extent.

techniques employed by various demographers, the story is always much the same. Ten thousand years ago, at the time the domestication of plants and animals was first taking place, the world was inhabited by perhaps 5 million people. By the beginning of the first millennium A.D., more certain supplies of food, improved technologies, and more efficient forms of political and urban organization had helped increase world population to perhaps as much as 275 million, although other estimates for this period are as low as 133 million. For the next ten centuries, until A.D. 1000, the total population of the world scarcely varied. Deficits in some places were countered by local surges of population in others, but the overall pattern was one of relative stagnation. Thereafter, from A.D. 1000 to about 1650, world population doubled in size, but it was only in the nineteenth century that the 1 billion mark was passed. Sometime between 1930 and 1940, world population had

again doubled. By 1970, more than 3.5 billion people were alive, and all of you who read this book may expect to share planet Earth with more than 6 billion neighbors if you survive until the year 2000.

This spectacular and ominous increase in population can be better appreciated in terms of the survival rates of various groups of people throughout history. Figure 3-5 shows the longevity, or average life expectancy, of eleven human groups in ancient and modern times. Illness, warfare, malnutrition, and famine were everyday occurrences for all our ancestors. The selection process was brutal and only the strongest survived.

Contrast the longevity of preindustrial societies with that of Americans in recent times. The results of the industrial and scientific revolutions of the nineteenth century are demonstrated here. Improvements not only in the means of production but also in the fields of

*Physical Geography: Environment and Man*

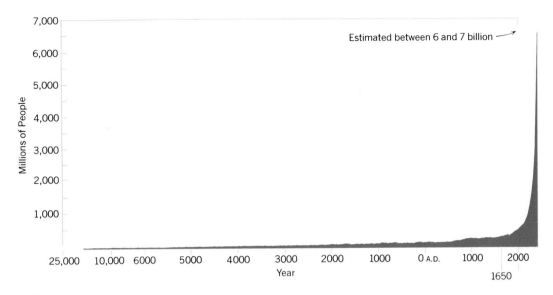

Estimated between 6 and 7 billion

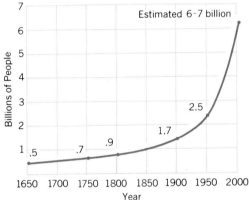

Estimated 6-7 billion

**Figure 3-4 World population growth** The long-time view of world population growth emphasizes how unusual current conditions are. The current rate of growth has no precedence, and it cannot continue very long. On a different scale, the smaller graph shows changes in the past three centuries with a forecast of the year 2000. (References: William Petersen, *Population,* 2d ed., Macmillan, New York, 1969; Glenn T. Trewartha, *A Geography of Population: World Patterns,* Wiley, New York, 1969; Edward S. Deevey, Jr., " The Human Population," *Scientific American,* vol. 203, September, 1960; John D. Durrand, "A Long-Range View of World Population Growth," *Annals of the American Academy of Political and Social Science,* vol. 367, 1967; *U.N. Demographic Yearbook 1970*).

agriculture, transportation, banking, trade, communications, and medicine all serve to extend the life-span.

One important element of these examples must not be overlooked. Just as the growth of world population has been sporadic through time until the last few centuries, so have the average length of life and the increase in numbers of people varied from place to place at any given period in history. This is illustrated in part by the variations in life expectancy at birth, as shown in Figure 3-6. We must also remember that local variations are hidden within data compiled for large political or census units. The survival rate in Appalachia is not the same as in New England. Moreover, those parts of the world with the fastest population growth rates (Figure 3-7) are often those with the lowest per capita incomes.

### The distribution of urbanization

In the same way that we may subdivide the world into geographical regions for which we show survival rates and other demographic characteristics, we may also want to talk about populations in terms of their urban and rural qualities. People, in changing their life-styles,

*Ways to Weight the World: A Preface to Man-Environment Systems*

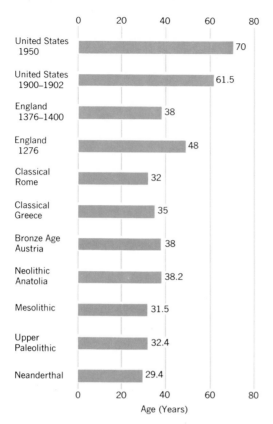

Figure 3-5   Longevity in ancient and modern times   Longevity, or average life expectancy, is a useful population statistic established by means of life tables. Life tables include records of the number of people alive in each age bracket, the age-specific death rate, and other data. The average number of years of expected life at birth may be derived from these data. Life insurance companies use them to determine life insurance risk rates. Longevity is usually calculated at birth. One of the main reasons for the longer life expectancy in modern times, compared with earlier periods, is the reduction in infant deaths. Considerable data are needed to calculate accurate life-expectancy rates. Since these data clearly are not available for ancient times, the rates quoted in the graph must be taken as gross approximations. (Edward S. Deevey, Jr., "The Human Population," *Scientific American*, September 1960, copyright © 1960 by Scientific American, Inc.)

also change their geographical locations. American cities have, in the last 70 years, sheltered a larger and larger proportion of the American population. In 1890 approximately 18 percent of the total inhabitants of the United States lived in its cities; by 1900, 40 percent; by 1930, the urban segment had increased to 56.2 percent; while in 1970, 69.9 percent were classified by the census as urban dwellers. The same trend is found everywhere in the world. Table 3-1 gives some indication of the magnitude of this change from rural to urban modes of life.

An example from the non-Western world illustrates these startling trends in the redistribution of population. During the 10 years from 1941 to 1951, more than 9 million people in India moved to the cities. Their destinations were the larger urban places; their origins, the poor and isolated hamlets and villages scattered across the countryside. This was approximately 3 percent of the 1941 rural population which generated the move and 20 percent of the original urban population in the same year.

Urbanization has become one of the most important issues of the last 150 years. World population in that time increased about 3.5 times, from 960 million in 1800 to 3.6 billion in 1970. During the same period, the total population of cities and urbanized areas with more than 100,000 inhabitants grew nearly 43 times, from 15.6 to 669.0 million. The distribution of

Figure 3-6   Life expectancy at birth, by nation, 1970   The values shown on this map are subject to the measurement errors described in the text and in the caption of Figure 3-5. Nevertheless, a clear spatial pattern is observable. Life expectancy is generally less in the underdeveloped countries of the world. The darkest areas indicate places where infant and child mortality rates are high. In these societies, the commonplace loss of children through death is in great contrast to the experience of people in more advanced economies.

Figure 3-7   Annual rate of population increase, 1971   The darker the area, the greater the birth rate. If the annual birth rate remained at a constant value of 2.0, it would take thirty-five years for the population to double. At a constant rate of 3.0, the population would double in twenty-three years.

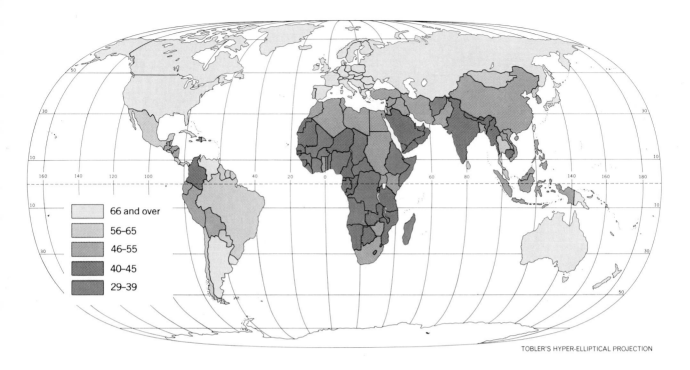

| | |
|---|---|
| | 66 and over |
| | 56–65 |
| | 46–55 |
| | 40–45 |
| | 29–39 |

TOBLER'S HYPER-ELLIPTICAL PROJECTION

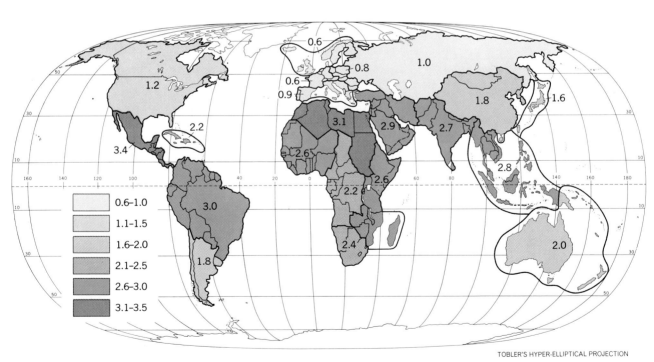

| | |
|---|---|
| | 0.6–1.0 |
| | 1.1–1.5 |
| | 1.6–2.0 |
| | 2.1–2.5 |
| | 2.6–3.0 |
| | 3.1–3.5 |

TOBLER'S HYPER-ELLIPTICAL PROJECTION

**Table 3–1  Development of Urbanism**

| Region | % of Total Regional Population in Cities over 100,000 | | | |
|---|---|---|---|---|
| | 1850 | 1900 | 1950 | 1970 |
| Africa | 0.2 | 1.1 | 5.2 | 10.2 |
| America | 3.0 | 12.8 | 22.6 | 24.7 |
| Asia | 1.7 | 2.1 | 7.5 | 10.1 |
| Europe and U.S.S.R. | 4.9 | 11.9 | 19.9 | 27.1 |
| Oceania | | 21.7 | 39.2 | 44.6 |

Source: 1850 to 1950: Gerald Breese, *Urbanization in Newly Developing Countries,* Prentice-Hall, Englewood Cliffs, N.J., 1966, p. 22. 1970 data: *U.N. Demographic Yearbook 1970, U.S. Bureau of the Census 1970, Statistical Abstract of Latin America 1969* (U.N. data are for latest available year, mainly 1965–1970).

urbanization by world regions is shown in Table 3-2 and Figure 3-8. The pattern of this distribution is similar to that of the distribution of per capita income. This suggests that it is more than coincidence that the developed countries of the world are the urbanized ones.

**Per capita income**  If numbers of people alone were the measure of human success, it would be difficult to understand world population distributions. Extreme concentrations of people exist both in northwest Europe and on the Gangetic Plain of India. England, with its

**Table 3–2  World Urban Population, 1970**

| Region | Total Population (millions) | Urban Population (millions) | % Urban |
|---|---|---|---|
| Northern Africa | 82.2 | 27.6 | 39 |
| Western Africa | 90.8 | 12.5 | 15 |
| Eastern Africa | 92.0 | 8.7 | 9 |
| Middle Africa | 28.2 | 5.1 | 18 |
| Southern Africa | 26.4 | 7.7 | 42 |
| Northern America | 199.3 | 140.0 | 70 |
| Middle America | 65.5 | 35.7 | 54 |
| Caribbean | 20.7 | 8.2 | 40 |
| Tropical South America | 121.4 | 60.1 | 49 |
| Temperate South America | 34.6 | 24.8 | 72 |
| Southwest Asia | 51.9 | 21.8 | 42 |
| Middle South Asia | 713.8 | 138.1 | 19 |
| Southeast Asia | 186.5 | 31.4 | 17 |
| Mainland East Asia | 658.0 | 99.4 | 15 |
| Island East Asia | 116.4 | 78.3 | 67 |
| Northern Europe | 80.1 | 56.6 | 71 |
| Western Europe | 145.5 | 80.8 | 56 |
| Eastern Europe | 102.4 | 53.1 | 52 |
| Southern Europe | 119.2 | 49.1 | 41 |
| U.S.S.R. | 241.7 | 136.0 | 56 |
| Oceania | 14.9 | 11.5 | 77 |

Source: *U.N. Demographic Yearbook 1970.*

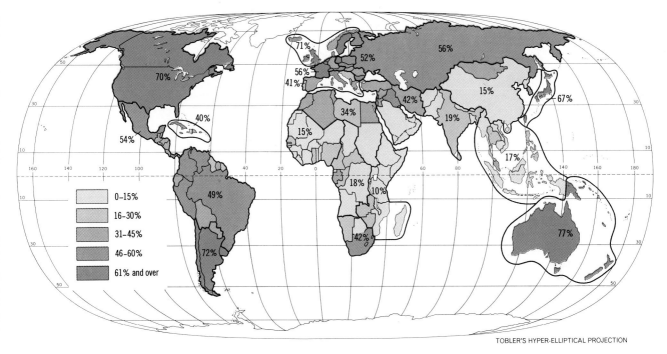

**Figure 3-8    Percent of urban population, by nation. 1970**    Definitions of urban areas vary from country to country and over time. The map shows the latest available U.N. statistics for countries and employs slightly different definitions of urban area as reported by each country. The countries are grouped into world regions, with their average percent of urban population shown. Variation of tones within regions shows something of the range in values which exists. Especially in the large countries subregions would also show great variation in percent of urban population. (Data: *U.N. Demographic Yearbook, 1970*)

coal fields and iron ore deposits, has only slightly fewer people per square mile than does Japan on its mineral-poor island. The material success of mankind must, therefore, be measured not only in terms of population density but also by the level of living that groups have attained. We, therefore, add to our considerations at this point *per capita income* as a measure of a groups's level of living (Figure 3-9).

Figure 3-10 brings together the four variables discussed above: level of living shown by per capita income, population density, available technology, and resource endowments. The diagram shows that countries with good re-

sources and high levels of technology can support dense populations at high per capita incomes. It does not follow, however, that all such national groups automatically will have high population densities. Nations with many technical skills but few resources have relatively small populations (Norway); or if they are densely settled, per capita incomes are considerable lower (Japan). On the low end of the technological and resource scale, incomes are universally small and are often matched by sparse populations.

We realize that this discussion for the time being overlooks the philosophical questions of

*Ways to Weight the World: A Preface to Man-Environment Systems*

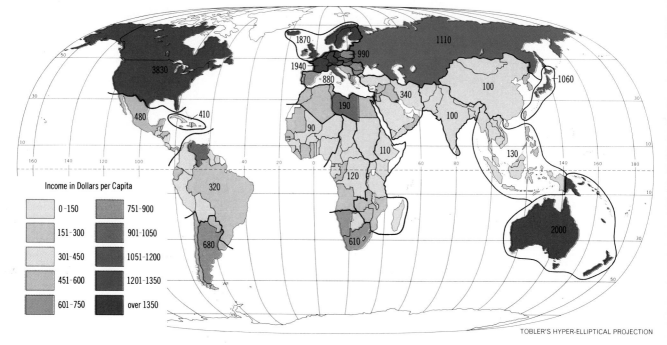

**Figure 3-9   World per capita national income in U.S. dollars, 1968**
(Population Reference Bureau, 1971; based on 1968 data supplied by the
International Bank for Reconstruction and Development)

*personal satisfaction* and *quality of life*, but the
definition of these terms is so complex that we
must begin with more mundane considerations
like per capita income.

We also recognize the importance of other
human factors that are difficult to measure.
Certainly, the *aspirations* of a people are basic to
their activities. Americans aspire to the lavish
consumption of goods; some groups spend their
time engaged in complex ritual and prayer; still
others treasure leisure. But where once it was
popular to point to more and more material
production as a measure of progress, now only
the dullest of technocrats defines satisfaction
and happiness solely in terms of goods and
gadgets.

**Population density and the organization of
space**   Egypt, India, and the United States,
among other nations, remain somewhat
anomalous when discussed in the above terms.
When census data are aggregated for entire
nations, the results often conceal smaller areas
within their boundaries having much higher or
much lower population densities. To appreciate
the problem of accurately determining population densities, let us consider the role of spatial
organization in the utilization of resources.

The relationship between technology, spatial
organization, and population densities is illustrated in part by Figure 3-11. Population densities measured in number of people per 100
square miles of territory are shown along the
ordinate.[1] Seven stages of human technological
development, with their attendant spatial characteristics, are ranked along the abscissa. These
range from the simplest ancient forms of hunt-

[1] *A useful suggestion:* In all graphs the two lines at right
angles to each other are referred to as the *y*, or vertical,
coordinate, and the *x*, or horizontal, coordinate. These lines
are also called the *ordinate* and the *abscissa.* Nearly everyone
has trouble remembering which is which. If you remember
that your lips form a vertical "O" shape when you say
ordinate, and a thin, flat line when you say abscissa, the
distinction can be easily remembered.

*Physical
Geography:
Environment
and Man*

ing, gathering, and collecting on the left to modern and highly developed life-styles like those in the United States and northwest Europe listed on the far right. The first column corresponds roughly to prehistoric Indian communities such as those that once inhabited the Tehuacan Valley in highland Mexico. Those wandering microbands, operating at simple technical levels, needed extremely large territories to support their migratory existence, dependent as they were upon nature's scattered distribution of nutrients in the form of wild plants and animals. Later groups employing more efficient predomesticate systems of plant collecting are shown in the same column. All these values represent estimates of populations using similar shifting patterns of spatial organization like those discussed in Chapter 10. The point labeled "Microbands" represents a population estimate for an early postglacial temporary campsite in what is now east-central England. The point marked "Food Gathering"

**Figure 3-10  Level of technology, resource availability, and per capita income**  Given a high level of technology, a nation can provide its population with high average per capita incomes. This is true even when natural resources are in short supply. This is also true whether population density is high or low. The location of the nations along the population density scale would vary greatly if only arable land were considered. The United Arab Republic (Egypt) is an example of this latter point. (Data: Population Reference Bureau, 1971)

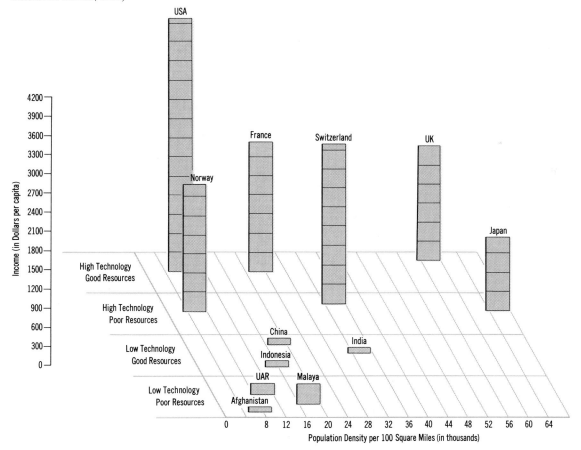

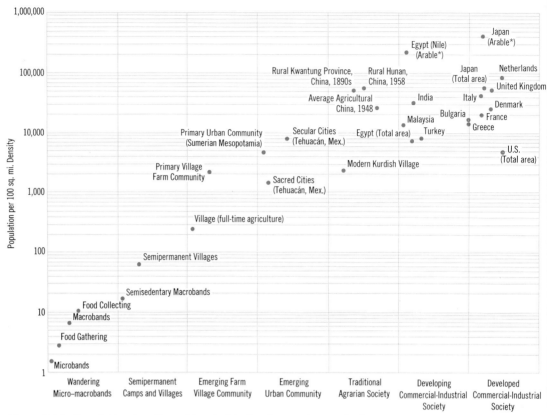

**Figure 3-11  Population density and level of technology and organization**
There has been an upward progression from prehistoric to modern times in
the population densities which mankind has been able to support. This is
closely linked to man's level of technology and organization of space. This
diagram should be considered with caution. No attempt has been made to
assign dates to the levels of technology ranked along the abscissa. Early
population estimates are "guesstimates" at best. Note, too, the shift in the
positions of Egypt and Japan depending on whether total areas or arable
areas are considered. (References: Edward S. Deevey, Jr., "The Human Population *Scientific American,* September 1960; N. Ginsburg, *Atlas of Economic
Development,* University of Chicago Press, Chicago, Ill., 1961; G. W. Skinner,
"Marketing and Social Structure in Rural China," *Journal of Asian Studies,*
vol. 24, no. 1, November 1964; R. Braidwood, and C. Reed, "The Achievement and Early Consequences of Food Production: A Consideration of the
Archaeological and Natural-Historical Evidence," *Cold Springs Harbor Symposium on Quantitative Biology,* 1957, also Bobbs-Merrill Reprint A-23;
R. MacNeish, "Ancient Mesoamerican Civilization," *Science,* vol. 143,
no. 3606, Feb. 7, 1964, also Bobbs-Merrill Reprint A-364)

is the best general estimate computed by Robert
Braidwood and Charles Reed for that way of
life.

The column designated "Emerging Farm

Community" represents a level of technology
higher than that of wandering hunters and
collectors. Food supplies are more readily available because of man's having acquired simple

agricultural skills. Settlements with permanent buildings begin to dot the landscape. Ancient towns and cities develop next, and urban culture becomes a way of life for a small proportion of the population. Now the city and the country-side surrounding it must be considered together as parts of a single ecosystem. This is followed by fully developed traditional societies like those in China. As groups gain more technical skills, nation-states emerge with most of their populations living in cities rather than on the land. Population densities associated with modern nations are shown in the last column.

We are careful not to give specific dates to any of the columns. The reason for this is that domestications and urban development took place at different times at different places throughout the world. For example, the use of domesticated plants begins about 5000 B.C. in the Tehuacan area, whereas in the Near East, Braidwood has identified domesticates occurring by 6500 B.C. at the village site of Jarmo. On the other hand, the simplest permanent settlements did not begin to appear in the Tehuacan area until about 3000 B.C. It follows that it is safer to keep our measuring units general and sequential rather than tying them to exact dates. The important thing to keep in mind is that every new state of technological development increases the complexity of the spatial organization of the society involved.

This does not mean that the quality of life necessarily improves. It is entirely possible for a nation to have many technical skills and rich resources and yet to create a nightmare state similar to that in George Orwell's 1984. Political control and economic power can be vested with the multitude or serve only an elite few. The latter is unfortunately the case in many developing countries as well as in some economically advanced nations. The result in the developing countries is a great and growing gap between the crowded poor and the wealthy few. In developed economies controlled by dictators, political organization, wealth, and technology are the means by which control is exerted. In such a society the quality of life for the common man can be unenviable as the result of his inability to control his own affairs.

**Problems of measuring population density**

Having observed the chart, we should pause for a moment to consider the nature of the data upon which it is based. Remember, *population density* means the number of people living within a fixed unit area; often, such figures are estimated. No one has yet invented a glass-bottomed time machine with which to return to the Tehuacan Valley or any other place and moment in history in order to take an actual population census. Such figures are "educated guesses" based upon archaeological evidence and current observations of similar situations. Nevertheless, when we collect estimates and "guesstimates" from several different sources and find them all falling within a relatively narrow range, as in this case, we can have some confidence that the picture we present is reasonable.

Estimating population density figures presents problems which are particularly spatial in character. For example, Braidwood and Reed's population density figure for the emerging urban community is based on estimates of the whole of ancient Sumer, including densely built-up cities and arid wasteland. The inclusion of wasteland is not some capricious whim of those prehistorians. Rather, there is little evidence with which to decide which areas were irrigated, which were pasture, and which lay waste so very long ago, and so there is little choice except to include the entire territory. The same thing is true for the Tehuacan figures which we have calculated from the author's reports.[2] For want of a better figure, we took the statement that the Tehuacan area totals 1,400 square miles and used it as the denominator in our calculations. But who knows the exact territories used and coveted by those ancient hunters, warriors, and farmers? A different number would have yielded larger or smaller densities, but again, lacking first-hand knowledge, we must go by *best estimates* if we are to go at all.

These observations remain true for today's censuses. Is the viable United States all the area

[2]Richard S. MacNeish, "Ancient Mesoamerican Civilization," *Science*, vol. 143, 1964, pp. 531–537.

*Ways to Weight the World: A Preface to Man-Environment Systems*

within its legal boundaries? Or should we exclude the wastelands of Alaska and Nevada and only consider the rich fields of Kansas and Iowa? And what about calculating urban densities? Should we use counties which may include both cities and their suburbs as statistical areas, or should we use only the areas within the legal boundaries of cities for computing urban population densities? Or should we compute densities using a neighborhood scale and include only residential lots in our calculations? Consider the figures shown in the right-hand columns of the chart. If the desert is included in our calculation of Egypt's population density, we arrive at a relatively low figure of 7,600 people per 100 square miles. On the other hand, if we consider only the arable and inhabitable portions of Egypt—that is, the valley of the Nile and its delta—the population density of Egypt climbs to an astounding 236,000 people per 100 square miles of farmland. (See the photograph on page 183 for a view of the two Egypts.) A comparable contrast exists between total area and cultivated area for Japan (63,000 versus 433,000 per 100 square miles).

There is no pat answer to the question of which areal units to use. The terminology of answers is determined by the terminology of the questions asked. And the questions asked are in large part dependent upon the nature of the systems under investigation. Up to this point in our discussion we have used ideas and terminology relating to systems in a largely intuitive manner. In the remainder of this book we will discuss larger and more complex processes and will rely even more on systems concepts. We, therefore, need to define some basic terms relating to these ideas. In following this line of reasoning, the first concept we must understand and define is *system* itself.

## General Systems Theory

Numbers of people, resources, technology, and level of living must be considered simultaneously if we are to understand the distribution of human population. To really make sense of man and his activities, the above four items form much too short a list. Moreover, simply listing things tells us nothing about relationships between the various elements, and nothing about the processes which link them together. To appreciate such processes we must be able to identify cause and effect relationships: If *A*, then *B*; given *B*, then *M* or *N*. . . . As geographers, we also want to see causal relationships in terms of their spatial and environmental characteristics.

Consider the above association of wealth, productivity, and population. At a given level of technology and resource availability a society within a fixed territory may have an unchanging per capita income and a stable population wherein births and deaths balance each other. A subsequent technological breakthrough might make more wealth (food, money to buy food, medicine, etc.) available to the group. At first, everyone would be better off, with more food to eat and money to spend. Given better conditions, the number of deaths per thousand people might diminish and the population increase correspondingly. But if the technological breakthrough was such that the resulting productivity could not keep pace with population growth, the per capita share of total wealth would again drop. Deaths might again increase because of renewed poverty, and a new condition of equilibrium would result. The only difference between the old and the new states would be that a larger number of people would be supported; but they would be no better off than before. A pessimistic view such as this was held by many early economists and led to a laissez faire philosophy which saw the poor getting poorer while the rich grew richer. Whether or not this situation is inevitable depends in large part on the interpretation given it by the people making the decisions and the kinds of analytical thinking they employ.

This can all be summarized: Resources plus technology yield a product, which in turn gives a population at a certain level of living. New technology increases productivity; increased production leads first to wealth and then to more people; more people's sharing a fixed product means lower per capita incomes. Lower incomes yield poverty; poverty yields a higher death rate. The circle is closed although the numbers involved are greater. We have here a set of things and their attributes which are linked together by a process that establishes their interdependence. However, a description such as the one just given can be awkward. It can also be difficult to repeat for another set of conditions or another time by a different person. It is important, therefore, that we establish a uniform technique of simultaneously viewing many variables and the processes which link them. To do this geographers are turning more and more to the idea of *general systems theory* and *systems analysis.*

### System and environment

The processes which characterize the world of man and nature can be thought of as chains or networks of related events and material things. *Any set of objects and the relations between them can be called a system.* Humans and human artifacts make up man-made systems. Natural phenomena and the processes which link them can be thought of as natural systems. The separation of man-made systems from natural systems is ultimately impossible. The presence of man has affected almost all natural systems with which he has come in contact. In turn, man is a biotic creature dependent upon the environment for all things: light, heat, food, shelter, the very air he breathes.

Let us reconsider the case of Egypt's population density. The question was whether to use the valley of the Nile or the entire territory within the nation's borders as the areal unit defining population density. To simplify things, let us assume that the population is entirely dependent on agricultural production. The critical area is defined then as those places where agriculture is possible, that is, the valley of the Nile and the delta. The total system includes the population supported by agriculture; the natural valley environment, including vegetation, soil, climate, and water; and various culturally defined pieces of equipment and artifacts, including everything from fertilizer to farm houses. The surrounding desert plays very little part in the operation of this farm system but cannot be ignored. At this point, we must make a careful distinction between the farm *system,* which is the *basic object of investigation,* and the surrounding desert, which constitutes the larger *environment,* that is, *those parts of the external world within which the system exists and which interact with it.* To be complete, we would also want to include as part of the environment the national and international economies within which the Egyptian agricultural system operates. But we will simplify the discussion by omitting those things. In agricultural terms, the proper areal unit for computing population density is the valley and delta where the system exists rather than the total country, which includes part of the desert environment surrounding the system.

### Closed, isolated, and open systems

Now let us consider the relationship between any system and the environment which surrounds it. Imagine a theoretical city and its suburbs and the workers who commute back and forth between the two places by electric trains. In this model, no people leave for places external to the system and none enter it from outside. The thing that is imported is energy in the form of electricity which provides the means by which travel takes place. Heat, another form of energy, escapes from the system back into the surrounding environment. This is a *closed system* very much like the sealed cooling system in an automobile, in which the same fluid circulates over and over and only energy to drive a pump and excess engine heat cross the system's boundary (Figure 3-12). The mass of the cooling fluid never varies. In other words, *a closed system is one which exchanges energy but never mass with its environment.*

"But," someone will ask, "isn't the cooling system an integral part of an automobile?" And, "Isn't it impossible to find a city and its suburbs completely isolated from the world market?" The answer is *yes* in both cases. But in our theoretical example both the automobile and the world market are not the objects of study, and by definition become the environments of the smaller systems. In other words,

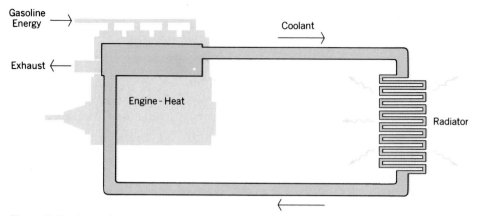

**Figure 3-12   An engine as a closed system**   The coolant recirculates in an engine. A closed system is one which exchanges energy but not mass with its environment.

*the environment of a system is all the systems larger than itself in which it is embedded.*

In fact, our use of a theoretical city and suburbs which exchange energy but no mass with their environment raises an important point. Reality contains some systems which for at least a limited period of time are physically discrete although energy flows across their boundaries. But it would be impossible to find a completely insulated real system which exchanged *neither mass nor energy* with its environment. Such systems can exist only in theory, as in the perfect thermodynamic models used in physics or in a total cosmology where the entire universe is treated as a single system. This type is called an *isolated* system.

On the other hand, if we consider the city and its suburbs or Egypt and the valley of the Nile as they exist in reality, we must picture them as kinetic and undergoing constant change. They grow and prosper, or decline; they exchange people, materials, goods, and ideas with the rest of the world that is their environment. Put another way, mass and energy constantly move in both directions across their boundaries. Systems of this type, which most closely describe reality, are referred to as *open systems*. Remember, Egypt's most important resource is the water of the Nile River, which originates outside the system as we have defined it here.

### Dynamic equilibrium

A steady supply of water is as essential to any city as it is to any farm system. Some cities take their water directly from rivers and during periods of drought must ration its use. A more dependable way to supply water to a city is to collect it in a reservoir for release when needed. (The city and the river can be thought of here as a partial system, since we do not consider where the river water comes from or where it goes after being used in the urban area.) In order to regulate the flow of water to the city as it is needed, the reservoir is used as a *storage device* to smooth out variations in the annual availability of water. Many systems include similar *stores* or *regulators* in order that the *input* to the system will balance the *output* that follows. (A budget and a bank account serve this purpose in the management of money. To paraphrase Mr. McCawber, ''Happiness is spending one cent less than you earn.'') A healthy city depends in part on its having enough water to meet its needs and keeping its needs within the amount of water available to it.

A city reservoir releases a varying flow of water to the city. In turn, it receives varying amounts of water from the rivers that feed it. When the rivers run low, the reservoir contributes more water than it receives. The flow to the city does not alter, but the level of the reservoir drops. When the rivers are in flood, the reservoir is refilled and excess water is allowed to bypass the reservoir. At that time, just enough water is taken in to balance the outflow at the city end without lowering or raising the level of the reservoir. If we measured the water level in the reservoir repeatedly and found no variation in its depth, we could not tell from that single set of measurements what the reservoir's inputs and outputs were. Such a condition, where a system temporarily is neither growing nor shrinking but is nevertheless in complete operation, is called a state of *dynamic equilibrium.* If the outlets and inlets to the reservoir were simultaneously closed, the system would exist in a condition of *static equilibrium.* But almost immediately evaporation and perhaps leakage would begin to lower the water level. Thus, static equilibrium is difficult to maintain, and what we most often observe is dynamic equilibrium. This is true of a person with very regular weight who must eat and drink every day in order to replace the energy he burns and the water he loses through perspiration and as a disposer of waste products.

## Systems in geologic and historic time

It is important to keep in mind that human systems operate over periods of time that are not even a wink in the eye of earth's geologic history. The demands made by a city upon its water supply vary from hour to hour, perhaps from minute to minute, and certainly from season to season within the year. But a molecule of water requires anywhere from 10 days to 150 years or longer to circulate from the atmosphere to the surface of the earth or beneath it and back to the atmosphere (Table 12-1). In Chapter 12 we will see how our insatiable demand for iron has within the span of 150 years exhausted rich ore deposits which took literally billions of years to accumulate. Within the biosphere, some insects and plants complete their life cycles in a matter of hours, while larger animals live from a year or two to more than 100, and our most elderly neighbors, the

bristle cone pine trees, live only 3,000 or 4,000 years. In contrast, the formation of the earth is estimated to have taken somewhat in excess of 4.5 billion years; life as we usually recognize it has existed for more than 600 million years; mammals have dominated earth environments for about 60 million years; and man's ancestors are now traced back to their simplest beginnings some 3 million years ago. Table 3-3 illustrates the scope of geologic time and indicates some of the major geologic and biotic events that have occurred along the way.

But these are empty and meaningless numbers. Which of us can really remember the span of his own life? Perhaps a few people can truly appreciate the geologic time scale—people like Colin Fletcher, who walked alone from end to end of the Grand Canyon deep within its walls. After immersing himself in the immensities of space and time represented by the canyon, he described his insights in a book entitled *The Man Who Walked Through Time,* but we must try to sensitize ourselves to the span of geologic time in a simpler way.

Imagine that a sheet of paper one-tenth of a centimeter thick represents one year of time, and that a 300-page book (that is, 150 sheets of paper) represents the century and a half that has passed since the first railroads and the real beginnings of the Industrial Revolution. If we take that book, and another, and another, and place them on an imaginary shelf side by side, we would need ten meters of books (32.8 feet) to represent the span of 10,000 years stretching back to the beginning of the Recent Epoch, which marks mankind's earliest attempts at domesticating plants and animals and creating the noösphere as we experience it. But that would be a small library compared to the kilometer of volumes (3,279 feet) necessary to return us to the beginning of the Pleistocene Epoch and the initial advance of recent continental glaciation. The Tertiary Period, also known as the Age of Mammals, began 63 million years ago, a shelf of books 63 kilometers in length (39.1 miles). The Mesozoic Era, or the Age of Reptiles, would begin 230 kilometers along our shelf; the Paleozoic Era, which began

## Table 3-3  Table of Geological History

| ERA | PERIOD | EPOCH | Absolute Age in Years Before Present | Major Geologic Events in United States Given in Order of Increasing Age | Distinctive Features of Plant and Animal Life | |
|---|---|---|---|---|---|---|
| CENOZOIC | QUATERNARY | Recent (HOLOCENE) | 10,000 | Minor changes in land forms by work of streams, waves, wind | Rise of civilizations | Age of Man |
| | | PLEISTOCENE | 1,000,000 | Four stages of spread of continental ice sheets and mountain glaciers | Development of man; extinction of large mammals | |
| | | | | *Cascadian orogeny: Cascade and Sierra Nevada ranges uplifted; volcanoes built* | | |
| | TERTIARY | PLIOCENE | | Marine sediments deposited on Atlantic and Gulf coastal plain; stream deposits spread over Great Plains and Rocky Mountain basins; thick marine sediments deposited in Pacific coastal region | Early evolution of man; dominance of elephants, horses, and large carnivores | Age of Mammals |
| | | MIOCENE | 13,000,000 | | Development of whales, bats, monkeys | |
| | | OLIGOCENE | 25,000,000 | | Rise of anthropoids | |
| | | EOCENE | 36,000,000 | | Development of primitive mammals; rise of grasses, cereals, fruits | |
| | | PALEOCENE | 58,000,000 / 63,000,000 | *Laramide orogeny: Rocky Mountains formed* | Earliest horses | |
| MESOZOIC | CRETACEOUS | | | Marine sediment deposition over Atlantic and Gulf coastal plain and in geosyncline of Rocky Mountain region | Extinction of dinosaurs; development of flowering plants | Age of Reptiles |
| | | | 135,000,000 | *Nevadian orogeny: Intrusion of batholith of Sierra Nevada region* | | |
| | JURASSIC | | | Marine sediment deposition in seas of Western United States; desert sands deposited in Colorado Plateau | Culmination of dinosaurs; first birds appear | |
| | | | 180,000,000 | *Palisadian disturbance: Block faulting in Eastern United States* | | |
| | TRIASSIC | | | Deposition of red beds in fault basins of Eastern United States and in shallow basins of Western United States | First dinosaurs; first primitive mammals; spread of cycads and conifers | |
| | | | 230,000,000 | *Appalachian orogeny: Folding of Paleozoic strata of Appalachian geosyncline* | | |
| PALEOZOIC | PERMIAN | | | Deposition of red shales and limestones in Southwestern United States; much salt and gypsum (glaciation of Southern Hemisphere continents) | Conifers abundant; reptiles developed; spread of insects and amphibians; trilobites become extinct. | Age of Amphibians |
| | CARBONIFEROUS — PENNSYLVANIAN | | 280,000,000 | Deposition of coal-bearing strata in Eastern and Central United States | Widespread forests of coal-forming spore-bearing plants; first reptiles; abundant insects | |
| | CARBONIFEROUS — MISSISSIPPIAN | | 310,000,000 | Deposition of limy, shaly sediments in widespread, shallow seas of Central and Eastern United States | Spread of sharks; culmination of crinoids | |
| | | | 345,000,000 | *Acadian orogeny: Folding and igneous rock intrusion in New England* | | |
| | DEVONIAN | | | Deposition of thick marine strata in geosynclines of Eastern and Western United States | First amphibians; many corals; earliest forests spread over lands | Age of Fishes |
| | SILURIAN | | 405,000,000 | | First land plants and air-breathing animals; development of fishes | |
| | | | 425,000,000 | *Taconian orogeny: Folding of rocks in Eastern United States, Nevada, and Utah* | | |
| | ORDOVICIAN | | | Deposition of thick marine strata in geosynclines of Eastern and Western United States | Life only in seas; spread of molluscs; culmination of trilobites | Age of Marine Invertebrates |
| | CAMBRIAN | | 500,000,000 / 600,000,000 | | Trilobites predominant; many marine invertebrates | |
| | Precambrian time; age goes back to over four billion years | | | Many periods of sediment deposition alternating with orogeny | Earliest known forms of life; few fossils known | |

Source: A. N Strahler, *Physical Geography*, 3d. ed., John Wiley & Sons, Inc., New York, 1969.

with the emergence of recognizable life in the form of marine invertebrates, stretches 600 kilometers from us into the distance. And earth itself formed in space about 4.5 billion years ago: a book shelf 4,500 kilometers (2,794.5 miles) long. Page by page, volume by volume, the years are written on earth's ledger. But most of the ecological systems in which modern man is a principal component occupy only a few paragraphs or a chapter or two. Yet with those few lines we are rewriting, for better or for worse, the sum of earth history. Here again the question of the "fit" between scales of operation becomes critical. Let us now return to our general description of systems and see what all this means.

## Control and feedback

The use of water in the city varies from hour to hour, from night to day, and from season to season. An interesting example of this is the manner in which a series of small water-demand peaks occur during early evening hours in areas where television is a major form of entertainment. Every hour and every half-hour a brief, intense period of water use indicates trips to kitchen and toilet by thousands of viewers taking simultaneous breaks during commercials and station announcements. On a longer time scale, summer obviously places peak annual demands on a city's water supply. It follows that any active system such as a city and its reservoirs must respond to several simultaneous and continuously changing sets of demands such as those shown diagrammatically in Figure 3-13. Long-range increases in the city's water needs, perhaps dependent upon long-term upward trends in urbanization or consumer life-style, further disturb or alter the system.

Such fluctuations in demand require a return flow of information to the men or machines that open and close the reservoir's intakes and outlets. As water lowers in the reservoir, a signal may be given by some floating switch mechanism, or visual observation may be employed. In either case, news of the deficit finds its way back to the intake, and the gates are opened.

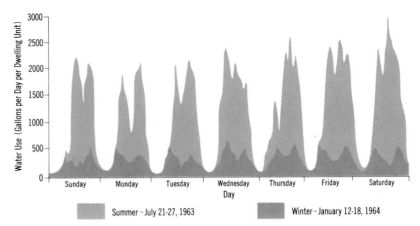

Figure 3-13 **Winter and summer pattern of daily water use—Creekside Acres, California** High summer water demand based largely on air conditioning, lawn watering, and swimming pool use is in sharp contrast to lower winter needs. The twin peaks for each day mark breakfast-time and dinner-time activities. Use in the early morning hours is least of all. (T.R. Detwyler, M. G. Marcus, and others, *Urbanization and Environment,* fig. 5-11, p. 121, after Linaweaver, 1965, © 1972 by Wadsworth Publishing Comapny, Inc., Belmont, California 94002. Reprinted by permission of the publisher, Duxbury Press.

Thus, a return track or *loop* has been activated in the system (Figure 3-14). In our human example we could say that as a person lives and does work, he uses up energy in the form of food. His stomach grows empty, and his immediate supplies of energy run low. His appetite increases, and he begins eating, largely because of the hunger message which has reached his brain, thereby triggering an activity which replenishes ebbing energy supplies in his body.

In both these examples *the presence of a loop means that the objects and events involved are mutually dependent upon each other.* An empty stomach generates appetite, which causes the person to eat, which satisfies his hunger, which reduces his appetite. The reservoir begins to empty, which causes the intake gates to be opened, which raises the level of the reservoir, which thereupon closes the intake. Since in each case information in one or another of many possible forms is fed back to a previous

*Ways to Weight the World: A Preface to Man-Environment Systems*

**53**

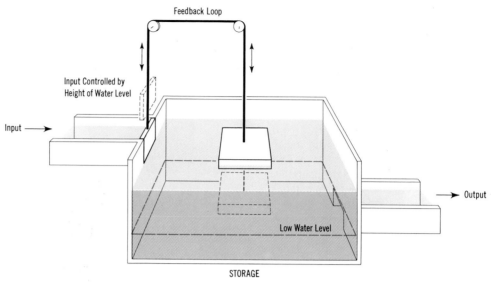

Figure 3-14 **A negative feedback or control loop** A deviation-correcting mechanism acts to bring the water level in the reservoir back to normal.

position in the system, this type of process is referred to as *feedback.*

In the examples just given the messages act to keep the system in equilibrium; that is, they dampen or reverse the developing condition. This form of control is called *negative feedback.* For example, action takes place to bring the water level in the reservoir back to normal. In other words, *negative feedback is deviation-correcting.*

In another case, the message which loops back can be a positive one. For example, when prospectors send the magic word GOLD! back to their friends from some remote place, those receiving the message may in turn tell their friends and relations, and they in turn will tell theirs until thousands of people get the message and a gold rush is on. In this case, a small event may mushroom into a big one. Simple models of city growth describe a similar situation. Two or three service establishments in a small settlement may lure additional customers, who in turn, by their presence, persuade a few more stores to open, which lure still more customers and workers, et cetera, et cetera, et cetera. In this way the equilibrium of the system is dis-

turbed and a new condition continues to grow. Just as the garbled account of one gossip may develop into a panic when passed on by other mouths, sometimes an act of little seeming consequence can bring about great changes, so that with time's passage it may become impossible to recognize the humble origins of major events. We recognize this when we say, "As the twig is bent so grows the tree," or "For want of a nail the shoe was lost. . . ." *Positive feedback is deviation-amplifying.* Only when a still newer set of messages reestablishes negative feedback—for example, "The city is a dangerous place to live," or "There's no more gold in them thar hills!"—can a new steady state or dynamic equilibrium be temporarily reached.

## *A structural classification of systems*

**Simple systems: Great Sand Dunes National Monument** The feedback just described implies information flow through the system. Information carries with it a strong implication of intelligence at work. But many sets of interacting things obviously lack conscious intelligence.

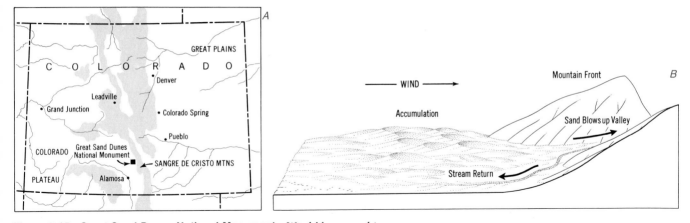

**Figure 3-15 Great Sand Dunes National Monument** Wind-blown sand is carried from the valley floor up the mountain slopes toward a low pass, where it is temporarily stored until water returns it to lower elevations. Very little escapes over the mountains. This may be considered a *closed system* because the sand is recycled again and again.

In fact, systems often include human participants who are so unaware of the roles they play that they become almost mindless robots in some larger scheme. The description which follows of the system at work within Great Sand Dunes National Monument, Colorado, illustrates how energy and material (mass) can move through a system independent of conscious information.

On the southwest flank of the Sangre de Cristo Mountains in semiarid south-central Colorado, giant sand dunes 500 to 600 feet high form part of a dynamic natural system more and more influenced by the works of man. In this area, just north of the New Mexico border, the flat, 40-mile-wide San Louis Valley lies between the San Juan Mountains to the southwest and the Sangre de Cristo Mountains to the northeast. In the valley center near the town of Alamosa (Figure 3-15A) is an area of prehistoric lake bottoms and natural levees, now dry, but formed in wetter times at the end of the Pleistocene glaciation. Prevailing winds blowing out of the southwest carry silt and sand from this ancient alluvial accumulation in a northeasterly direction toward the barrier of the Sangre de Cristo Mountains. As the sand-bearing winds rise over these mountains (Figure 3-15B), their energy is insufficient to lift the burden they carry above the peaks and crests. Their load drops, as it has for thousands of years, at the foot and on the flanks of the mountains. This has resulted in an accumulation of giant transverse dunes now set aside and recognized as Great Sand Dunes National Monument. Sand from these dunes is carried eastward into the mountains up Medano Pass, which acts as a funnel for the strongest gusts. The occasional rain that falls on the mountains forms intermittent streams which return the sand to alluvial fans on the valley floor. Little or no sand escapes beyond the mountains to the east. Only the wind and its energy continue on. The sand recycles again and again.

Wind, sand, dunes, slopes, rain, and streams are the objects which taken together make up this particular natural system. The San Luis Valley and the two mountain ranges form the environment in which the system is embedded.

These physical objects taken by themselves form a set with little or no indication of the processes which bind them together. Such sets constitute the simplest view of systems, and as Kenneth Boulding says, make up "the geography and anatomy of the universe."[3] They represent the *frameworks* for all the other more complex views of systems we may take.

In the last few thousand years the Great Sand Dunes system has been for all practical purposes a *closed* one. Wind-borne energy passes through it, but its mass in the form of sand remains almost constant. The path through the system taken by the energy as well as the path taken by the sand itself presents a somewhat more complex view of the same system. The dry lake bottoms contribute sand to accumulation dunes just on the windward side of the great transverse dunes for which the monument is famous (Figure 3-15B). The great transverse dunes in their turn are the source of the sand found in a set of climbing dunes which endlessly advance up the slopes of the Sangre de Cristo range, only to be dropped by the wind and returned by the streams to the valley below. Once again in the valley, the returned sand is subject to further wind erosion and the cycle repeats. The mapped path of energy and mass through this system describes a series of *cascades* with the output of one part or subsystem becoming the input of the next. In combination, the framework of objects and the cascades of energy and mass form a kind of *clockwork system* which operates without requiring information feedback. Systems organized at this clockwork level exist everywhere about us. In the next chapter we will discuss movements of the earth and sun in space as though they make up a gigantic clockwork system.

The above terms sufficiently describe the system operating at Great Sand Dunes National Monument. A temporary equilibrium exists, though perhaps slightly more sand accumulates each year with progressive wind erosion of the valley floor. No conscious control mecha-

[3]Kenneth E. Boulding, "General Systems Theory—The Skeleton of Science," *Modern Systems Research for the Behavioral Scientist*, Walter Buckley (ed.), Aldine, Chicago, 1968, pp. 3–10.

nism is necessary for the existence of the dunes. We will round out our typology of systems with more complex ones in a moment, but before saying goodbye to the Great Sand Dunes let us consider what may happen to those fragile features in the near future.

The equilibrium of the system depends upon a steady supply of sand passing through the great dunes from the valley floor to the mountains and back. One way in which blowing sand dunes are stabilized is to moisten them and to encourage vegetation to grow and fix their surfaces. The dry valley bottom near Alamosa and the alluvial outwash fans along the foot of the mountains provide sand because of their aridity. But man everywhere works to readjust nature and, to his way of thinking, improve it. This could very well happen in the San Luis Valley if extensive irrigation projects are constructed outside the national monument but well within the critical sand source area. If this happens, the sands there will be stabilized, as will the materials returned by the streams to the valley floor. With their source of supply cut off, the Great Sands would gradually dwindle away as more and more of their materials became stabilized. Conversely, overgrazing of the valley floor might possibly lead to more erosion, more sand, and bigger dunes. In either case, man would serve an almost mindless role in the destruction or enlargement of the natural feature.

**Complex systems** We have already mentioned many systems more complex than simple clockwork ones. They include some control mechanism or *thermostat* that operates to keep them in equilibrium, much as that name implies. The floating switch at the city reservoir which opened the intake valve when water levels dropped and closed it when they approached normal is typical of this type of *thermostat system.*

*Self-maintaining systems* that are able to repair themselves come next. Living and nonliving structures are usually separated from each other at this point. However, simplicity is still a determining characteristic, and living cells are perhaps the best example.

*Plants,* self-maintaining and self-reproducing systems lacking mobility, consciousness, and intelligence, as we ordinarily recognize those properties, are more complex forms of self-maintaining systems. *Animals,* with mobility, intelligence, consciousness, and a variety of sense mechanisms, constitute living systems beyond the plant level. *Animals, plants,* and the *environment* which supports them form *ecosystems. Man* must be given an even higher status as a system because of his superior intelligence, his ability to abstract and symbolize, his accumulated culture, and his tools. Groups of men form *social systems,* for obvious reasons more complex than single human beings, and these in turn interact with all other systems which make up the environment of man to form the most complicated structure of all, *human ecosystems.*

## Hierarchies of Systems

The above list suggests that the more complex forms, such as ecosystems and human ecosystems, are in fact hierarchically arranged and include all the other simpler systems at lower levels. The problem when using a systems approach is to choose the proper scale. In fact, we might now want to take exception to the saying, "The proper study of mankind is man." It might be better to say, "The proper study of mankind is the human ecosystem." But in any event, no one can hope to look at everything there is to study. Having distinguished a system from the larger ones which form its environment, we must still decide the lower limits of the study. How deeply do we wish to penetrate into the subsystems which make up the object of our study?

### Black, gray, and white boxes

One way to simplify the problem is to look only at the inputs and outputs of those subsystems rather than to examine their internal workings. We may take the internal operations of such things for granted, or if we are unable to understand all there is to know about some smaller part of a system, we may simply accept the results of it. Such unanalyzed subsystems may be thought of as *black boxes* through which things pass and in which the conditions of things are changed, although we don't know how.

The use of black boxes is reasonable when trying to understand the larger systems in which they are embedded. Naturally, black boxes should be treated with caution if the system to which they belong begins to change and drift into new levels. This is what is happening in the current ecologic crisis, where some unpleasant surprises are beginning to show up. The side effects of some insecticides on higher animal life illustrate the unforeseen consequences of our acts.

If the processes inside subsystems are partially understood, we can view them as translucent if not transparent containers and call them *gray boxes*. Finally, when a subsystem is understood and mapped to our complete satisfaction, we may refer to it as a *white box*. No secrets there!

### Spatial hierarchies as scale transformers

We have already suggested that systems operate at many different scales. This fact becomes critically important when we return to our consideration of food chains and food webs and to how energy is transferred through such systems. The sun is the source of energy upon which all life ultimately depends. Plants, and only plants, are able through the process of photosynthesis and the attendant use of chlorophyll to convert solar energy into forms that animals can use. We will return to this process in greater detail when vegetation is considered in Chapter 9. For the moment, let us keep in mind how dependent we are upon the primary producers described in Chapter 1, for it is the energy made available by plants that ultimately keeps all humankind alive.

The solar energy that enters earth's ecosystems cannot be increased beyond its original amount, nor can it be diminished. This is in accordance with the first law of thermodynamics, which states: Energy can be neither

created nor destroyed. Thus, radiant energy through photosynthesis by green plants is converted to some form of chemical energy, usually a carbohydrate, which in turn is oxidized by animals, thereupon releasing energy to perform work. Heat will be lost to the environment at each trophic level through the respiration and oxidation which keeps both producers and consumers alive and functioning. When energy is transferred in this manner from producers to consumers, upward from one trophic level to another, less and less is available for each succeeding life form. This is according to the second law of thermodynamics, which states: Energy will tend to distribute itself equally throughout any closed system. In other words, with each transfer work will be done and some energy in the form of heat will be lost to the surrounding environment. Other energy may be stored temporarily in the materials of which the producers and consumers are composed. Sometimes energy thus stored will remain out of circulation for long periods of time, as when vegetation accumulated in swamps during the Carboniferous Period (see Table 3-3) and remained in the earth's crust as coal for 280 million years until now, when it is being released through combustion in furnaces and engines.

These processes can be viewed in a variety of ways. Perhaps we can think of them most simply in terms of the mass of food produced and consumed at each trophic level. Figure 3-16 shows that 10,000 pounds of phytoplankton (simple aquatic plant life) must be produced in the sea in order that some fisherman or his child can gain one pound in weight. The pyramidal shape of this diagram suggests the term by which such complex food chains and webs are known. *Food pyramids* can be used to discuss the *biomass*, or total amount of organic matter involved within a specified area, or they can show the actual amounts of energy transferred through the food chain. In either case, the measurement of such quantities is difficult and time-consuming. One of the few available studies describing the energy levels of a food pyramid is that done by H. T. Odum for Silver

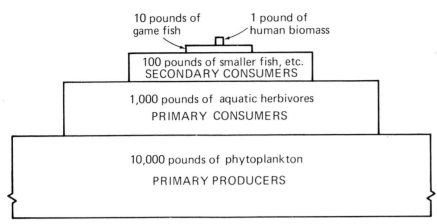

10 pounds of game fish

1 pound of human biomass

100 pounds of smaller fish, etc.
SECONDARY CONSUMERS

1,000 pounds of aquatic herbivores
PRIMARY CONSUMERS

10,000 pounds of phytoplankton

PRIMARY PRODUCERS

**Figure 3-16 Food pyramid** Simplified biomass pyramid with man at the top feeding only on second-order carnivores. In this case, the sea must produce 10,000 pounds of organic matter through photosynthesis in order for a man to gain one pound in weight. Productivity ratios are (to say the least) schematic but give some idea of the relative magnitudes involved. If the figure were drawn to scale, it would run off the page. (From Ned Greenwood and J. M. B. Edwards, *Human Environments and Natural Systems*, © 1973 by Wadsworth Publishing Company, Inc., Belmont, California 94002. Reprinted by permission of the publisher, Duxbury Press.)

Springs, Florida. We illustrate his findings in two different ways in Figures 3-17 and 3-18. In the former, the kilocalories of energy produced at each level are shown divided into those lost through respiration/oxidation, those retained within the biomass to be passed on to the next trophic level, and those entering a subsidiary system in which *decomposers*, mainly bacteria, ultimately reduce waste materials to inorganic substances and heat. The second of these figures (3-18) takes a more spatial view by showing the amount of territory required for the same number of kilocalories to be generated at each level. While one square meter of primary producers generates more than 20,000 kilocalories, 991 square meters of territory would be required for the same amount to be produced by a top carnivore.

Seen in this way, it comes as no surprise that the larger the animal the greater must be its feeding territory and the smaller its population in numbers. This also means that carnivores must travel more—that is, must cover more ground in order to live. Food pyramids, hier-

**Figure 3-17 A pyramid of energy flow** Solar energy transformed by photosynthesis by *primary producers* is shown distributed to respiratory loss, decomposers, and that portion passed on to higher levels in the food chain. The retained portion of *primary producers* is consumed by the *herbivores*, and is again utilized in respiratory loss and decomposers, with a portion passed on to *carnivores*. A secondary or *top carnivore* utilizes only a small fraction of the solar energy that entered the system. (Data from H. T. Odum, "Trophic Structure and Productivity of Silver Springs, Florida," *Ecological Monographs*, vol. 27, pp. 55–112)

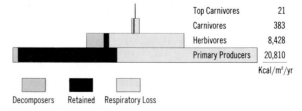

| | Kcal/m²/yr |
|---|---|
| Top Carnivores | 21 |
| Carnivores | 383 |
| Herbivores | 8,428 |
| Primary Producers | 20,810 |

Decomposers    Retained    Respiratory Loss

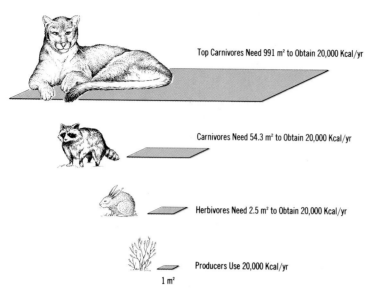

Top Carnivores Need 991 m² to Obtain 20,000 Kcal/yr

Carnivores Need 54.3 m² to Obtain 20,000 Kcal/yr

Herbivores Need 2.5 m² to Obtain 20,000 Kcal/yr

Producers Use 20,000 Kcal/yr

1 m²

**Figure 3-18   Area required to yield 20,000 Kcal/yr by level of food hierarchy.**   Using the proportions shown in Figure 3-17, the amount of area required to make a given amount of energy available at each level of a food pyramid is shown.

archies, systems, scale, connectivity: how does all this apply to human spatial organization?

Hierarchies as systems become important to us in the gathering, processing, distribution, and use of resources of all kinds. This is such an important point that we will return to it again and again, but a single example will suffice for now. The following simplified food system—sun and plants – fields – farms – graineries – mills – bakeries – stores – homes – individuals – heat and waste—can be shown as a complete tree (Figure 3-19). Tiny increments of solar energy and soil nutrients are concentrated by each wheat plant into kernels of grain. Grain is harvested and concentrated in grain elevators for shipment to flour mills. Flour in turn moves to bakeries from which bread is shipped to numerous retail outlets and finally to individual homes. Once eaten, the energy is released; everything else that remains is waste. The energy is used to perform work, which creates heat as its final by-product. The waste, either garbage or sewage, eventually breaks down into

its component parts, releasing a final increment of heat energy back into the earth environment. Thus the whole system represents a coming together and a dissipating of energy through space and time.

**Figure 3-19   Energy pathways in the wheat-to-bread food chain**   Many systems display a dumbell, or "complete tree," spatial network. Energy and material are assembled, change of state occurs, long transport hauls are completed, and products are distributed for use and, finally, returned to the environment after use.

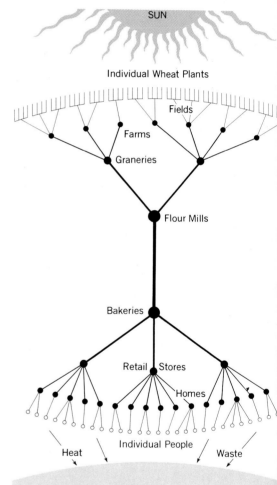

SUN

Individual Wheat Plants

Fields

Farms

Graneries

Flour Mills

Bakeries

Retail  Stores

Homes

Individual People

Heat

Waste

This forms a basic dumbbell-shaped system of concentration and dispersion similar in form to a real tree with both roots and branches. Consider how leaves concentrate the sun's energy for long trips to the tips of root tendrils, and roots in turn absorb moisture from the soil to help maintain the tree's metabolism. On the next warm day lie down on the grass near some tree, and in your prone and sideways view see both ends of the tree, one in air and the other in soil. Imagine the system in its entirety and enjoy seeing birds and worms as similar, each species living in and around the energy system that is the tree, and each an energy system organized in its own way along hierarchial lines (Figure 3-20).

Returning to human considerations, this is one reason why we have talked of urbanization in the preceding pages. Different-sized settlements, each typified by different types of activities or by different stages in the same system of activities, are an efficient way to fill space and take advantage of the resources distributed across the face of the globe. Thus our wheat – bread – energy system in Figure 3-19 could be shown in part as a farm – village – town – city – neighborhood – house settlement system. Neat and formal arrangements of space are often lost or undecipherable, but in any system, either natural or social, the connectivity which binds its many parts together can be shown as treelike series of paths or links connecting together hierarchically arranged focal points. At such foci, energy, materials, or ideas are concentrated, perhaps changed into another form, and sent along to higher or lower centers for further concentration or distribution. *Trees* of this kind are found everywhere and ultimately can be thought of as communication and transportation systems which organize and hold together all human and natural phenomena. Solar radiation provides the energy for all such systems,

**Figure 3-20  A complete tree system**  The tree is a system which transfers energy and material between two volumetric domains via a hierarchical network.

and it is this basic relationship between earth and sun to which we turn in the next chapter.

*Ways to Weight the World: A Preface to Man-Environment Systems*

**61**

Two examples of human adjustments to the basic clockwork flux of solar energy are school graduation and June brides. As the school year draws to an end, graduation and a new life loom. To some students this means freedom; to some marriage. But no matter how the year's seniors may feel about their situation, June graduation and June brides are part of a long and honored tradition tied to the passage of the seasons and to man's adjustment to the annual cycle of solar energy availability.

When most of the population lived on farms or in small towns serving the agricultural community, the need for field labor was greatest during the summer months. Schools were dismissed so that students could help on the farms. At the same time, winter was past and the quagmire roads had dried out. Families and friends could get together, and since early harvests or the promise of a good growing season were as good as money in the farmer's pockets, it was a natural time for weddings.

With the shift into an urban age the need for large, seasonally available numbers of farm workers diminished. But institutions are slow to change, and schools still let out in late May or June. Freedom from school reinforces the old tradition of June farm weddings, and a spate of campus marriages still marks that time of year. Of course, we are talking about the Northern Hemisphere, and of course some schools are experimenting with new time schedules which pay little attention to the agricultural cycle.

## Energy and Man

The human use of energy and the amount of energy provided by the sun offer curious contrasts in magnitude. The energy produced in engines and furnaces has made profound changes in man's life-style. Yet, in caloric equivalents, fuels and other sources of energy used by man in 1965, for example, produced energy equal to only about one one-hundredth of the solar energy arriving at the top of the earth's atmosphere each day! (The earth, in turn, intercepts only about one part in two billion of the sun's total energy output.) For this reason alone it is important to distinguish among world energy systems and view each at its proper scale.

Man differs from all other life-forms in that he employs engines and furnaces to gain access to nonliving sources of energy. Engines convert fuels to electrical and mechanical energy; furnaces produce heat. Nathaniel Guyol summarizes the magnitude of such energy conversion:

*World consumption of energy from all sources, including non-commercial fuels, was equivalent, in 1965, to approximately 6 billion tons of coal (or 4 billion tons of oil).*

*To supply this energy, the world consumed 2.7 billion tons of coal and lignite; 1.5 billion tons of crude petroleum; 25 trillion cubic feet of natural gas; 900 billion kilowatt hours of hydro-, nuclear, and geothermal electricity; about 1 billion tons of non-commercial fuels; and something under 100 million tons of peat, oil shale, and natural gasoline.[1]*

One item that we should note is that our consumption of energy is constantly increasing. In 1965, energy use from commercial fuel sources was the equivalent of 5.3 billion metric tons of coal. In 1971, this energy use reached a coal equivalence of 7.3 billion metric tons, an increase of nearly 38 percent in six years. Another significant thing about the above quotation is that only two words in it pertain to types of energy derived from sources other than the sun: *nuclear* and *geothermal.* These two energy sources derive directly or indirectly from radioactive materials and are assuming greater significance with every decade. Their contribution to man, however, is as yet insignificant when compared with other sources. All fossil fuels are simply stores of ancient sunlight; hydroenergy, whether in rivers or waves, derives directly or indirectly from the sun. In the former case, ancient plants and animals stored solar energy in their bodies as part of normal growth processes; in turn, their remains form deposits of coal, petroleum, and natural gas. In the latter case, solar energy through evaporation and precipitation lifts water high onto the land, thus providing the kinetic energy of running water.

## The Energy Balance of the Earth

In discussing the energy relationship of the earth to the sun we will talk about *heat,* which is one form of energy. Other forms of energy include kinetic energy, the energy of an object in motion; potential energy, the energy inherent in an object due to its relative position; work, or force moved through distance (gram

[1]Nathaniel Guyol, *Energy in the Perspective of Geography,* Prentice-Hall, Inc., Englewood Cliffs, N. J., 1971, p. 5.

centimeters/second$^2$ × centimeters = erg); chemical energy; nuclear and electromagnetic energy; and the energy of mass, described by the familiar $E = mc^2$. Heat is measured or perceived in terms of *temperature,* which is a measure of the kinetic energy of the molecules that make up the heated substance. Temperature can refer to the degree of molecular motion in gases, liquids, and solids. This temperature, particularly air temperature, is sometimes referred to as *sensible heat.*

The relationship between the temperature of the earth and the radiant energy of the sun represents a system in dynamic equilibrium. We can imagine a simple model of this system that consists of an intensely radiating sun and a solid ball representing the earth suspended in the vacuum of space some distance away. The sun radiates energy in all directions, but only a small portion of that energy will be absorbed by the earth. The amount of energy intercepted by the earth will be proportional to the intensity of the sun's radiation and the area of the earth's cross section, and inversely proportional to the square of the distance between the sun and the planet.

As the radiant energy of the sun strikes the surface of the sphere, some will be reflected back into space; the rest will be absorbed. It is convenient to use the term *insolation* (from the phrase *in*coming *sol*ar radi*ation*) when referring to the interception of solar energy by any surface. The ability of a surface to reflect radiation is called its *albedo.* A snow field has high albedo; plowed ground a very low albedo. On the average, the earth reflects about 36 percent of the insolation it receives from the sun.

If the sun were to increase its radiation, the earth would heat up until increased reradiation once more balanced the inflow of energy; a new state of equilibrium would be attained at a higher temperature. However, as long as the elements of the system remain constant, reradiation and reflection from the sphere balance the inflowing energy and the sphere maintains itself at a steady temperature. Fortunately for us, the fusion process which converts hydrogen to helium within the sun is remarkably regular in its production of energy. The unvarying rate

*Solar Energy and Man: Earth as a Heat Engine*

of flow of radiant energy that results is called the *solar constant*. It is measured as the amount of energy received on a square centimeter of surface held outside the atmosphere at the average distance of the earth's orbit from the sun (93 million miles) and at right angles to the sun's rays. Under such conditions 2 gram calories per square centimeter per minute are received. Since 1 gram calorie per square centimeter is a measure of heat energy called a *langley*, the solar constant is the equivalent of 2 langleys per minute.

In this example, insolation (short-wave radiant energy) is absorbed and transformed into heat (kinetic energy) and then again transformed into long-wave radiation which returns to space. These changes illustrate our need to understand the first law of thermodynamics when examining energy flows through earth systems. As stated in Chapter 3: Energy can be neither created nor destroyed. Thus, if the earth has its temperature in equilibrium, neither increasing nor decreasing, the outflow of energy from the system must equal that which enters it. However, the total energy input to a system may be transformed into more than one form, and we must use careful accounting procedures to find where all of it has gone. Also, in some cases, energy may be temporarily stored within a system as potential energy—for example, in waters ponded high above sea level or as chemical energy in the form of plant and animal materials.

### The spatial and temporal distribution of solar energy

The model of the sun and earth-sphere described above has slight similarity to the complexities of the real system. The seasonal march of temperatures distributes heat unequally across the surface of our planet. The spinning globe warms each place by day and cools it at night. Low-lying areas are warmer on the average than the tops of mountains; cloudy regions receive less energy than those with open skies. Many mechanisms exist at a global scale which distribute and redistribute energy across the surface of the earth. In its grandest and most simple form, this process can be described by the second law of thermodynamics which can be restated: Of itself, the flow of heat is always from warmer to cooler areas. As with the first law of thermodynamics, the second is sometimes difficult to see clearly and immediately in action. The words *of itself* are a key to understanding this. Energy may sometimes move from cooler to warmer areas—the continued chilling of a refrigerator is an example of this—but this reverse flow is possible only with inputs of additional energy from outside the system in question.

The daily amount of insolation at a point on the earth's surface depends upon the number of hours of daylight and the angle at which the sun's rays strike the earth. The great distance of the sun from the earth results for all practical purposes in all the sun's rays approaching the planet parallel to one another. Consider three shafts of sunlight parallel to each other and of equal diameter in space (Figure 4-1). Let one ray strike the earth at the equator, another at 30° latitude, and the third at 60°. Note in Figure 4-1 that all three cross sections in space are equal

**Figure 4-1 Intensity of sunlight as a function of latitude** Three shafts of sunlight strike the earth centered upon 0°, 30°, and 60° latitude, respectively. Their cross sections have the same area in space, but the area each illuminates on the global surface differs from the others. Assuming a steady and equal flow of radiant energy along each shaft, insolation per unit area will be most intense at the equator and least nearest the poles. This is because of the increase in size of the illuminated areas from low to high latitudes.

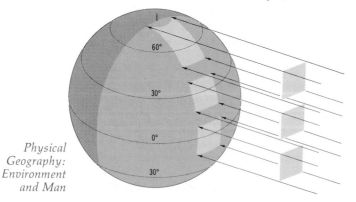

and that the energy per unit area is the same for each. Now note that where the vertical ray strikes the equator the energy arriving at the earth is distributed over an area the same as that of the shaft's cross section in space. At 30° the same amount of energy is distributed over a larger area, thus reducing its intensity. The third shaft hitting nearest the pole is thinly spread over a still larger surface. This effect is increased still further by the length of the path that solar energy must take to reach the surface once it has entered the earth's atmosphere. At the equator, the vertical ray penetrates the least amount of atmosphere; near the poles the path is much greater. Such distances are called the *optical air path* of sunlight.

Spatial differences in the seasonal distribution of energy will be accounted for in the next section, but for the moment let us consider the shorter days and lower angle of the sun during winter months followed by the long days and high sun of summer. This reaches its extreme with six months of darkness and six months of daylight at both the North and South Poles. Figure 4-2 shows the intensity of insolation for the entire year along a meridian of longitude from pole to pole. The elevation of the surface indicates the amount of energy received. The flat-floored basins at the extreme north and south show those months when no sunlight reaches the poles. The values shown are measured outside the atmosphere in order to avoid variations resulting from differences in optical air paths, as well as differences in atmospheric turbidity, density, etc. Global subsystems within the atmosphere and oceans constantly act according to the second law of thermodynamics to even out the differences in energy availability by moving energy from the equator toward the poles. Before considering such systems let us for a moment look more closely at the celestial mechanism through which such differences in the distribution of insolation occur.

## The earth, the sun, and the seasonal cycle

The earth-sun system produces cyclical and seasonal changes which serve as the pulse as well as the inspiration of mankind. For our

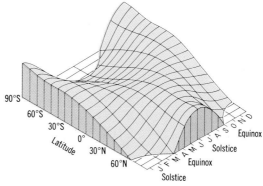

**Figure 4-2 The effect of both latitude and season of year on the intensity of insolation** At any given latitude and date, the relative amount of energy received is proportional to the height of the surface point above the flat base of the block. It should be noted that insolation along only one meridian is shown as a continuous surface for every day of the year. The flat areas in the northern corners and south-central part of the diagram represent the periods of Arctic and Antarctic night. This diagram is drawn looking from north to south. (Diagram by W. M. Davis, Reproduced by permission from A. N. Strahler, *Physical Geography.* Copyright 1951, © 1960, 1969 by John Wiley & Sons, Inc.)

purposes we can think of this system as consisting of only two elements.

We begin with a radiant body, the sun, which may be considered stationary in space, although in reference to other star systems it is not. A spherical earth, *rotating on its axis, revolves* around the sun. The earth's path of revolution is called its *orbit.* The plane determined by the sun and the orbit of the earth is called the *plane of the ecliptic.* The earth's axis of rotation is not vertical to the plane of the ecliptic but rather is inclined to it at an angle of 66½°. All these terms are illustrated in Figure 4-3. Though the axis of the earth has a slight wobble, the effects of which are unimportant here, its north end always points in the same direction with reference to the heavens outside the solar system: approximately toward Polaris, the North Star.

During a 12-month period the earth travels once around the sun, following an unvarying path which defines and is traced upon the plane of the ecliptic. The earth is closest to the sun

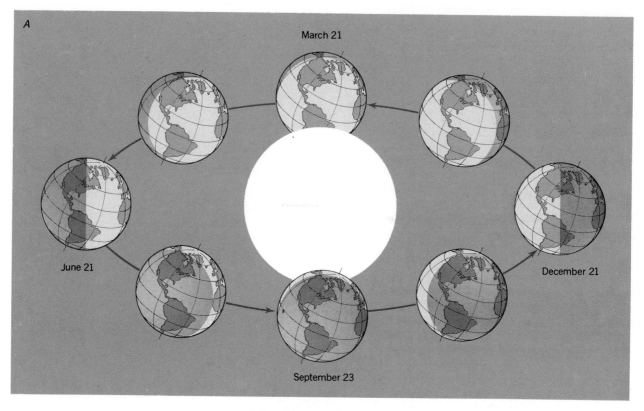

A

March 21

June 21

December 21

September 23

Earth Viewed from the Sun at the Time of

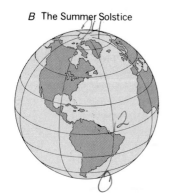

B The Summer Solstice

The Vernal and Autumnal Equinoxes

The Winter Solstice

(91½ million miles) during winter in the Northern Hemisphere. Six months later it is farthest from the sun (94 million miles). The elliptical shape of this orbit is insignificant in terms of the effect it has on the amount of insolation received when compared with differences caused by the tilt of the earth's axis and the position of the vertical ray of the sun north or south of the equator.

The *summer solstice* (June 21 or June 22, depending upon the particular year) is a good time and place to begin tracing the annual trip of the earth around the sun (Figure 4-3A). At that time, the north end of the axis points most

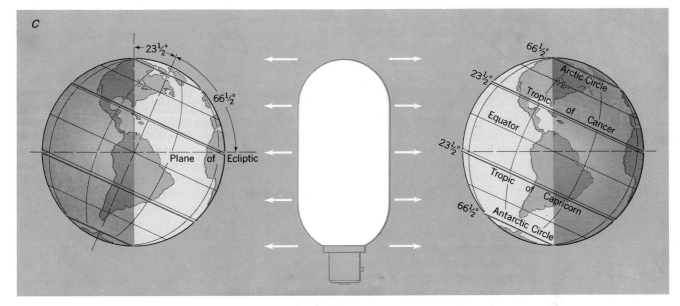

**Figure 4-3  The inclination of the earth's axis and the changing seasons**  The north pole of the earth always points the same direction toward the star Polaris (*A*). The inclination of the axis from the plane of the ecliptic accounts for the north-south march of the vertical rays of the sun and therefore, of the seasons (*B* and *C*).

nearly toward the sun. It is the longest day north of the equator, and the vertical ray of the sun has moved as far north as the Tropic of Cancer, 23½° north latitude.

As the earth continues on its journey, the north end of its axis continues to point toward far Polaris. Thus, to people in the Northern Hemisphere it seems that the sun is moving lower in the sky each noon relative to their position on earth. A careful examination of Figure 4-3 will show, however, that the movement of the vertical ray of the sun toward the equator results from the special geometry of earth-sun relationships in space and not from some change in the orientation of the earth's axis. Three months later (September 22 or 23) the axis of the earth points neither toward nor away from the sun but is at right angles to it. At this time, day and night are of equal length everywhere in the world, and a ray of sunlight is just tangent to both the North and South Poles. This is the *autumnal equinox.* Three months later, December 22 or December 23 marks the

*winter solstice* with the north end of the earth's axis pointing away from the sun and the vertical ray striking at the Tropic of Capricorn, 23½° south latitude. In ancient times this was a period of feasting and sacrifice to the sun god in order that he might be lured back from his wintry home. Three months later, the *vernal equinox,* on March 20 or March 21, again sees the earth's axis at right angles to the sun, and again the conditions that held during the autumnal equinox prevail. Finally, with the passage of another three months the earth has returned to a position at the summer solstice and its cycle is completed, only to begin once more.

To understand the effect that this annual cycle has upon the receipt of energy at the surface of the earth, consider the earth as a sphere lighted from a single source, the sun. Given these conditions one-half the sphere will be lighted, the other dark. The lighted half is called the *circle of illumination.* Now, at the winter solstice the circle of illumination on the

*Solar Energy and Man: Earth as a Heat Engine*

earth extends only to 66½° north latitude and the North Pole area is in darkness (Figure 4-3C). This is the shortest day of the year in the Northern Hemisphere. Six months later, during the summer solstice, the circle of illumination has shifted northward and the South Pole area at latitudes higher than 66½° is in darkness (Figure 4-3C). The line marking the extreme extent of darkness in the north is called the *Arctic Circle,* in the south, the *Antarctic Circle.* The absence of sunlight in polar regions for long periods of time obviously accounts in large part for their extreme cold.

Meanwhile, the vertical ray of the sun, that is, the ray following the shortest optical air path, has moved from 23½° south latitude on December 22 to 23½° north latitude on June 21. As we have already seen, insolation is greatest where the vertical ray strikes. Since the vertical ray moves with the seasons, the greatest amount of incoming radiant energy per unit area per minute shifts north and south during the year and crosses the equator twice, once at each of the equinoxes (Figure 4-3B). This cyclical progress accounts for the northward movement of higher temperatures in our summertime and the reverse effect during our winter.

Although insolation is at its maximum during the summer solstice, ground and air temperatures reach their maximum about one month later. This lag occurs because the ground continues to reradiate energy, heating the air above it in addition to the daily increment of direct solar radiation. In much the same way, daily temperatures are highest three to four hours after noon, as the earth returns some of the radiation it has absorbed. The cold months are similar. January is usually colder than December north of the Tropic of Cancer because the ground for another month continues to lose more heat through reradiation than it gains from increasing insolation. These relationships are, of course, symmetrical but opposite south of the equator, with summer there being our winter, with Christmas at the beach and July ski trips.

### The terrestrial heat engine

During the summer the polar regions receive 24 hours of sunlight and accumulate more insolation in one day's time than does the equatorial zone, where day and night are approximately 12 hours long every day of the year. However, during the rest of the year the long polar night lowers the average annual amount of insolation received in the Arctic and Antarctic. When all factors are taken into consideration—differences in the angles of the sun's rays, differences in the length of day and night, and so forth—we find that the area near the equator has an annual heat surplus while that near the poles has an overall heat deficit. Figure 4-4 shows incoming shortwave solar energy and outgoing long-wave radiation plotted according to latitude. Near the equator the area between the two lines represents heat surplus; near the poles the area shows heat loss. Although they have different shapes, the two areas are the same size, thus illustrating the overall spatial balance and equi-

**Figure 4-4  The latitudinal energy balance of the earth between incoming shortwave solar radiation and outgoing longwave radiation from the earth**  Note the predominance of incoming radiation at low latitudes and the excess of outgoing radiation near the poles. The deficit and surplus areas shown on the graph have the same areas but different shapes. If they did not equal each other, the planet earth would either overheat or become colder. The movement of energy from surplus to deficit areas drives the giant atmospheric and hydrologic heat engine described in the text. (R. E. Newell, "The Circulation of the Upper Atmosphere," *Scientific American,* vol. 210, no. 3, March 1964, p. 69, copyright © 1964 by Scientific American Inc.)

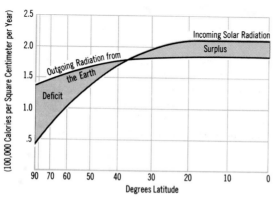

librium of the heat budget of the earth. One important aspect of this equilibrium condition is the latitudinal separation of the surplus and deficit areas. Energy in very large amounts must move from place to place in order to balance deficits and surpluses. The mechanism through which this occurs is a global subsystem of winds and ocean currents which can be thought of as a gigantic heat engine constantly in motion and constantly fueled by the sun.

## Budgeting the earth's energy cycle

In order to examine the components of this engine, we must understand how energy is transferred from the sun to the earth's surface, and from the earth's surface to the atmosphere and back to space. This flow is shown by the vertical cross section through the atmosphere in Figure 4-5. The horizontal patterns made by its moving parts can be seen in the maps of the winds and currents that follow.

Scale is an important consideration in the energy cycle of the earth. An ant may walk sedately beneath grass whipped by a mighty wind, and minnows may flash along the ocean bed as giant waves break above them. Only the occasional gust or eddy disturbs such creatures at home in the flux of the elements. Man's relationship to the flow of energy in which he lives is the same. Although tornadoes and hurricanes may endanger us, the energy in the fiercest winds is still only a fraction of that in which the earth is bathed each day. For example, the kinetic energy in a tornado with winds in excess of 300 miles per hour is something in the neighborhood of $7 \times 10^{10}$ calories, or only about $1/2$ of 1 percent of a summer day's insolation on a square mile of the earth.

Though the amount of radiation varies from place to place and from season to season, the solar energy intercepted by the earth equals approximately $3.67 \times 10^{21}$ calories per day. The general paths of the vast amount of energy arriving at the earth are shown in Figure 4-5. Of the insolation reaching the outer edge of the atmosphere (100 percent), 41 percent penetrates the protective mantle of air, clouds, and dust and is absorbed by the surface of the earth itself.

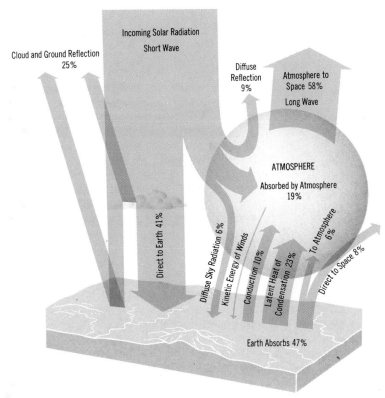

**Figure 4-5  Energy balance of the earth**
Incoming shortwave solar radiation is equal to 100 percent. Outgoing radiation also equals 100 percent. The paths taken by component parts of the total are simple (e.g., direct reflection back into space from clouds) or complex (e.g., diffuse sky radiation, absorbed by the earth, is thereafter released into the atmosphere as latent heat of condensation, and finally is reradiated into space). Careful examination will show that the subtotals as well as totals on this diagram are in balance. Such symmetry is, in part, the result of the simplification by the authors of much more complex and incompletely understood processes.

Another 25 percent is reflected directly back to space from the earth and clouds. The remaining 34 percent is either absorbed or diffused by the atmosphere. Of that, 15 percent is diffused, from which 9 percent is reflected back to space, while 6 percent reaches the earth. The atmosphere temporarily retains the other 19 percent.

The path followed by the 47 percent which is absorbed by the earth before it returns to outer space is a complex one. For the moment, we will simplify things by treating the earth as a black box in the energy system. Since 47 percent of the total insolation is absorbed by black box earth, an equivalent amount must return to the atmosphere and eventually to space in order to maintain equilibrium. *Conduction* accounts for 10 percent of this returning energy. This is simply the direct heating of cooler air as it comes in contact with the warmer surfaces of the earth. The warm earth also emits long-wave radiation which accounts for another 14 percent, of which 8 percent goes directly to space while 6 percent is temporarily absorbed by the atmosphere. The remaining 23 percent returns to the atmosphere in the form of *latent heat.* This requires further explanation.

### Fusion, vaporization, condensation, and sublimation

When an object either absorbs radiation or is otherwise heated, energy is imparted to the molecules making up that object. The speed of the molecules is increased, and this may be thought of as a form of kinetic energy. A molecule of water in the form of ice will have the least kinetic energy; water in fluid form will have more energy. The extra increment of energy needed to change ice to liquid is called the *latent heat of fusion;* the amount needed to change liquid water to water vapor is called the *latent heat of vaporization.* Conversely, when water vapor condenses to a liquid, its molecules move more slowly. The increment of energy originally used to change the liquid to vapor must go somewhere when vapor is condensed. When water vapor in the atmosphere condenses and becomes rain, the energy released heats the atmosphere. This amount is called the *latent heat of condensation.* In the same manner, there is an increment of energy released by each molecule when liquid water freezes into ice. This accounts in large part for warmer temperatures observed during snow storms. Citrus growers, when threatened by frosts of brief duration, sometimes take advantage of the release of latent heat and may actually spray their orchards with water. As the water freezes on contact with the cold air, its latent heat is released, thus warming the surroundings. If the frost is not too severe, a little ice coating the buds and twigs may actually serve as an insulating overcoat.

The frost that forms fantastic patterns on winter windowpanes results when water vapor turns directly to ice without passing through a liquid state. Conversely, frozen water can vaporize without first liquefying if the air is very dry. This process, as it proceeds in either direction, is known as *sublimation.* The energy taken up by sublimating molecules of water as they move faster and faster is called the *latent heat of sublimation.* The amounts of energy involved in vaporization, condensation, fusion, and sublimation can significantly alter surrounding air temperatures. Figure 4-6 shows these transformations and the caloric energy absorbed or given up as a gram of water converts from one state to another. Since the details of these processes are complicated and depend upon surrounding air temperatures, humidities, and pressures, the caloric values can vary. Those given in Figure 4-6 are for normal, sea-level conditions with ice at 0° C (32° F) and water vapor at 100° C (212° F).

### The atmospheric reservoir

The atmosphere also can be conveniently thought of as a black box in this system. Conduction, latent heat, and long-wave radiation return 39 percent of the energy from the earth to the atmosphere where it is added to 19 percent directly absorbed from the sun's rays. The heated atmosphere and the heated earth exchange energy back and forth many times in many ways. The net result of this exchange has already been mentioned as the 14 percent which is emitted from earth as long-wave radiation. Another small amount—less than 1 percent—also returns temporarily to the earth's surface as the wind energy which helps propel the currents of the world's large water bodies. Finally, the atmosphere reradiates all the energy it receives from all sources. This amounts to 58 percent of the original insolation. All these paths can be identified and cross-checked by carefully adding up the percentages shown at various points in Figure 4-5.

One word of caution is necessary. No one, as yet, really knows the exact percentages of energy at any given point in the system. Also, these amounts vary from season to season and from day to day depending upon cloud cover, the angle of the sun's rays, the amount of dust in the atmosphere, the reflectivity of the particular portion of the earth's surface under consideration, and many other things. Thus, these comments should be taken as indications of the amounts involved rather than as absolute values.

## Composition, Temperature, and Pressure of the Atmosphere

Fifty-eight percent of all solar energy intercepted by the earth finds its way into the heat reservoir of our atmosphere for varying lengths of stay. When radiant energy is transformed into heat within the atmosphere, it is unevenly distributed because of differences in latitude, land and water distribution, cloud cover, and a variety of other factors. Air masses of different temperatures are the result, and the presence and movement of such air masses in both vertical and horizontal directions as well as the interaction between unlike portions of the atmosphere are responsible in large part for the short-term effects we call *weather* and the long-term conditions we know as *climate.* We have already mentioned how the movement of air helps to balance the heat surpluses near the equator against the deficit areas near the poles. The composition and characteristics of the atmosphere are integral components of the earth heat engine.

The atmosphere contains pure dry air, water vapor, and atmospheric solids including dust particles. By *pure dry air* we mean a mixture of gases which has changed significantly since the beginnings of geologic time, but which has remained more or less constant in composition within man's span on earth. This mixture consists by volume of 78 percent nitrogen, 21 percent oxygen, 0.9 percent argon, 0.03 percent carbon dioxide, and traces of a variety of other gases. Dust from windy deserts or plowed fields may color a sunset or blot out the sun, just as smoke particles from forest fires or factories may also serve to filter the sunlight from above. All kinds of matter are classified as atmospheric

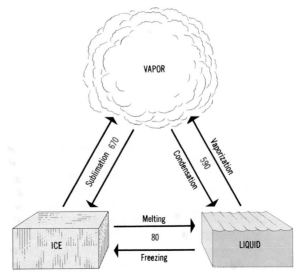

**Figure 4-6  Energies associated with conversion of water from one state to another**  Measures are approximate calories per gram.

solids, including airborne salt crystals from ocean spray and radioactive materials from hydrogen bomb tests. Water vapor is most conspicuous in its associated form of liquid cloud droplets, but high humidities on any warm day also indicate the presence of water molecules in the atmosphere.

### Air pollutants

Any discussion of the atmosphere from a human point of view must recognize the presence of dangerous gases and particles of matter in the air we breathe. There has always been air pollution. Long before the ancestors of Homo sapiens scuffled up clouds of dust while dancing or lighted their first cooking fires, volcanos belched clouds of gas and dust, forest fires started by lightning billowed smoke, and dust storms hid the land. But the natural systems involved provided the means, such as cleansing rain, of getting rid of pollutants. Most important, the scales at which those systems operated incorporated long spans of time, and the rate at which pollutants were produced was usually slow enough or irregular enough to let

*Solar Energy and Man: Earth as a Heat Engine*

71

the purifying processes catch up. Man has changed all that. Our technological systems and processes move at accelerated rates; we produce all manner of organic and inorganic compounds which are nearly indestructible by natural processes; and our organization of earth space concentrates materials in quantities which overload the carrying capacities of local systems.

The sources of air pollution can be considered in a spatial context. Point sources include the continuous plumes from factory chimneys and incinerators. At regional scales, burning dumps, factory complexes, cities, forest fires, plowed fields, and numerous other area sources of pollution exist. Line sources are less frequently encountered. Major highways would be one such source, the swathes from spray planes another. Point sources are easiest to identify and to control. Laws requiring smoke traps on chimneys are a good example of this. The increase in magnitude that comes with cleaning up areal sources of pollution presents enormous problems. So, too, the variety of pollution sources found in a typical city requires a variety of responses which makes the clean-up task more difficult.

The size of the areas that air pollution can involve is illustrated by Figure 4–7, which shows the distribution of air pollution in California during the period 1961–1963. In general, air pollution can be rated in increasing severity by whether it reduces visibility of the landscape, damages vegetation, or irritates the eyes. The sources of the air pollution shown on this map correspond with those areas marked by eye irritation—that is, the major metropolitan centers. Another way of viewing this map is to consider human population distribution within the state. About 70 percent of all Californians live in eye-irritation areas; 80 percent of the population lives within the plant-damage zone; and 97 percent seldom gets a chance on a clear day to see forever. We will return to these matters in the chapters that follow. For the moment, let us briefly consider the major air pollutants.

The exact number of gases created by human technology is unknown, but there are at least six significantly dangerous inorganic ones as well as numerous organic compounds that enter the atmosphere. Carbon dioxide results from burning organic fuels. This gas represents one of the question marks in the modern earth ecosystem. The recent rapid increase in the amount of carbon dioxide in the atmosphere has apparently not harmed any life forms. On the

Cloud formation of a low-pressure system near the Straits of Gibraltar as seen from the National Aeronautics and Space Administration's Gemini X spacecraft. The curving limb of the earth that marks the horizon is veiled by a thin layer of white, the thickness of which represents more than 90 percent of the atmosphere upon which we depend. (NASA photo)

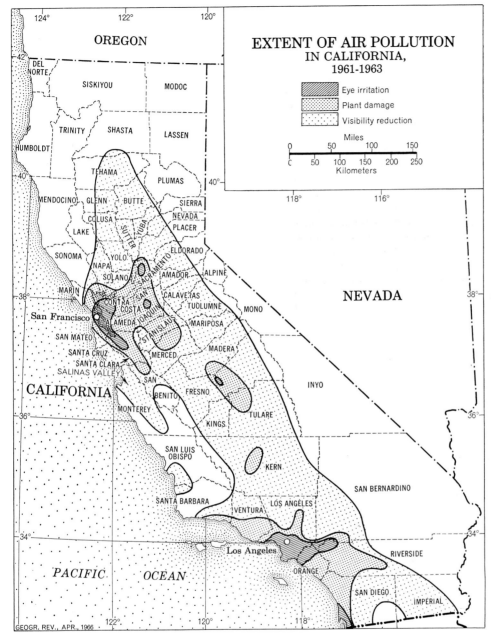

**Figure 4-7 Extent of air pollution in California, 1961–1963** (Philip A. Leighton, "Geographical Aspects of Air Pollution," *Geographical Review,* vol. 56, April 1966, p. 166, copyrighted by the American Geographical Society of New York)

other hand, there is a serious and unresolved question that such increases might help change the heat balance of the entire earth, perhaps even allowing the polar ice caps to melt by preventing the return of long-wave radiation to outer space. This phenomenon is sometimes referred to as the *greenhouse effect*. Short-wave incoming solar radiation can penetrate the at-

*Solar Energy and Man: Earth as a Heat Engine*

**73**

mosphere with comparative ease. That portion which strikes the earth's surface warms the ground, which, in turn, emits radiation of longer wavelengths. (A general rule is that the hotter the emiting surface the shorter the wavelength, and vice versa.) Part of this reradiation is absorbed by carbon dioxide, water vapor, and ozone. Thus, heat can be stored in the atmosphere, much as the glass panes of a greenhouse permit radiant energy to enter but prevent the warmer inner air from mixing with colder air outside.

As already mentioned, *ozone* is another gas which could help change the global heat balance. While deadly in heavy concentrations, relatively little ozone is produced by man. High-flying aircraft are thought to increase the ozone content of the atmosphere, but much more needs to be learned about these matters.

*Carbon monoxide* and a group of *oxides of nitrogen* result primarily from automotive combustion. The former can be toxic to humans at concentrations of more than 100 parts per million if exposure continues for any length of time. Nitric oxide and nitrogen dioxide are the major constituents of autombile smog. Reactive hydrocarbons from automobile exhausts and petroleum refineries undergo photochemical reactions in the presence of sunlight which change nitric oxide to nitrogen dioxide. The latter gas is responsible for the whiskey-colored air shrouding many cities and can cause damage to plant and animal life.

*Hydrogen sulfide* is poisonous, but even at nonlethal concentrations its odor of rotten eggs can make life miserable. Stagnant waters full of sewage and industrial wastes are a major source. *Sulfur dioxide* is a byproduct from the burning of coal and oil. Lethal smogs like the one which killed 4,000 Londoners in 1952 are a mixture of sulfur dioxide, sulfuric acid (which forms when water is added in the form of mist, fog, or rain), and other chemicals. Fortunately, London has largely solved its smog problem by banning the burning of soft coal and by enforcing antipollution laws.

*Organic gases* include ethylene, a vegetation-damaging simple hydrocarbon coming from industrial sources and automobile exhausts. In-

cinerators and stockyards furnish aldehydes such as formaldehyde, which makes the eyes smart and the throat burn on smoggy days. We mention these few gases, but it would take volumes to exhaust the subject.

In addition to gases, man also adds solid or particulate matter to the atmosphere. Cities are famous for their soot, which is formed from microscopic particles of carbon. Many city folk unwillingly sport soot-speckled collars as badges of their urban life-style. But rural people also contribute to atmospheric pollution with plowed land that can at times produce disasters like the famous "Dust Bowl" of the 1930s. In parts of the world where fields are burned to clear them, the sky can be darkened for days before the planting season. Modern insecticides, defoliants, herbicides, and radioactive fallout add to the deadly list. And then, of course, there is everything else that pulverizes and goes to waste. Millions of tons of rubber worn off countless auto tires; paper, plastic, even glass and rust can end up floating in the air as tiny particles. Table 4-1 shows the amount of air pollutants by kind and source produced in the United States in 1966. Tomorrow morning, jump out of bed, throw open your window and take a deep breath—if you dare.

## The vertical distribution of the atmosphere

The vertical distribution of the atmosphere is of critical importance. Here at the bottom of our airy envelope the atmosphere may seem to extend above us forever, but in reality we exist within a thin gauze of gas clinging to the much larger globe. This becomes evident in the many photographs from satellites which show the thin layer of atmosphere in cross section along the limb of the earth's horizon. In fact, almost 97 percent of the mass of the entire atmosphere is within 18 miles (29 kilometers) of the earth's surface. If you have ever climbed a mountain above 18,000 feet (5.5 kilometers) or flown that high in an airplane, you have had one-half of the atmosphere below you. The other half extends outward approximately 21,000 miles (33,800+ kilometers) to the *exosphere*, the absolute limit of earth's atmosphere and the point where

**Table 4-1   Amounts of Air Pollutants, by Kind and Source, Produced in the United States, 1966**

| Source | Carbon Monoxide | Particulates | Hydro-carbons | Nitrogen Oxides | Sulfur Oxides* | Total |
|---|---|---|---|---|---|---|
| | *(Millions of Tons Per Year)* | | | | | |
| Transportation | 64.5 | 1.2 | 17.6 | 7.6 | 0.4 | 91.3 |
| Fuel combustion in stationary sources | 1.9 | 9.2 | .7 | 6.7 | 22.9 | 41.4 |
| Industrial processes | 10.7 | 7.6 | 3.5 | .2 | 7.2 | 29.2 |
| Solid waste disposal | 7.6 | 1.0 | 1.5 | .5 | .1 | 10.7 |
| Miscellaneous | 9.7 | 2.9 | 6.0 | .5 | .6 | 19.7 |
| Total | 94.4 | 21.9 | 29.3 | 15.5 | 31.2 | 192.3 |
| Forest fires | 7.2 | 6.7 | 2.2 | 1.2 | (†) | 17.3 |
| Total | 101.6 | 28.6 | 31.5 | 16.7 | 31.2 | 209.6 |

*For the year 1967.
†Negligible.
Source: U.S. Department of Health, Education, and Welfare, 1970, p. 9.

a gas molecule can, with its own velocity, escape earth's gravitational field never to return.

The atmosphere presses upon the earth beneath. The weight of a column of air from sea level extending to the exosphere is on the average 14.7 pounds per square inch. This will support a 29.92-inch (76-centimeter) column of mercury in a barometer. At an elevation of 11 miles (17.7 kilometers) above sea level the weight of the remaining column of air is one-tenth as much; at 20 miles (32.2 kilometers), one-hundredth as much; at 70 miles (112.7 kilometers), one one-hundred-thousandth as much. We are speaking here of average conditions.

Sensible heat recorded as air temperature changes regularly with elevation. In the lower atmosphere there is a steady decrease in temperature equal to 3.6°F with every 1,000 feet of increased elevation (6.5°C per 1,000 meters). Changes of this kind are called the *lapse rate,* and the above rate of change is called the *normal lapse rate.* Atmospheric temperatures under normal lapse-rate conditions decrease regularly to an elevation of 6 or 7 miles (10 kilometers), where a low of −60 to −70°F is reached. This elevation marks the upper limit of the *troposphere*—that is, the portion of the atmosphere nearest the earth and which contains most of the water vapor, most particulate matter, and where most weather disturbances take place. However, beyond this elevation temperatures may stabilize, increase, or decrease depending upon the zone within the atmosphere. The portion above the troposphere to a height of about 30 miles (48 kilometers) is known as the *stratosphere.* Beyond that the *mesosphere* and *thermosphere* extend outward into space. The division between the troposphere and the stratosphere is called the *tropopause.* Variation in the elevation of the tropopause accounts for the varying heights to which the normal lapse rate extends (Figure 4-8). Almost but not quite all of the earth's weather occurs within the troposphere, and those of you who have flown in a commercial flight above 10 kilometers (33,000 feet) may have noticed how smooth your trip

*Solar Energy and Man: Earth as a Heat Engine*

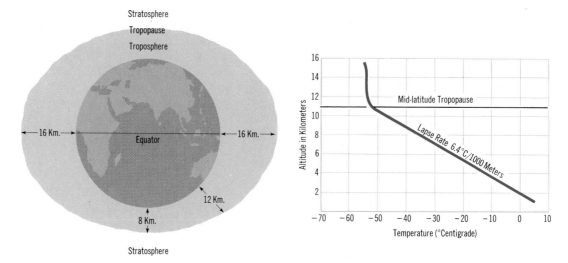

**Figure 4-8  The tropopause and the lapse rate**  The tropopause is the elevation at which atmospheric temperature ceases to fall in a regular fashion with increasing altitude. It is highest over the equator and lowest over the poles. The normal lapse rate is the rate of decrease in free-air temperature when passing from a lower to a higher elevation.

was, with all the clouds and air turbulence in a layer below you. You were undoubtedly winging along just above the tropopause, taking advantage of the relative uneventful conditions of the stratosphere.

The gaseous envelope, in which earth is contained like a space age parcel en route to some unknown destination, is unevenly heated horizontally as well as vertically. Air temperature varies from place to place, particularly with changes in latitude and elevation but also because of cloud cover and whether land or sea lies below. Water is a much better insulator than land, and water bodies heat more slowly under direct solar radiation than does land. Water also cools more slowly and gives up less energy than does an equal area of land surface. Water's slower rate of temperature change results from

(1) the continual mixing of warm and cool waters by waves and currents, (2) evaporation and the removal of large amounts of energy as the latent heat of vaporization, (3) the fact that sunlight can penetrate water to considerable depths while earth and stone are opaque and absorb most incoming energy within 1 or 2 inches of the surface, and (4) the greater specific heat of water, which requires about five times as much energy as an equivalent volume of land to increase its temperature the same amount.

The result of all this is that land is hotter by day and colder by night as well as considerably warmer in the summer and colder in winter than water. Variations in air temperature can be mapped at global as well as neighborhood scales. Figure 4-9 shows the global variation in

**Figure 4-9  Global variation in temperature, (A) January and (B) July**  Variations in temperature can be mapped at global scales. The isotherms on these maps indicate the distribution of energy at the earth's surface in the coldest and hottest months in each hemisphere. Note how the isotherms bend poleward over the continents in the summer hemisphere, while low temperatures reach farther toward the equator on the earth's winter half. Sharp bends in the isotherms along certain coasts show the presence of cold and warm offshore currents.

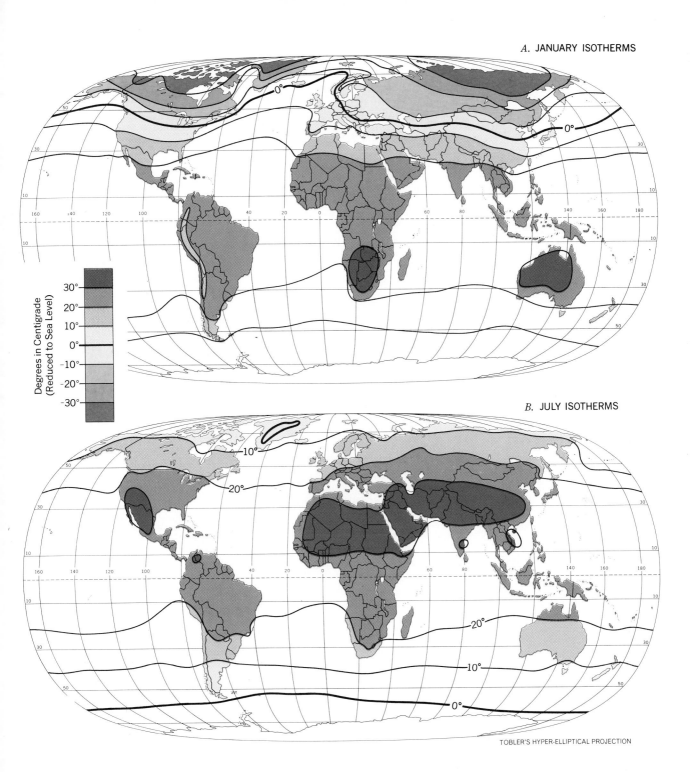

A. JANUARY ISOTHERMS

0°

0°

Degrees in Centigrade
(Reduced to Sea Level)

30°
20°
10°
0°
-10°
-20°
-30°

B. JULY ISOTHERMS

10°

20°

20°

10°

0°

TOBLER'S HYPER-ELLIPTICAL PROJECTION

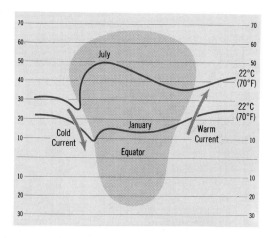

**Figure 4-10 Latitudinal shifts in an isotherm from July to January as the result of oceanic and continental influences (hypothetical Northern Hemisphere continent)** Cool offshore currents move lower temperatures southward along the west coast of the continent. Low January temperatures reach farther equatorward in the center of the continent. Conversely, warm July temperatures extend farther north in the center of the continent than on its edges. These variations result from the relatively poor insulating quality of the land, which warms and cools more quickly than does water.

temperature for July and January. The lines on this map connect points having equal temperatures and are called *isotherms.* They give an indication of the distribution of energy at the earth's surface for those times. Note how the isotherms tend to bend poleward over the continents in the summer hemisphere while low temperatures reach farther toward the equator on the earth's winter half. This differential shifting of temperatures over land and water with the seasons is shown diagrammatically in Figure 4-10.

Differential heating of the atmosphere on an earth scale results in large, distinct masses of cold and warm air. Cold air is dense and heavy compared with warm air and tends to sink earthward, while thinner, warm air rises. These differences are reflected as variations in air pressure measured by a barometer. Barometric pressure varies both above and below the 29.92 inches (67 centimeters) of mercury, which you will recall equals the weight of a column of air

under standard conditions. Air pressure, like heat, tends to adjust inequalities within the system by movements from high-pressure to low-pressure areas. This endless effort to stabilize barometric variations results in winds which, in general, move down the *barometric slope* or *pressure gradient.*

## The world of wind

Let us now look at both a vertical and horizontal view of the atmosphere and see how, by knowing a very few facts about the distribution of energy and the movement of air masses, it is possible to sketch a model of the global wind system. This knowledge is useful, for it is this wind system which in large part drives the ocean's currents, accounts for some important long-range climatic characteristics, and helps bring daily changes in the weather.

The direct rays of the sun fall upon the equator only twice each year. At the time of the equinoxes (March 15 and September 15) the heating of the earth is essentially symmetrically distributed between the Northern and Southern Hemispheres. The rest of the time more insolation strikes the earth on one side of the equator or the other. This makes the conditions at the equinox a good starting point for discussing the world wind system.

Despite annual variations in the symmetry of insolation, we may think of the total amount of energy earth receives as being greatest at the equator and least at the poles. In this simplified model we would expect to find the atmosphere at the equator heated the most and that at the poles least of all. This results in warm, ascending air girdling the earth with an equatorial belt of low pressure (Figure 4-11). But the air cannot rise indefinitely, and as it cools at greater elevations, it spreads horizontally to the north and south. Meanwhile, the air above the poles, being coldest, is most dense and settles earthward. The pressure gradient thus created should extend from a high at the poles to a low at the equator, and the atmosphere, acting to readjust these inequalities, should move accordingly. Dense, cold air should slip equatorward near the earth's surface, while rising, warmer air should move north and south as it gradually

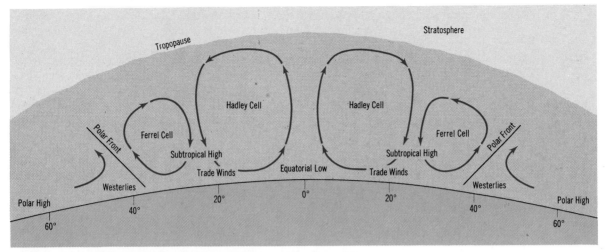

**Figure 4-11  Hypothetical airflow in the troposphere at the time of the equinox**  The equinox has been chosen in order to give hypothetical symmetry on both sides of the equator. Vertical components of the system are high and low-pressure belts. Horizontal air movement across the earth's surface results in the world wind system discussed and named in the text. The actual system is less well defined and less symmetrical than this diagram shows. (*Vertically exaggerated and not shown to scale.*)

cools. The result should be twin cells of circulating air, one north and the other south of the equator. This system was first suggested in 1735 by the English meteorologist George Hadley. However, the existence of such a gigantic cell in each hemisphere did not fit the observed global distribution of climatic features.

An American meteorologist, William Ferrel, modified Hadley's scheme in the nineteenth century to include three such cells on each side of the equator. As we shall see in the pages ahead, his suggestion of an equatorial band of warm, rising air, a second belt of cool, dense, descending air at 30° latitude north and south, and another band of relatively warm air ascending at 60° latitude more closely matches the distribution of world climates (Figure 6-1) and the pattern of world winds (Figure 4-12). In memory of these men the two equatorial cells are called *Hadley cells* and the two mid-latitude cells are called *Ferrel cells*. The spatial asymmetry of insolation north and south of the equator during most of the year results in a rather lopsided system of pressure belts. Only one Hadley cell develops fully for most of the earth's

**Figure 4-12  General world wind and pressure pattern**  This diagram shows a highly abstract and hypothetical pattern of world pressure belts and the winds which equalize their different barometric conditions. The polar easterlies are less well developed than other parts of the system, but the basic pattern prevails in a much more complicated form. Compare this with the satellite photo of the Southern Hemisphere for striking similarities.

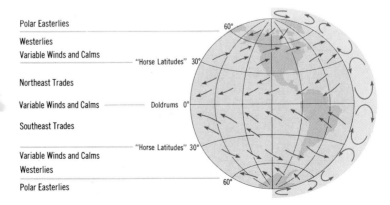

Polar Easterlies — 60°

Westerlies

Variable Winds and Calms — "Horse Latitudes" 30°

Northeast Trades

Variable Winds and Calms — Doldrums 0°

Southeast Trades

"Horse Latitudes" 30°

Variable Winds and Calms

Westerlies

Polar Easterlies — 60°

12-month trip around the sun, that in the winter hemisphere; the one in the summer hemisphere dwindles until it almost disappears. Nevertheless, the energy within a fully developed Hadley cell is enormous. The mass of air circulated annually by such a cell would equal about $378 \times 10^{15}$ metric tons, approximately one-seventeenth the mass of the earth! And yet the energy required to do this represents, in turn, only a fraction of the total insolation in which the earth is continually bathed.

Given the alternating belts of high and low pressure we have just described, we could also predict that winds would blow along the surface of the earth, adjusting differences between them. On a stationary globe such winds would move directly north and south. But the earth rotates, and its rotation produces an effect upon these winds, as it does on all moving objects. This effect, more specifically known as the *coriolis effect,* is stated in *Ferrel's law: A gas, fluid, or object moving in the Northern Hemisphere tends to be deflected to the right of its path of motion, no matter what compass direction that path may have. Moving objects are similarly deflected to the left in the Southern Hemisphere.* This effect is greater near the poles than near the equator. Various results occur from the coriolis effect. For example, in the Northern Hemisphere, at high latitudes, driftwood accumulates on the right bank of rivers more than on the left, and the right-hand rails on railways carrying one-way traffic are reported to wear out more quickly than those on the left side. This same deflection affects the winds blowing between belts of high and low pressure. The resulting pattern looks something like that in Figure 4-12.

At this point we must take note of man's desire to order the world in which he lives by giving names to everything. Figure 4-12 summarizes and labels the seven pressure belts and six major winds making up the global components of heat engine earth. Note the general symmetry of the pattern on either side of the equator; also note that winds are often named after the direction *from which they blow.*

At the equator the major vector of air movement is vertically ascending. Air near the surface is often calm or disturbed by random breezes blowing every which way. Here are the *doldrums,* one of the two areas of calms at sea dreaded by all sailors before the age of steam. Many a sailing ship has been becalmed for days, weeks, perhaps even months, near the equator; and the term *doldrums,* now in general use, means any period in our lives when we just can't get started or go anywhere. The convergence of twin wind systems to the north and south add to the general condition within the doldrums. These are the *northeast* and *southeast* trade winds, which represent the lower horizontal components of the tropical Hadley cells. Within the area of the trade winds the motion of the atmosphere is relatively predictable, with a steady flow of air converging at the equator. This is called the *zone of intertropical convergence,* which is the technical name given to the doldrums. The trade winds themselves are so named because of the role they played in the early exploration and settlement of the New World. Shipmasters soon learned to sail south-easterly from Europe to America in order to take advantage of these steady winds. However, poleward of the trade winds is another belt of intermittent calms and breezes. In this case, descending cool air at 30° north and south latitude makes the ocean beneath a trap for sailing ships. Here is the *subtropical high-pressure belt,* sometimes called the *horse latitudes,* about which strange stories are told. It is said that when the Spanish first ventured to cross the Atlantic, their galleons often were becalmed in those latitudes. Days passed, and the ships' supplies of fresh water soon became more precious than gold, too precious indeed to give to the horses on board. And so the poor beasts were jettisoned, left to drown, their bloated bodies terrible signposts of thirst warning other ships to beware a similar fate.

Undoubtedly, a more correct explanation of this term comes from the English verb "to horse," meaning to move without ability to reach a given position or direction. This is exactly what happens to a boat under sail power in that zone. As British naval terminology has it, it is horsed.

Still farther poleward the flow of air from

northwest to southeast in the middle latitudes provides a regular wind system known as the "brave westerlies" in the Northern Hemisphere and the "roaring forties" in the Southern. These gales continually move masses of wet marine and dry continental air from west to east around the world. In the days of sail they provided a quick, rough journey back to Europe from the New World, particularly in the summer. But in the winter months, and almost all year around in the Southern Hemisphere, where fewer continental land masses block their passage, the continuous rush of air pushes giant ocean waves before it. Thus the ships involved in the "triangular trade," which linked colonial America with the Eastern Hemisphere, would move from Europe to the Caribbean and Africa on the northeast trades and then northwest to the Carolina coast on a starboard tack. Once there, they would lay over in the port of Charleston until spring, when the westerlies would blow them back to Europe. The winter trip was not worth the risk and rough weather. Charleston thus became an important port in colonial America, but with the age of steam, ships no longer wintered over on the Carolina coast and the town became the quiet place it is today.

Poleward of the westerlies, the cold of higher latitudes acts to offset the regularity of winds and pressure belts described in our simplified model. Nevertheless, a pressure belt of *relatively* warmer air ascends at 60° north and south as the *subpolar low-pressure belt,* while at the poles frigid, descending air creates two powerful *polar high-pressure areas.* The winds that circle the poles equatorward of the polar high-pressure systems are called the *polar easterlies* but are less well understood, in part because of the lack of comprehensive and continuous weather observations in those latitudes.

## High- and low-pressure cells

While the above summarizes the world wind and pressure systems, there remains another basic, if secondary, characteristic of the world pressure system that we should note. The Northern Hemisphere, with its continental mas-

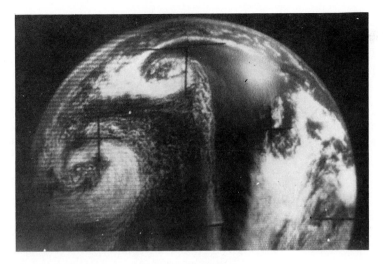

Two large, low-pressure cells over the North Atlantic. The counterclockwise movement of the cells is apparent from the cloud pattern. Storms and weather fronts are associated with the passage of this type of cell. (Tiros IX, 100th orbit, Jan. 30, 1965; NASA photo 65-H-136)

ses alternating with the Atlantic and Pacific Oceans, has a well-defined set of high- and low-pressure cells associated with temperature differences on the land and water. Particularly in its winter, the Northern Hemisphere develops two giant high-pressure cells, or *anticyclones,* one over central Asia which is called the *Siberian High,* the other over North America called the *Canadian High.* The pressure of the descending, cold air at the center of these anticyclones forces winds outward toward their edges. The coriolis effect deflects these winds to the right, so that the entire cell rotates in a clockwise direction. At the same time, since the oceans bordering the continents are relatively much warmer in winter than the land, warm, moist, and ascending air creates an *Aleutian Low* over the North Pacific and an *Icelandic Low* over the North Atlantic. These are termed *cyclones* and have a general counter-clockwise movement of winds toward their centers (photo above).

In the Northern Hemisphere in the summer months the location of high- and low-pressure cells is reversed, with low pressure developing

*Solar Energy and Man: Earth as a Heat Engine*

Apollo 17 (72H-157B), December 1972. The Southern Hemisphere appears with Antarctica a prominent white patch at the bottom. Africa and the Arabian peninsula are clearly seen. This photograph illustrates the pressure belts and wind systems discussed in the text. The sequence, beginning in Antarctica, is (a) the polar high with relatively clear, dry air, (b) the polar easterlies along the circumpolar low, shown as a band of broken irregular clouds, (c) a series of well-developed cells marking the path of westerly winds, (d) the open skies of southwest Africa, indicating the southern subtropical low-pressure zone, (e) linear cloud elements over the Indian Ocean indicating the southeast trade winds, (f) the equatorial low-pressure belt marked by scattered clouds formed by convectional air currents, (g) a less clearly seen area of northeast trade winds, and (h) the subtropical high-pressure belt in the Northern Hemisphere. (NASA)

over the hotter lands. However, cells of either kind are better defined in the winter months than in the summer. This is in part because the entire system tends to shift poleward during the high-sun period. The westerlies referred to above give impetus to the high- and low-pressure cells and help to drive them continually eastward around the world. Thus, in winter, the middle latitudes experience a continual parade of alternating high- and low-pressure systems which in large part account for the changeable weather with which most of us are familiar.

In the Southern Hemisphere the lack of large land masses leads to the development of more beltlike pressure systems, although depending on the season, smaller highs and lows do form over the continents. However, the northern winds and pressure cells are better known to mankind because most of the world's population lives north of the equator in the middle latitudes.

## Eddies in the World Ocean

Anyone who has experienced the battering drive of wave and wind as a hurricane comes ashore or who has read Joseph Conrad's or Jack London's descriptions of storms at sea knows something of the energy exchange between the atmosphere and the waters of the globe. The winds described in the previous section impart a fraction of their energy to the water surfaces over which they pass. Little is known about the actual frictional mechanism by means of which the energy transfer takes place, but the overall effect is important in directing the currents of the world ocean.

The direct effect of the wind on the water is simple to understand although difficult to test outside the laboratory. The surface layer of the ocean can be thought of as a stack of nearly frictionless sheets, one below the other. The top sheets slides over the others in almost the same direction that the wind pushes. However, the water is deflected slightly by the coriolis effect, just as it deflects the winds themselves. In the Northern Hemisphere, each succeeding downward sheet of water is deflected a little more to the right of its original path of motion. How-

**Figure 4-13 Ekman spiral in the Northern Hemisphere** Wind moving across the surface of water imparts energy to the water and causes it to move. The coriolis effect deflects the movement of the water (to the right in this example). Since water is essentially frictionless, the water molecules at each successive depth are deflected away from the path of movement of those directly above. This results in a spiral movement imparted to the water to a shallow depth. The average direction of this moving surface layer is at an angle to the original wind direction.

ever, the wind's energy cannot directly reach the deeper waters beneath the surface layer, and the spiral of motion is relatively shallow (Figure 4-13). This directional spiral in the surface waters is known as an *Ekman spiral,* named after V. W. Ekman, a Swedish oceanographer, who suggested its presence. The entire surface layer, with its main movement at an angle to the force of the wind, is called the *Ekman layer.* Movements within this layer in combination with other factors such as continental barriers to the west and east tend to direct the surface currents of the oceans toward the centers of the major basins in the midlatitudes (Figure 4-14). This results in a piling

*Solar Energy and Man: Earth as a Heat Engine*

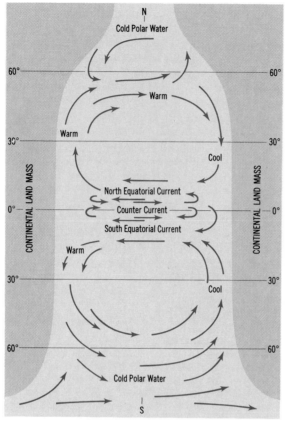

**Figure 4-14  Schematic ocean currents**
Two gyres are shown moving warm and cool
water into latitudes where otherwise it would not
be found. The equatorial currents are pushed
along largely by the trade winds; their east-west
direction results from the coriolis effect. The
currents, in turn, are deflected by the shores of the
continents and moved farther to the right in the
Northern Hemisphere and to the left in the
Southern Hemisphere by the continuing coriolis
effect. The westerlies drive the currents eastward
until they are again deflected and the gyre closes
upon itself. Cool water thus moves equatorward
along the western shores of the continents, while
warm water moves poleward on the eastern shores.
The west-to-east movement of the water and winds
is enhanced in the Southern Hemisphere by the
absence of land barriers at high latitudes.

waters beneath. These deeper waters must slide
out in response to the pressure from above. The
entire system operates much like an atmospheric high-pressure cell or anticyclone. The outward moving waters are deflected to the right in
the Northern Hemisphere and to the left in the
Southern.

The confining shape of the ocean basin further deflects the moving waters. In the North
Atlantic, for example, the westward flow encounters the American continent and is turned
first northward and then to the east. The east-
flowing current comes up against the shores of
Europe and is forced largely southward and
subsequently back to the west, thus completing
the circuit (Figure 4-15). These giant eddies in
the world ocean are called *gyres,* the most
famous of which is the one just mentioned with
the Sargasso Sea at its center.

The energy carried in these gyres—or more
accurately, variations in the energy contained in
them from one part to another—warms or chills
the shores of the world ocean. This will become
apparent as we follow the complete path of a
subtropical gyre. However, the circulation of
oceanic waters throughout the entire volume of
the seas is little understood, and the patterns
shown in Figures 4-14 and 4-15 relate only
to the surface layer.

Beginning with the limb or portion of the
gyre nearest the equator, the steady movement
of the trade winds imparts energy to the waters
beneath and sets them in motion. The coriolis
effect deflects the moving water at approximately 45° to the right or left of the wind's direction,
depending upon the hemisphere. The result is
an east-west flow of water called the *North and
South Equatorial Currents.* Between these, and
directly at the equator, is a lesser, reverse flow,
the *Equatorial Counter-current,* which helps to
equalize the return flow of water to the east.
The Equatorial Currents are warmed by the
high-sun insolation of the tropics. As they are
deflected by the continents on the western
edges of the ocean basins, they carry relatively
warm water poleward along the east side of the
continents. This water is a mixed blessing, for
being warm, it is a major source of energy for
those continental shores, while the moisture-

up of water, which while slight—less than 1
meter higher than at the edges of the basin—
further results in a downward pressure on the

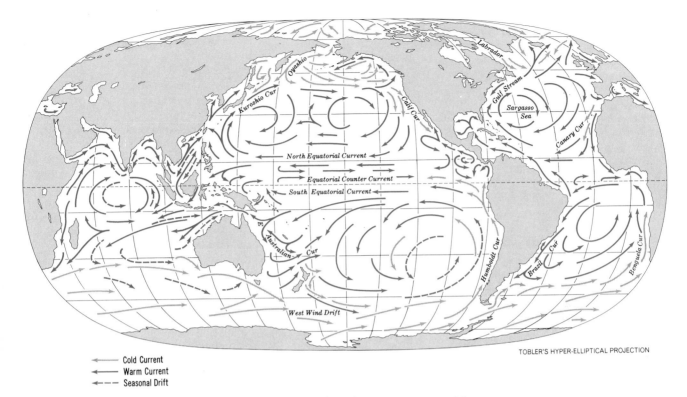

Cold Current
Warm Current
Seasonal Drift

TOBLER'S HYPER-ELLIPTICAL PROJECTION

**Figure 4-15  World ocean currents**  The complicated pattern shown here is an extension of the simplified version shown in the preceding figure (4-14). Warm and cool waters moving latitudinally north and south are important elements in the formation of climates utilized by man. The pattern is essentially symmetrical north and south of the equator and in both the Atlantic and Pacific Oceans.

laden air above it provides precipitation for the eastern coastlands. At the same time, however, the warm water contains smaller amounts of absorbed gases such as oxygen and carbon dioxide which are necessary for maintaining marine life. The warm, blue waters of the Gulf Stream and the Mediterranean may evoke the praise of poets, but they are relatively sterile, and it is the colder oceans, rich and green with microscopic plant and animal life, which sustain the fish which, in turn, help to feed the world.

The northern and southern extensions of the equatorial flow are perhaps the best known of all the world's currents. The *Gulf Stream* along the eastern United States was described as long ago as two centuries by Benjamin Franklin. A similar current in the Pacific, the *Japan,* or

*Kuroshio, Current,* is nearly as famous. Lesser known, but important in the Southern Hemisphere, are their counterparts, the *Brazil Current* and the *East Australia Current.* Poleward of 40° latitude, the warm waters begin to contact cold currents flowing from the Arctic, the *Labrador Current* of the North Atlantic, the *Oyashio* of the North Pacific, and a general upwelling of antarctic waters in the Southern Hemisphere. The colder, gas- and nutrient-filled waters of the polar seas mix with the warmer waters and are forced to the surface, where they support rich harvests of sea life.

In these latitudes the surface currents are pushed smartly along by the westerlies; they take the name *West Wind Drift*, called the *North Atlantic Drift* in the Atlantic Ocean. The West

*Solar Energy and Man: Earth as a Heat Engine*

Wind Drift contains more energy than would be normally expected at such latitudes. By the time the eastern shores of the ocean basins are reached, the difference is significant. England and Scandinavia, British Columbia and Alaska, and southern Chile all benefit from the warmth thus provided. For example, January temperatures at 60° north latitude average minus 10°F on the shores of Labrador, while at the same time southern Norway and the Shetland Islands experience average temperatures of 40°F. Of course, the shores of Labrador are further plunged in cold by southward-flowing arctic waters. In general, the temperature of the ocean affects temperatures on the adjacent coasts. This is shown by the poleward shift of isotherms where warm currents flow offshore and by the equatorward shift where cool or cold currents are found (Figure 4-10).

With their deflection equatorward along the eastern basin shores, the waters of the subtropical gyres once more enter relatively warmer latitudes. By comparison they are now cooler than their new environment, having lost much of their energy along the way. In turn, they have acquired more absorbed gases and more nutrients and are mixed in part with polar waters. The effect is shown by the cool, life-laden *California* and *Humboldt* (or *Peru*) *Currents* of the Pacific and their Atlantic counterparts, the *Canary* and *Benguela Currents.* At this point, the gyre closes upon itself like the dragon of immortality biting its own tail.

Though we close this chapter with a notion of immortality and endless process, we must avoid the notion that these liquid components of heat engine earth operate entirely outside the range of man's impact. Navies may float like tiny chips upon their surface, but oil spills from ships far at sea can be carried with disastrous effects to our shores. Battles have been fought over the control of fishing at certain critical spots such as the shallow waters off Iceland. And just as the green glass fishnet floats lost by Japanese fishermen wash ashore in the Pacific Northwest after transpacific voyages, so too can radioactive materials from island bomb tests find their way to populated shores. Nor does man's tampering end with acts such as these. Megalomaniacal schemes and dreams of some planners have included the damming of the Bering Straits, which would warm the eastern shores of Siberia by cutting off the frigid Oyashio Current. Others have talked of blocking the Red Sea near Aden in order to generate hydroelectric power by pouring the waters of the Indian Ocean into the empty basin left by the evaporation of the waters trapped behind the dam. These ideas may be the pipe dreams of bureaucrats, but with environmental control becoming more and more a reality and a political issue, we cannot entirely overlook them. We will return to considerations such as these, and more, in later chapters. For now, we have described in the briefest detail the major atmospheric and oceanic transfers of energy that help to maintain the surface of the earth in a habitable equilibrium. Our task in the next four chapters is to enlarge the scale at which we view the world and to consider the interaction between the atmosphere; the hydrosphere, or waters of the world; and the lithosphere, or rocky portions of the earth's surface. These interactions can be thought of as subsystems of energy and moisture flow, which sculpture and reshape the earth's everchanging face as well as create world patterns of weather and climate.

# 5 | ENERGY AND THE EARTH: THE HYDROLOGIC CYCLE

In the year 1927, Gutzon Borglum, an American sculptor with a vast dream and endless ambition, began carving massive figures of American Presidents from the granite of South Dakota's Mount Rushmore. Fifty-three months later, when lack of funds ended his project, four great faces fit for giants 465 feet tall stood out upon the mountainside. To reveal those craggy features 400,000 tons of granite were removed from the peak in just over four years. Natural erosion processes before man appeared on earth removed an equivalent amount from the continents into the sea every 24 minutes. The activities of man have so increased the rate of erosion on earth that today the same amount is moved by the world's rivers into the world's oceans in just 10 minutes.

Two important subsystems of planet earth are highlighted by the above comments. Enormous quantities of water must move endlessly across the earth in order to transport such vast tonnages of eroded materials. Also, the continents must be constantly renewed; for at the present rate of erosion, all the land surfaces with their hills and mountains would be reduced to sea level in just 34 million years. Yet there is no geologic evidence—that is, no proof exists—that the continents have ever worn completely away.

The endless cycling of water from the seas to the land and back again, *the hydrologic cycle*, will receive special attention in this chapter. Chapter 6 considers how energy and moisture in varying combinations create earth environments or climates conducive to different forms of life. Chapter 7 looks at the rocky stuff of which the earth is made and shows how it is moved from place to place and the transformations it undergoes during this *lithologic*, or *rock*, cycle. The following chapter describes how the earth's surface is sculptured into a variety of landforms. In every case, the hydrologic and lithologic cycles are intertwined to various degrees, but the time span of the hydrologic cycle is measured in years, while that of the lithologic cycle is measured in eons.

## The Physical Properties of Water

Before discussing the geographic movement of water, it is worthwhile commenting on this most interesting of the chemical compounds found on earth. In the outermost 3 miles (5 kilometers) of the earth (not counting the atmosphere) water is three times more plentiful than all other substances combined and six times more abundant than feldspar, the next most common compound. Water is almost alone in inorganic nature as a liquid, and is the only substance found on earth in all three states: solid, liquid, and vapor. Its heat of vaporization is the greatest of all substances,

*Energy and the Earth: The Hydrologic Cycle*

Mount Rushmore Memorial, Black Hills, South Dakota  The man-made talus slope beneath the monument, while huge, is only a tiny part of the materials eroded by human activity. (U.S. Dept. of Agriculture photo)

and its surface tension is the highest of all fluids. It is one of the best solvents and can dissolve a wider range of materials than any other. Water has many other special characteristics, but we will mention only one more. Like most substances, water contracts while cooling. Unlike all other materials, shrinkage ceases at 4°C (39.5°F) for fresh water. At lower temperatures expansion again occurs, and ice, which forms at 0°C (32°F), has a density 0.92 that of water. Because of dissolved salts, sea water freezes at a lower temperature than fresh water does, about −2°C (28.4°F). Sea water, unlike fresh water, contracts until the freezing point is reached. As freezing occurs, however, the dissolved materials are essentially excluded and sea ice has the same composition as ice formed from fresh water. Two important conditions follow from these properties. Marine waters near the poles become denser when chilled and sink to the bottom of the ocean. This allows warmer waters to take their place, thus helping to induce oceanic circulation. The vertical component of this circulation ventilates the deeper basins, while the horizontal component helps

equalize the global distribution of energy. The second condition is perhaps even more important. If ice were denser and heavier than water—a condition true for almost all other substances—it would sink to the bottom of the lakes and seas. Once there it would accumulate under the insulating layers of water above it. Gradually the basins of the world would fill with ice, and life on earth would be extremely difficult if not impossible under such circumstances. Kurt Vonnegut in his novel *Cat's Cradle* describes an imaginary ice that freezes at a temperature higher than that of the human body. The result is a dying and soon dead planet. If ice were heavier than water, the effect would be the same even if the process were somewhat slower. While we need not worry about such a predicament, it is useful to keep in mind the unique properties of water as we discuss its role in the earth's environmental systems.

Geographic and temporal changes in the location of water are characterized by many different processes, all of which become the means for channeling energy from place to place. Twenty-three percent of the sun's energy intercepted by the earth finds its way from the heated surface of the globe back into the atmosphere by means of the latent heat of vaporization. This and other increments of energy which pass through the atmospheric reservoir act as a source for power for the winds of which we have already written. Wind, precipitation, changes in temperature, and many other similar phenomena combine into a set of short-term atmospheric conditions we all know as *weather*. Weather, in turn, acts upon the surface of the earth in various ways. The resulting breakdown of rocky materials, *weathering*, helps to create conditions suitable for the formation of soils and the growth of vegetation. The cumulative, long-term effects of weather are characterized as different types of *climate* most easily identified by the vegetation associated with them. Meanwhile, weathered materials and soils are constantly being removed and transported by moving water. The removal of materials by water, *erosion*, continuously planes down the earth's surface. The redeposition of these materials, *sedimentation*, continuously fills in the low-lying portions of the earth. At still another scale, water moves endlessly through the bodies of plants and animals, sustaining them and keeping them alive and healthy. In combination, all the processes and paths which water follows become so complex that in order to understand them we need simple models with which to order our thinking. The idea of the hydrologic cycle offers us such a place to begin.

## The Hydrologic Cycle

Earth is the wet planet. More than 295 million cubic miles of water in the form of ice, liquid, and vapor lie awash the globe. Why are we constantly cautioned that humankind is rapidly "running out of water?" If we draw up a budget of the world's waters, part of the problem becomes immediately apparent. Almost 97 percent of all water is the salty stuff that fills the oceans. Only 3.3 percent, or just under 10 million cubic miles, is fresh or salt-free enough to drink. But not all the fresh water on earth is available for immediate use. Three-fourths of it (2.46+ percent) is locked up in glaciers and ice caps. Another 0.43 percent is stored within the surface layer of the earth as groundwater more than 2,500 feet deep. Groundwater stored nearer the surface accounts for another 0.36 percent. Most amazing is the relatively small amount found in rivers and lakes (0.01 percent), that within the soil and available to plants (0.002 percent), and that in the atmosphere itself (0.001 percent). These proportions are shown in Figure 5-1.

If the moisture in the atmosphere, rivers and lakes, and the soil were not in constant flux, there indeed would not be enough water to meet all the demands placed upon such a small store. Fortunately, the solar energy which drives the earth system constantly recycles water and allows its reuse. Thus, while the atmosphere contains only an estimated 3,400 cubic miles of water in the form of vapor, an estimated 95,000 cubic miles of water is evaporated from the oceans, lakes, rivers, and moist lands each year. Obviously, what goes up must come down, and 95,000 cubic miles of precipita-

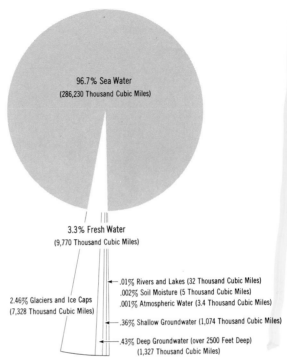

96.7% Sea Water
(286,230 Thousand Cubic Miles)

3.3% Fresh Water
(9,770 Thousand Cubic Miles)

2.46% Glaciers and Ice Caps
(7,328 Thousand Cubic Miles)

.01% Rivers and Lakes (32 Thousand Cubic Miles)
.002% Soil Moisture (5 Thousand Cubic Miles)
.001% Atmospheric Water (3.4 Thousand Cubic Miles)

.36% Shallow Groundwater (1,074 Thousand Cubic Miles)

.43% Deep Groundwater (over 2500 Feet Deep)
(1,327 Thousand Cubic Miles)

**Figure 5-1  Distribution of water on earth**  So great is the proportion of the world's water that is stored within the sea that two scales have been used in this diagram. Of the 3.3 percent of the total that is fresh water, less than 1.0 percent remains unfrozen. Though earth has been described as a "wet" planet, very little fresh water is available for use. (Data from J. E. Van Riper, *Man's Physical World*, McGraw-Hill, New York, 1971, p. 174)

tion balance the equation. It is the paths taken by this constantly recycling water that concern us here.

The oceans are the source of almost all the water moving through the hydrologic cycle, and there is strong evidence that the accumulation of oceanic waters is very old. The seas have lapped at the continents from at least early Paleozoic times more than 400 million years ago.

There are varying opinions on the amount of water evaporated from the ocean surface each year, but one good estimate places this at about 80,000 cubic miles. Once in the atmosphere, this water follows one of a number of possible routes or cycles until it returns to the sea (Figure

*Energy and
the Earth:
The Hydrologic
Cycle*

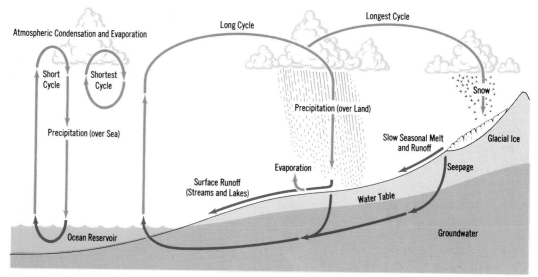

**Figure 5-2 The hydrologic cycle and some of its variations** A molecule of water moving through the atmosphere and the oceans and across the surface of the land can be caught up in hydrologic cycles of various duration. The short cycles may be simply from the ocean to the atmosphere and back again or from the atmosphere to a falling drop of rain which evaporates before reaching the ground. Cycles of intermediate length would move the same molecule from the sea to the atmosphere, over the land in rain clouds, and eventually back to the sea in streams of groundwater. If the water were locked up in glacial ice, years might pass before melting released it for a return trip to the sea.

5-2). The shortest cycle of all does not involve the ocean directly and consists of the condensation of rain and its immediate evaporation in the atmosphere without reaching the earth's surface. Nevertheless, this *shortest cycle* still depends upon the ocean's having provided the atmosphere with water at some earlier time. A more significant *short cycle* takes place when condensation occurs and rain falls upon the surface of the sea. Some of the water vapor provided by the oceans is blown inland and enters a *long cycle*. In this case, rain falls upon the land, where part of it immediately evaporates. Much of the water that returns to the atmosphere above the continents is moved from one place to another by winds and can be precipitated and evaporated over and over again before reaching the sea. Some precipitation soaks into the earth to fill the twin reservoirs of soil moisture and groundwater, and some runs back into the sea in the form of streams and

rivers. Along the way water is temporarily stored in lakes and also, in tiny but critical quantities, in plants and animals. A still *longer cycle* involves precipitation in the form of snow and ice. In this cycle, years may pass before a particular molecule of water finds its way slowly from the head of a glacier to the point where melting occurs. After that, the return to the sea is quickly accomplished. The longest cycle incorporates all the shorter cycles within it, as each succeedingly shorter cycle includes those still smaller. Thus, in addition to the water evaporated from the oceans, another 15,000 cubic miles is evaporated from rivers, lakes, and moist lands. Since precipitation must balance this evaporation, we find 24,000 cubic miles of water precipitated each year on the land and another 71,000 onto the oceans. Moreover, since more water precipitates upon the land than evaporates from it, a final 9,000 cubic miles eventually reaches the sea as liquid runoff. The final picture is one of endless cycles within cycles, some of which we will now try to describe.

## The latitudinal balance sheet

Energy is unequally distributed across the globe from the equator to the poles. Land and water are also unequally distributed latitudinally upon the earth (Figure 5-3). Moreover, the surface area of the earth itself varies with latitudinal position. While the inequalities of land and water area are self-evident from the accompanying figure, the statement about surface area and latitude needs a little more explanation. For example, a band of latitude measured from the equator (0°) to 5° north latitude is approximately 25,000 miles (40,250 kilometers) in circumference on its southern edge. A similar 5° latitudinal band measured between 60° and 65° latitude is only 12,500 miles (20,125 kilometers) in circumference along its equatorward side.

All else being equal, insolation upon the earth's surface results in water evaporation proportional to the temperature of the atmospheric cover. We would, therefore, expect to find evaporation from the seas and wetlands to be greatest near the equator and least near the poles. For the moment, let us consider only the total amount of evaporation and its inevitable parallel, precipitation, as these two processes are distributed latitudinally across the earth. Figure 5-4 shows the distribution of excess evaporation and excess precipitation north and south of the equator by 10° bands. By excess we mean whichever amount is greater when all that has gone up (evaporation) is subtracted from all that has come down (precipitation). At some latitudes much more moisture is lost to the atmosphere than is returned; at others just the opposite is true. Only between 30° and 40° south latitude is the ledger in balance. A distinction must also be made between the continents and the oceans, for the seas generally contribute and receive more water than do the continents.

Now consider the general pattern of moisture debits and credits. Between 10° and 40° north latitude and 0° and 30° south latitude the oceans

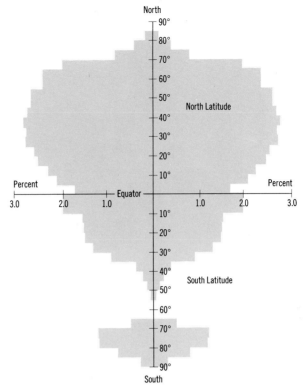

**Figure 5-3  Distribution of land on earth by latitude**  The total land mass in each five-degree band of latitude north and south of the equator is shown as a percentage of the total. The data have been grouped symmetrically on either side of the central axis in order to suggest the shield shape which we have assigned the hypothetical continent discussed throughout the book. The diminishing of the total surface area of the globe near the poles is clearly evident in the south polar region, which has only a small portion of the total land area, although that area is considered to be completely above sea level. (Based on data: Waldo R. Tobler, *Geographical Coordinate Computations Part II,* Finite Map Projection Distortions, Technical Report No. 3, University of Michigan, Dept. of Geography, December 1964, under contract with the Office of Naval Research, Geog. Branch Contract Nonr. 1224(48), Task No. 389-37)

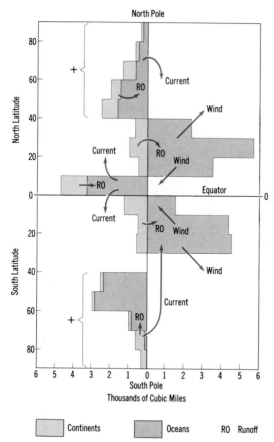

**Figure 5-4 The water balance of the earth by latitudinal zones** The left side of the diagram shows the areas of the world where precipitation exceeds evaporation; the right side shows the areas where evaporation exceeds precipitation. An excess of liquid water is found poleward of 40° latitude in both hemispheres. At lower latitudes, evaporation (water vapor) exceeds precipitation (liquid water). Only in the Southern Hemisphere between 30° and 40° latitude do the two balance each other. In order to redress these inequalities, water in both its liquid and vapor forms must move long distances latitudinally. The arrows indicate the direction and type of transporting agent important in maintaining a worldwide water balance. (After J. E. Van Riper, *Man's Physical World*, McGraw-Hill, New York, 1971, p. 177; taken from Wuest)

give up enormous quantities of water which enter the atmosphere in vapor form. (Remember that the quantities given in this figure refer only to the absolute differences between evaporation and precipitation. Thus, the total shown for the entire world is about 22,000 cubic miles, compared with 95,000 cubic miles mentioned earlier. The difference between the two figures represents those waters which are recycled several times in the same immediate area within a single year.) Poleward of 40° in both hemispheres precipitation exceeds evaporation. This is clearly a function of less solar energy's being available at those latitudes. The absolute amount of either excess evaporation or excess precipitation is also less, because of the smaller surface areas involved, as discussed above. The greatest evaporation occurs near 30° north and south latitude in the belts of subtropical high pressure.

It is obvious, though, that the subtropical seas do not dry up and that the higher latitudes are not necessarily drowned in ice or water. A constant readjustment or geographic redistributing of water vapor and liquid keeps the ledger of the earth in balance. Water vapor from areas of excess evaporation is carried by the wind into areas of excess precipitation. Much of the precipitation simply falls upon the surface of more northerly and southerly seas, but another portion descends upon the land. The excess precipitation, in turn, runs off as rivers or sometimes as glaciers which reach the sea. At higher latitudes excess precipitation which falls upon the surface of the sea eventually finds its way equatorward in the form of deep sea currents. Where excess precipitation also occurs just north of the equator, runoff from the land and rain upon the ocean's surface is carried by currents toward the subtropical high-pressure belts. Once there, it again evaporates. We should not imagine, though, that these trips are direct or without interruption. The same molecule of water may travel between earth and high heaven a number of times along the way. In fact, it is estimated that the average length of time that water in vapor form remains in the atmosphere is only about nine or ten days.

What we are talking about is the total picture, and as with all views of totality much detail is lost in the telling. Nevertheless, even these simple statements about evaporation and precipitation can be combined with our knowledge of the world's wind and ocean current systems to give us a better view of the total earth ecosystem. Having done this let us move closer and look at certain hydrologic processes at an even larger scale.

## Humidity and condensation

The words *evaporation* and *precipitation* go hand in hand. We have already considered evaporation in some detail but have not yet looked closely at the causes and conditions of precipitation. Precipitation is water in either solid or liquid form which falls from the atmosphere, occurring mainly as rain, snow, hail, or sleet. Its presence depends upon the *condensation* of water vapor when the temperature of the atmosphere reaches the *dew point*. This term refers to the temperature at which the carrying capacity of the air matches the actual amount of water vapor contained within it. We mean by this that warm air has greater capacity to hold water vapor than does air which is cold. For every temperature there is a fixed and definite weight of water vapor which can be held in a given volume of air. When this amount is present, the air is said to be *saturated*. The proportion of water vapor present to the total carrying capacity of the air is called the *relative humidity* and is measured as a percentage. Thus, for example, at 30°C (86°F) a cubic meter of air can hold up to 30 grams of water vapor. If at that temperature only 15 grams are contained in a cubic meter, we describe the relative humidity as 50 percent. When the air is saturated and contains 30 grams, the relative humidity is 100 percent. If the temperature of this same cubic meter of air were to drop to approximately 17°C (62°F), its capacity to contain water vapor would be halved and it would become saturated by only 15 grams of water vapor (Figure 5-5). When saturation is reached, any excess moisture condenses, and precipitation results. The

term *absolute humidity* refers to the amount of water actually present in a volume of air and is measured by *weight* rather than as a percentage.

All of this discussion assumes that the volume of air involved remains constant. But a cubic meter of air when moved to higher elevations becomes considerably larger, while one moved nearer sea level is compressed and becomes smaller in volume. As a result, the measurement of absolute humidity presents some problem. To overcome this difficulty, the *specific humidity* of the air is often used by meteorologists and geographers. This is the *weight* of water vapor contained within a given *weight* of air including the water vapor itself. Very similar to this measure is the *mixing ratio*, which is the weight of water vapor compared with the dry weight of the air which contains it. The advantage of both these measures is that they remain constant regardless of changes in volume as long as the unit of air involved neither gives up nor receives additional moisture.

What all this means in terms of world ge-

Figure 5-5 **Water-holding capacity of air as a function of temperature** (After A. N. Strahler, *Introduction to Physical Geography,* John Wiley, New York, 1965, p. 74)

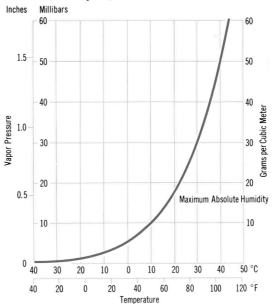

ography is that at a global scale the cold air of the poles contains far less moisture than does the warm air of lower latitudes. At the same time, the relative humidity of the atmosphere is lowest near the subtropical high-pressure belts and greater both at the equator and in polar latitudes. Not only is air more nearly saturated near the equator (relative humidity), it also contains more available moisture in absolute terms (specific humidity or mixing ratio). As we will see, this has important consequences for human occupance in those general areas.

## Types of Precipitation

The processes by means of which precipitation occurs in all its forms are more complicated than might at first be supposed. In this discussion it will suffice to say that large masses of moist air must be cooled below the dew point in order for precipitation to take place. Nighttime cooling by long-wave energy reradiation into space cannot accomplish this. Instead, it is necessary for warm, moist air at lower elevations to be lifted to greater heights where suitable cooling can occur. The discussion that follows deals essentially with the lifting processes which accomplish this.

### The adiabatic rate

The general decrease in temperature as one travels away from the earth's surface has already been mentioned. This is known as the *lapse rate*, normally 6.4°C. per thousand meters change in elevation, and is associated with the fact that air at lower elevations is more dense than air at greater altitudes. Dense air is susceptible to heating through conduction from warm ground surfaces and from solar radiation. Also, the increased presence at lower elevations of materials such as water droplets and $CO_2$, both ready absorbers of radiation, helps produce higher temperatures nearer sea level.

The lapse rate should not be confused with a second type of temperature change within the atmosphere. As a given volume of air rises away from the earth's surface, its temperature decreases even when it loses no energy to the space beyond its own limits. This is because as the air expands, the molecules contained within it have more and more room in which to move, and strike each other with less and less frequency, thus generating less heat. This steady decrease of air pressure, and its parallel decrease in sensible heat, is called the *adiabatic rate*. More specifically, if the temperature of the air, cooling as it rises, has not fallen below the dew point, no precipitation will occur. Air temperature change while in this condition is known as the *dry adiabatic rate* and is equal to approximately 10°C for every 1,000 meters or 5.5°F for every 1,000 feet change in elevation. If the dew point is reached, precipitation follows and heat is released into the atmosphere as water vapor changes to liquid (the latent heat of condensation). This slows the adiabatic rate to about 6°C per 1,000 meters or 3.2°F per 1,000 feet, a condition known as the *wet adiabatic rate*. As precipitation continues, a point may be reached where the air is no longer saturated and precipitation will then cease. Now let us examine the three major ways in which moist air masses are lifted to elevations where cooling and precipitation can occur.

### Convectional rainfall

Imagine a warm day somewhere in the tropics or on the American Great Plains. At 9 A.M. the sky is clear but the air may already seem heavy and humid. As the sun rises hour by hour toward noon, the temperature climbs to 80° and then to nearer 90°F. By twelve o'clock small, puffy cloudlets fill the sky overhead and along the horizon. The day may be quite still, but little cat's-paws of breeze begin to stir the grass and trees. Actually, much of the movement of the air is vertically upward from the baking earth. This convectional flow of air, by itself, would seldom reach more than a few thousand feet from the surface, but if the air is nearly saturated, it may require only a short vertical rise to bring it to elevations where the temperature has decreased to the dew point. Condensation will follow, forming small amounts of rain which may never reach the earth. The latent heat released by the initial condensation will add

Apollo 9 photo. This nearly vertical view of thunderclouds over the Amazon Basin, Brazil, was made on the eightieth revolution of the Apollo 9 spaceflight. The energy released by this single storm exceeds that of any weapon devised by man, and yet is so distributed that the selva beneath benefits from its effect. (NASA)

sufficient energy to the system to actually accelerate the vertical movement of the moist air to greater altitudes. By 3 or 4 P.M. cumulus clouds will begin to boil slowly upward. Somewhere near 10,000 feet (3,200 meters) the dry adiabatic rate will have reduced the temperature of the main air mass sufficiently for much heavier condensation to occur. Even though the wet adiabatic rate at still higher elevations will be somewhat slower, the cumulus clouds will begin to develop great white domes which push higher and higher. Soon mountains of white cumulonimbus clouds rear 3 to 4 miles into the sky, their bases filled with supersaturated air that may actually be below the freezing point (Figure 5-6A). Humidities of more than 100 percent under these conditions can exist if there are no suitable nuclei around which condensation can occur. The vertical winds which drive the moist air upward quickly push it to elevations where low temperatures permit ice crystals to form. Thus, the tops of thunderheads may be white with ice rather than water droplets. These crystals, falling earthward, enter the lower cloud and provide the solid bits necessary for condensation of the supersaturated air. Vi-

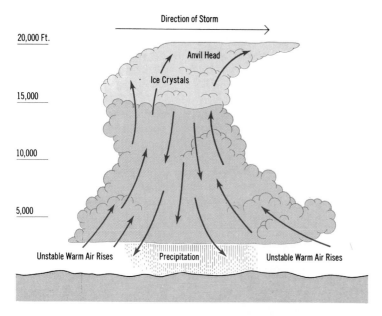

20,000 Ft.

Direction of Storm

Anvil Head

Ice Crystals

15,000

10,000

5,000

Unstable Warm Air Rises          Precipitation          Unstable Warm Air Rises

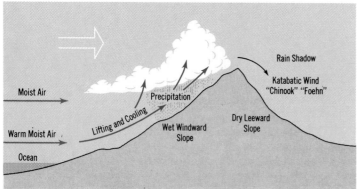

Moist Air

Warm Moist Air

Ocean

Lifting and Cooling

Precipitation

Wet Windward Slope

Dry Leeward Slope

Rain Shadow

Katabatic Wind "Chinook" "Foehn"

**Figure 5-6  Convectional and orographic rainfall. *A.* Anatomy of an altocumulus cloud with convectional rainfall. *B.* Orographic rainfall**
Moist air must be cooled below the dew point in order for precipitation to take place. Two occasions on which this happens are when convectional air currents lift air masses to altitudes where lower temperatures prevail, and when winds force moist air up mountain slopes into similar conditions. Altocumulus clouds or thunderheads (*A*) are frequently associated with the first method; orographic rainfall on the windward slopes of mountains accounts for the second (*B*). (*A* after Joseph E. Van Riper, *Man's Physical World,* Copyright 1962, 1971 McGraw-Hill Book Co., New York, fig. 6-31, p. 232)

*Physical Geography: Environment and Man*

olent rains of brief duration follow, adding additional latent heat to the seething cloud mass.

Some particles of ice and water may make several trips up and down through the cloud before their torrential plunge to earth. Layer on layer of water may freeze around the original bits until hail accompanies the thunderstorm. Then, as the sun begins to set, energy is removed from the system and its movements begin to ease. The anvil-shaped masses of the mature cumulonimbus start to break up, the accompanying lightning and thunder grow less frequent, and sometime in the night's dark hours the clouds disappear until another day of sun refuels the system. This convectional process is the most dramatic of the three types of rainfall and accounts for the high average annual precipitation in tropical areas (Figure 5-7). Convectional rainfall is also important in mid-latitude continental locations where fields and forests may heat unevenly with localized convection cells forming as a result. This accounts for the bumpiness of low-level airplane flights on summer days in those areas. These vertical cells, or *thermals*, may be the delight of glider pilots, but they also breed altocumulus clouds dangerous to any kind of flight. Even the open skies near thunderheads can hide enough *clear air turbulence* (CAT) to tear an airliner apart. Nevertheless, heavy summer rainfall over mid-latitude farm regions is a major benefit of such air currents and clouds.

### Orographic rainfall

The basic requirement for all types of rainfall is the lifting of moisture-laden air to elevations where prevailing temperatures can cool it below the dew point. Obviously, such air masses also must have sufficient water vapor and condensation nuclei within them to bring about precipitation. In areas where prevailing winds encounter hills and mountains, the horizontally moving air is pushed up and over those barriers. Somewhere up the windward slope rain or snow occurs in large quantities; thus the name for this type of rain derives from the Greek word *oreos,* meaning "mountain" (Figure 5-6*B*).

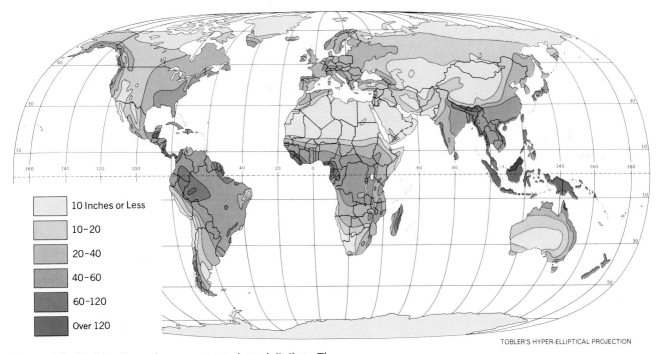

TOBLER'S HYPER-ELLIPTICAL PROJECTION

**Figure 5-7  World pattern of average annual precipitation**  The occurrence of precipitation on a worldwide basis provides good examples of the climatic controls discussed in the text. Parallel bands of different rainfall intensities trend north-south along the Pacific coast of both North and South America. This reflects the barrier effect of mountain ranges. Deserts (less than 10 inches or 25 centimeters rain) mark much of the zone of subtropical high pressure. The west coasts of the continents at higher latitudes are well watered by storms brought on by westerly winds. These are only a few of the numerous examples shown by this map.

Heavy annual rainfall resulting from these conditions is typical of the western margins of the continents between 40° and 60° latitude. This is particularly true in the Western Hemisphere, where the Cascade Mountains in North America and the Andes in South America thrust up major barriers into the path of westerly winds. Another smaller but effective barrier, the Olympic Mountains just west of the Cascades in Washington State, receives more than 180 inches of rain annually. When such moisture turns to snow in the winter months, the result can be spectacular. There was a record-breaking accumulation of more than 1,000 inches of snow on Mount Rainier in the Cascades in 1971–1972. Even where mountain elevations are not as

great, slight lifting through orographic processes can result in heavy rain. The western British Isles and parts of Norway and northwestern Spain are noted for their wet weather, although their rocky heights are considerably lower than the Cascades or Andes.

Another part of the world where orographic lifting is important is on the subcontinent of India. The monsoon winds, moving in from the Bay of Bengal, encounter the foothills of the Himalayas, particularly the Khasi Hills of Assam Province just north of the Bangladesh border. A weather station located there in the town of Cherapunji consistently records incredible amounts of rain. The most spectacular quantity was 905 inches of rain measured there

*Energy and
the Earth:
The Hydrologic
Cycle*

in 1861; during that year, in the month of July alone, 366 inches fell.

The lee sides of mountain barriers can be as dry as their windward sides are wet. Once air has been pushed to elevations where it gives up its moisture, the winds may continue over the mountains and down the far side. The descent serves to compress and heat the air at the dry adiabatic rate of 10°C per 1,000 meters (5.5°F per 1,000 feet). As the air warms, its ability to hold moisture increases and the land in its path is dried out. These drying, descending winds are technically called *katabatic winds*, but wherever they are found they have colorful local names. The *chinook* which blows down the eastern slopes of the Rocky Mountains and the *foehn* sweeping out of the Alps are two famous examples.

The overall effect of orographic processes is to create belts of heavy rainfall on the windward sides of mountains and *rain shadows* and deserts on the lee sides. Where mountain ranges are aligned at right angles to the winds, as they are along the west coasts of North and South America, a pattern of wet and dry lands develops paralleling the coastline (Figure 5-7). In other places, mountains may be oriented in the same direction that the winds blow. This is the case in Europe, where much moisture-laden air carried in from the sea by the westerlies penetrates beyond the hills of the British Isles and southern Scandinavia. The main mountain barrier in Europe is the Alpine system running from the Pyrenees in the west to the Carpathians and Pontic Mountains in the east. Much of Europe's wet weather is kept north of this impressive barrier. The result is a relatively large area along the Mediterranean Sea which is warmer and drier than might otherwise be expected without the sheltering mountains. Here, too, we find a distinctive spatial pattern of rainfall (Figure 5-7) and a parallel set of human activities marking mankind's utilization of this special distribution of resources.

### Cyclonic systems and frontal rainfall

Entire continents, or better yet, whole hemispheres, are the arena where the processes producing frontal rainfall take place. Precipitation of this origin occurs when major masses of warm and cold air encounter each other with subsequent lifting and mixing of warm, moist air with that which is dry and cold. To understand these gigantic meetings, we must consider that the lower atmosphere is divided into two major types of air. The first is the colder air of higher latitudes, viewed by meteorologists as a series of *polar air masses*. The second is the warmer air of lower latitudes made up of *tropical air masses*.[1]

Polar and tropical air masses are further subdivided according to the water vapor they contain, which is, in turn, a function of their continental or marine origins. These various masses of air correspond closely to the low- and high-pressure cells described in the preceding chapter. Thus, we find three polar air masses alternating across the northern part of the Western Hemisphere (Figure 5-8). The North Pacific is the source of cool, moist, maritime polar air. (The notation for this is mP.) Northern Canada produces cold, dry, continental polar air (cP). The North Atlantic is again the source of mP air masses. Far to the south, similar tropical air masses also form. Warm, moist, maritime air masses (mT) develop over the pacific southwest of California, over the Gulf of Mexico, and over the Atlantic southeast of the United States. A fourth and smaller continental tropical (cT) air mass noted for its dry air and high temperatures occurs above the American southwest in the summer.

In their totality, these tropical and polar air masses remain remarkably distinct from one another. Their zone of contact is called the *polar front* and stretches around the world in the middle latitudes. High-pressure cells associated with this polar air mass rotate in a clockwise direction, resulting in a flow of cold winds from east to west along the polar front. The tropical counterpart of this flow moves from west to east as the result of a basic counterclockwise movement. As the warm and cold air masses come into contact with each other, a series of eddies

[1]It should be noted in passing that these terms are something of an oversimplification since their major source areas are in the *sub*artic and *sub*tropical areas of the world.

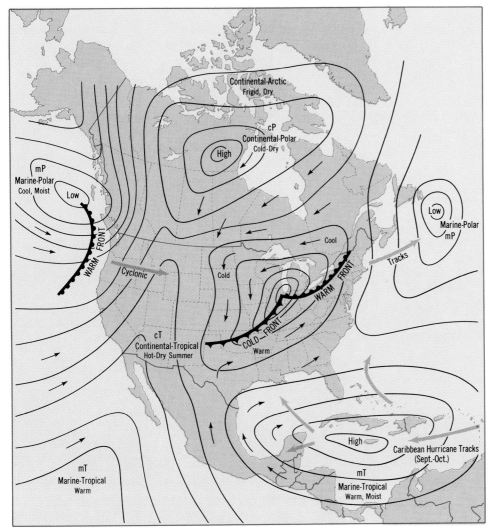

**Figure 5-8  Typical North American weather system and air masses**

The pattern shown here represents an arrangement of the basic air mass
components described in the text. Two low-pressure cells are crossing North
America, bringing frontal systems and the storms associated with them.
Rain and perhaps violent squall lines will occur where the continental polar
air over northern Canada is being pumped southward into the low-pressure
cell over the Eastern United States. The zone of contact between tropical
and polar air masses would mark the polar front.

occur along the front formed by their opposing
flows (Figure 5-9). The dense polar air tends to
push into the warmer, lighter tropical air as a
rather steep or blunt-nosed intrusion, a *cold*
*front*. The tropical air takes the form of a
thinner wedge which slides up over the colder
air against which it moves. This advance is
known as a *warm front*; both are shown in

*Energy and
the Earth:
The Hydrologic
Cycle*

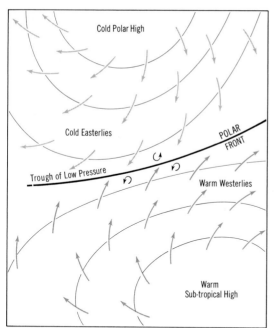

**Figure 5-9 Eddies in a trough of low pressure between two air masses**

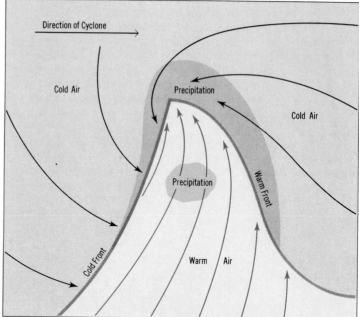

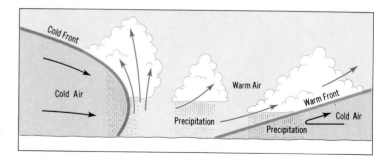

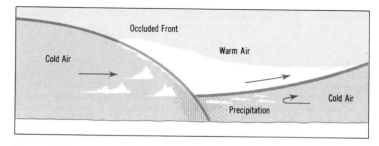

Figure 5-10*A* and *B*. As the warm air is deflected over the polar air, a prong of warm air can be caught between the two limbs of the polar mass. Since a cold front generally moves more rapidly than a warm front does, the advancing cold air sometimes overtakes and pushes under the warmer air. The result is that warm air is trapped above the colder air as the advancing limb of the cold front catches up with the warm front. When this occurs, the trapped overriding warm air forms what is called an *occluded front* (Figure 5-10*C*). As major masses of cold and warm air advance across the land, their zone of contact is marked by precipitation and the release of energy through the latent heat of condensation. Two mechanisms, therefore, are working to lift moisture-bearing air to higher elevations. The first is a wedging effect which drives warm air over cold air; the second is a version of the convection processes described in the preceding paragraphs. The two in combination produce significant amounts of precipita-

**Figure 5-10 Low-pressure cell with associated frontal systems** Cyclones in the Northern Hemisphere rotate in counterclockwise direction. Cold fronts may advance more rapidly than warm fronts and, overrunning them, form occluded fronts. (After Joseph E. Van Riper, *Man's Physical World,* copyright 1962, 1971, McGraw-Hill Book Company, New York, fig. 6-24, p. 224.)

tion and sometimes, particularly along the advancing edge of a cold front, spectacular and violent storms.

Squall-line storms are a frequent occurrence accompanying the advance of cold fronts across the Great Plains and the Midwest (Figure 5-11). These storms are marked by violent gusts of wind, heavy rain of brief duration, thunder, lightning, and tornadoes. These latter winds with their funnel-shaped clouds and shattering violence travel in association with the most turbulent portions of cumulonimbus buildups. They advance in the general direction of the squall line's movement at speeds from 25 to 40 miles per hour; but within their funnels, winds may sometimes exceed 500 miles per hour.

Because the central United States has virtually no natural barriers running in an east-west direction, polar and tropical air masses meet in unrestricted battle more frequently here than anyplace else in the world. The advancing limbs of high-pressure cells, that is, the cold fronts which we have just mentioned, are moving in a counterclockwise, or southwest to northeast, direction by the time they have reached the Midwest. Thus, the most frequent path of squalls and tornadoes is toward the northeast in this part of the world. Moving eastward into densely settled areas, tornadoes have become part of American folklore and fiction. There would be no Dorothy and no Toto and no Judy singing "Somewhere Over the Rainbow" without the famous Kansas "twisters." In reality, and on a far grimmer note, the death and destruction from these storms is tremendous. Figure 5-12 shows the occurrence of tornadoes by one-degree squares in the United States from 1953 to 1969. Contrary to popular belief, although tornadoes are generally more frequent in the Great Plains states, such as Kansas and Texas, their destructive potential to human life is far greater in Illinois, Indiana, Michigan, and some parts of the Eastern United States. This is

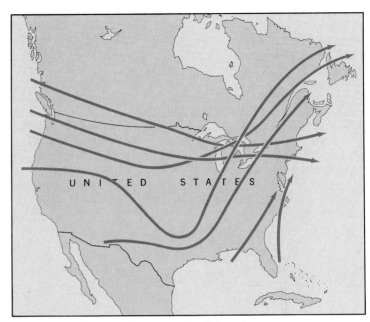

**Figure 5-11  Storm tracks across North America**
The mid-latitudes experience a steady progression of high- and low-pressure cells which move from west to east. The frontal systems associated with these cells produce the storms whose "tracks" can be traced across the continent. In the summertime, tropical air masses move far to the north and the frontal systems and their storms pass most frequently across Canada. In the winter, polar air masses push far south, bringing storms into the Central and even Southern United States. It should be noted that this figure includes two tracks of hurricanes, which are associated with another set of conditions.

A tornado which struck Tracy, Minnesota, in June 1968. Nine persons were killed and 100 injured. Note the narrow base which creates a *line* of destruction. If the area across which it moved had been more densely populated, the tragedy could have been even greater. (Wide World photos)

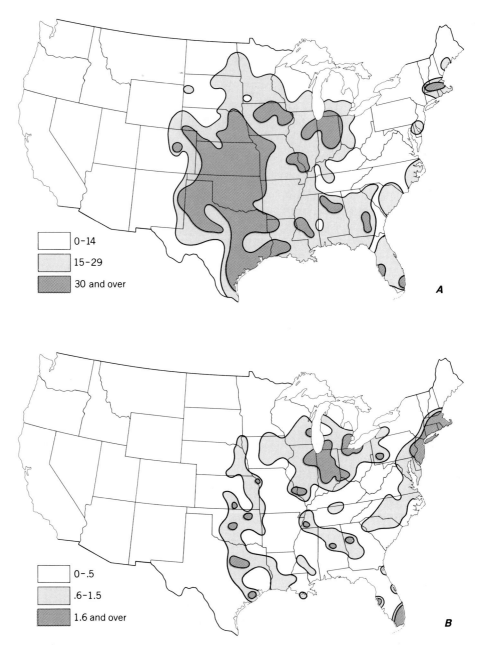

Figure 5-12 **A.** Occurrence of tornadoes by one-degree squares, 1953-1969. Distribution of tornadoes reported during the seventeen-year period cited. Note the dominance of the southern Great Plains and Texas Gulf Coast **B.** Potential casualties from tornadoes per square mile. This map has been weighted in two ways: the area of each one-degree square actually swept by tornadoes during the seventeen-year period was carefully estimated; the 1970 population of each square was considered to be evenly distributed

a function of population density. Figure 5-12*B* shows the *potential* casualties from tornadoes per square mile. Note the northeastward shift of the high-risk areas when the distribution of population is taken into account.

Less dramatic but just as familiar is the slower advance of a warm front.

> *Mares' tails*
> *And mackerel scales*
> *Bring lofty ships*
> *With lowered sails.*

This traditional jingle describes the first messengers announcing a warm front weather change. Because it overrides rather than wedges beneath the cold air mass before it, a warm front first reaches any given location not at ground level but high overhead (Figure 5-13). The first clouds formed along its leading edge are high-altitude *cirrus*, known to most "folk meteorologists" as *mares' tails*. Just behind them, and somewhat lower, will come either a thin veil of *cirrostratus* clouds that put a "ring around the sun," or sometimes the evenly distributed and puffy *cirrocumulus* clouds called *mackerel scales*. Still farther behind the leading edge of warm air, and lower still, will come *altostratus* clouds. At last, reaching down to the earth's surface, will come *stratus* and rain-heavy *nimbostratus* layers. This is the second major type of rain production associated with frontal or cyclonic systems. The somewhat more complicated precipitation mechanics of an occluded front will not be treated here. It is enough to say that an occluded front again involves a combination of wedging and convectional turbulence.

Total destruction caused by the Xenia, Ohio, tornado, April 4, 1974. The twister moved directly across a thickly settled portion of the town, killing 35 people and injuring scores of others. Note the abrupt transition between flattened buildings and others still standing. As suburban sprawl covers larger and larger areas, scenes like this will become more common. (Wide World photos)

## Water on the Ground

Once moisture has changed state from vapor to liquid and fallen to the earth, its most direct route back to the sea involves infiltration into the soil and the runoff of any excess. Let us, for the moment, assume that frozen precipitation melts and joins rainwater with no further complications. The accumulated moisture either soaks into permeable surfaces or runs off those which are impermeable. *Permeability* is a measure of how easily or quickly moisture can travel through a substance and depends on the connectivity between existing air spaces within it. *Porosity* describes the amount of empty space within a substance, that is, the enclosed volume not taken up by solid particles. Most rock, unless it is broken or shattered or unless it has solution cavities within it, may be considered impermeable. Rain falling upon it will simply

*Energy and the Earth: The Hydrologic Cycle*

**103**

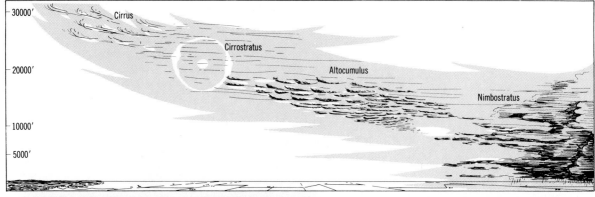

**Figure 5-13 Cloud sequence of an advancing warm front** As the warm, moist air overrides the cold air before it, clouds form along the zone of contact. An observer looking directly overhead at the passing clouds would first see cirrus clouds composed of ice crystals at very great heights. The sequence of clouds that follows the cirrus approaches nearer and nearer the ground, becoming more and more dense, until nimbostratus clouds bring drizzling rain. Cirrus (A) and altocumulus (B) clouds are typical of advancing warm fronts. Their positions in the sequence of cloud types can be seen below. (Photographs courtesy of NOAA.)

flow elsewhere as surface *runoff*. In most cases rock is neither permeable nor porous. Soil, on the other hand, incorporates varying degrees of these two qualities depending upon its own characteristics.[2] Loose sand is highly permeable, as every child who ever tried to fill a lake in his sandbox knows. Sand is also porous, with its porosity depending upon the size and shape of the grains. On the other end of the continuum of soil textures are colloidal and coarser clays. These are composed of microscopic platelets of a number of different minerals. When the platelets are all oriented in the same direction, clay is nearly impermeable; but, surprisingly, its porosity still is very great. If you pour a cup of water on a lump of freshly dug clay, it will run off. But the same lump of clay, which may appear almost dry at first, often will become soaking

[2]Soil is a complex substance needing special consideration. We ask the reader's indulgence in our use of the term at this point before we define and discuss it in a later chapter.

wet after continued kneading. The manipulation of the clay between your hands will disturb the orientation of the platelets of which it is composed, thus freeing the water trapped between them.

Soil may be compared to a sponge in its ability to absorb a certain amount of water but no more. Hold a dry sponge under a faucet. At first its hard surface will shed water, but as its fibers soften, more and more is absorbed. Finally, when every cavity and pore space is filled, water will once more flow directly from the surface of the sponge. The baked soils of summer and those of all arid regions are usually slow to absorb the rain that falls upon them. (Loose sands, which are not true soils, are an exception to this.) If the downpour is of brief duration, very little water can soak in locally. Instead, a film of water may run off downslope, gaining volume and depth as it goes. This *sheet flooding* eventually accumulates in river valleys and often creates *flash floods* of brief duration. Desert thunderstorms are a common source of flash flooding, and many an unwary traveler camped in some cool arroyo or wadi far downstream from a rainy area has drowned in a unexpected flood.

Where the soil is more moist and porous and where precipitation is slow and steady rather than torrential, water soaks readily into the surface. If enough water is available, it will seep downward until solid rock or other impermeable material is encountered. The water begins to build up from that level and soon fills the pore space of the overlying materials. The stored water thus accumulated in this *saturated zone* is called *groundwater*, and its upper surface, that is, the top of the saturated zone, is referred to as the *water table*. Above the water table the soil is partially filled and may serve as a container for more moisture or may become a zone of upward water migration if the surface is dry and the energy in the air above it is great enough.

Groundwater moves slowly through the soil and upper layers of the earth wherever permeability permits. The water table tends to follow the contour of the land surface above it. Nevertheless, this water is under the influence of gravity as much as any surface water. Just as normal air pressure represents the accumulated weight of a column of air reaching to the exosphere, the hydraulic pressure exerted on groundwater varies from place to place depending upon the water above it or upslope. This accounts for the flow of groundwater which is always down the *hydraulic gradient*.

As a general rule, the water table is farthest from the surface on the tops of hills. Farther downslope the water table may actually intersect the sides of valleys, where it appears as a

Figure 5-14 Groundwater The permeable sandstone mantle of soil and underlying rock serves as a reservoir for underground water. The top of the completely saturated zone is called the *water table*. Wells may penetrate to it, and the water table may also intersect the sides of valleys, creating springs. In general, the water table roughly parallels the contour of the surface above it.

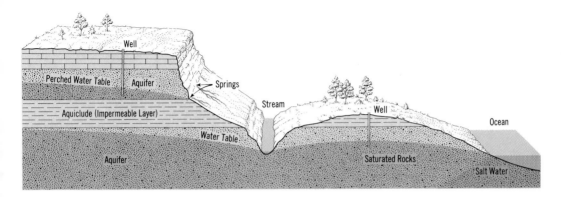

series of springs (Figure 5-14). In this and other ways water may reach the surface and return to the saturated zone several times on its trip to the sea. Once there, some water actually travels underground beyond the margins of the land and wells up as freshwater springs on the ocean floor on the continental shelf. In general, though, the contact or interface between groundwater and sea water is static. Freshwater springs at sea are an unusual feature.

Water may also penetrate beneath the unconsolidated upper covering of earth materials into the rock layers or strata beneath. Certain strata, such as the famous St. Petersburg sandstone that underlies much of the Great Plains, carry large quantities of water. Such water-rich layers, or *aquifers*, are important sources for irrigation and city water systems in many parts of the world. Other strata may be impermeable and dense and are known as *aquicludes*. Aquicludes prevent the passage of water and form effective traps or barriers. Sometimes a locally occurring aquiclude may hold a lens of groundwater above the normal water table. *Perched water tables* such as these offer short-term benefits to small populations who use them, but

Figure 5-15 **Effluent and influent streams** In humid regions a stream may pick up water from the surrounding groundwater table. In arid regions the stream may give up water to the groundwater table. (After A. N. Strahler, *Introduction to Physical Geography*, John Wiley, New York, 1965, p. 126.)

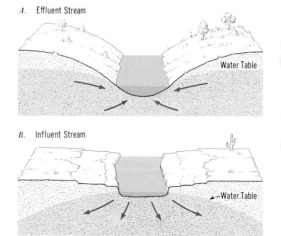

A. Effluent Stream

Water Table

B. Influent Stream

Water Table

are often easily exhausted with intensive use.

In any case, when the storage capacity of the soil (field capacity) is filled, additional precipitation finds its way to streams through one of two avenues. Groundwater may move down the hydraulic gradient to intersect streams which carry it away. If the rate of accumulation of precipitation exceeds the rate of groundwater removal, underground flow will still continue, but in addition, the remainder of the surplus water will become direct surface runoff as streams and rivers.

Lakes may form as the result of damming or in natural and man-made basins. In terms of the hydrologic cycle, one of the most important aspects of rivers and lakes is the evaporation of water from their surfaces. The development pattern for a river in humid regions is to grow progressively larger as it nears the sea. Each tributary adds its bit to the total volume of water flowing from the land, and if energy demands of the atmosphere are not too great, precipitation can keep pace with or exceed evaporation. In arid climates, particularly at low latitudes, where the dry air of the subtropical high-pressure belt dries out the land beneath, rivers may actually grow smaller as they near the sea. Typical of such streams are the Nile in North Africa and the Orange River which rises in Lesotho and flows westward across the Republic of South Africa. In both cases the sources of these rivers are at high elevations in better-watered regions. Their downstream portions at lower elevations cross deserts where no tributaries enter. *Exotic streams* such as these give up enormous quantities of moisture to the air and in the case of the Orange River may actually have a dry streambed downstream while water still flows near the source. Another way in which exotic streams lose large quantities of water is through *influent flow* which gives up water to sands and gravels which they may cross. Just the opposite occurs in humid regions where groundwater continuously enters streams increasing their flow (Figure 5-15).

## Water in Vegetation

The above description has treated all water as simply a flowing mineral substance without

*Physical Geography: Environment and Man*

106

reference to its place in life processes of plants and animals. We must now come closer to the earth and choose a still larger scale on which to view the hydrologic cycle. It is in its use within life forms that the true miracle of water occurs.

"Willow weep for me," once sang a popular entertainer. Anyone who stands beneath a willow tree on a hot day will be bathed in a fine mist dropping from its leaves. Desert streams lined with vegetation flow on the surface at night during periods of low energy demand. In the daytime the plants along the bank can draw so much water from the streambed that if the streams are small, they may actually go temporarily dry. Water is essential for plant growth and for plant health. It constitutes a major portion of active plant tissue; it is a reagent in the photosynthetic process, and a solvent which serves as a carrier transporting salts, sugars, and other dissolved materials throughout the plant. Just as important is water's role in maintaining the energy balance of every living thing, including vegetation. When air temperatures are high and radiation is intense, the body heat of a plant may increase to the point where chemical and other processes within it can be seriously and perhaps permanently impaired. Vegetation overcomes this problem by releasing water in vapor form from small openings, or *stomata*, on leaf surfaces. This process is known as *transpiration* and in many ways is the equivalent of perspiration in animals. The change in state of the water from liquid to vapor removes excess heat from the plant tissue. If, for example, a leaf 10 square centimeters in area (1.55 square inches) gives up only 0.005 gram of water each minute, this can produce an energy loss of 3 calories and a lowering of tissue temperature by as much as 15°C (59°F). The need for water is real and immediate. For example, a single corn plant can transpire as much as 204 liters (54 gallons) of water during its growing season. An acre of corn containing 6,000 plants would need 1,225,000 liters (324,000 gallons). A similar study has found that, in addition, an amount equal to somewhat more than 40 percent of the water used in transpiration *evaporates* from a corn plant. In other words, an additional 83 liters (22 gallons) of water would

pass from the surface of the plant to the atmosphere.

The combined demands placed upon available water supplies by both evaporation and transpiration have been given the term *evapotranspiration*. In high-sun periods the *potential evapotranspiration* possible if water were available in unlimited quantities can actually exceed the *actual evapotranspiration* that takes place. When the difference between these two quantities is negative, a *water deficit* occurs. If the deficit is great enough, plants will be forced to give up part of the water normally stored in their tissues. Depending upon the type of plant, permanent wilting will occur after from 25 to 75 percent of their water content is lost. Obviously, the water needs of plants will vary from season to season and from region to region. We will discuss these matters in greater detail in Chapter 6. For the moment, it is only necessary to add that man often intercedes and attempts to provide needed moisture for his thirsty crops through irrigation.

*Evapotranspiration: A national balance sheet*

The complexity of the hydrologic cycle with its hierarchy of shorter cycles nested within ones of longer and longer duration prevents our making exact statements about it. It is useful, however, to attempt to balance the ledger of precipitation and evapotranspiration for a large area like the United States. Figure 5-16 presents such an effort in graphic form. The source of all but a small portion of the moisture is assumed to be the atmospheric reservoir. In this case, only the estimated amount of precipitation has been taken into consideration. Though precipitation appears in the diagram to fall and evaporate only once, the same molecule of water might, in reality, make several cycles.

Nearly 5 billion acre-feet of water are involved. By *acre-foot* we mean the amount of water necessary to cover 1 acre to the depth of 1 foot, a total of 43,560 cubic feet, or 326,700 gallons. The first debit in our national ledger is somewhat more than 3.3 billion acre-feet which

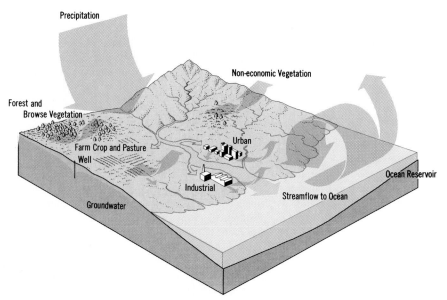

**Figure 5-16  United States water balance (hydrologic cycle)**  It is estimated that nearly 5,000 million acre-feet of water are precipitated on the United States each year. If this constitutes 100 percent of the total (actually, an additional 6 million acre-feet of water are mined from aquifers), then an equal amount must eventually find its way back into the oceans and the atmosphere in order to maintain the system in reasonable equilibrium. Much of the total is evaporated from farmland and noneconomic vegetation back into the atmosphere. A small amount is used by municipalities and industry. The water which runs off into the sea eventually evaporates and returns to the land as precipitation. See Table 15-1 for the amounts indicated by each arrow. (Based on data in "Water Resources." Report to the Committee on Natural Resources of the National Academy of Sciences, *National Research Council Publication 1000-B,* Washington, D. C., 1962)

are evaporated and transpired from all the non-irrigated land in the United States. Note that this constitutes 71 percent of the total, nearly half of which (39 percent) is used by plants which are in some way utilized by man and his animals. The remaining 29 percent runs off the land or falls upon lakes and streams and ends up as stream flow. At this point water stored in aquifers in previous years, perhaps as long ago as the last ice age, is pumped out for a variety of human uses including irrigation. This water is mined just as we might mine petroleum or solid minerals. Once such water is removed, there is little or no chance of its being replaced within our lifetime.

Of the 1,380 million acre-feet found in Ameri-

can streams, about 345 million are withdrawn for human use for varying lengths of time. Most of this is utilized in irrigating farm crops; a surprisingly small amount is permanently removed for industrial or municipal purposes. Some water flows directly to the sea; the remainder, minus only about 100 million acre-feet, is returned to the river system. Finally, the ocean reservoir returns the balance to the atmosphere. For all practical purposes, all the precipitation that falls finds its way back, or perhaps even somewhat more returns than fell if we wish to include mined water. Table 5-1 and Figure 5-17 help to clarify this discussion. The same water movements are described as those shown in Figure 5-16, but the use of the

**Table 5-1  U.S. Water Balance (estimated)**

| | Millions of Acre-feet | | % of Total Precipitation |
|---|---|---|---|
| Input: Annual precipitation over entire U.S. | 4,750 | | 99+ |
| groundwater | 10 | | + |
| | 4,760 | | 100 |
| To: Nonirrigated land (and returned to atmosphere)* | 3,380 | | 71 |
| To: Stream flow | 1,380 | | 29 |
| | | | 100 |
| Withdrawn: | 345 | 7 | |
| For: Irrigation | 159 | 3 | |
| Industry | 159 | 3 | |
| Municipal | 27 | 1 | |
| Output: Returned to atmosphere from land | 3,480 | | 73 |
| From: Irrigation | 95 | 2 | |
| Industry | 3 | + | |
| Municipal | 2 | + | |
| From: Nonirrigated land* | 3,380 | 71 | |
| Farm crop and pasture | 1,100 | 23 | |
| Forest and browse vegetation | 750 | 16 | |
| Noneconomic vegetation | 1,530 | 32 | |
| Total stream flow reaching ocean reservoir | 1,280 | | 27 |
| | | | 100 |
| Returned to streams | 245 | 5 | |
| From: Irrigation | 64 | 1 | |
| Industry | 156 | 3− | |
| Municipal | 25 | 1− | |
| Not withdrawn | 1,035 | 22 | |

Note: Water may be held for varying lengths of time in the various parts of the cycle.
Source: Based on data from Abel Wolman, "Water Resources a Report to the Committee on Natural Resources of the National Academy of Science," *National Research Council, Publication 1000-B*, Washington, D.C., 1962.

graphic form in Figure 5-17 shows how this technique can help us to conceptualize complex processes.

In view of the above data, we can once again raise the question, *if water is essentially available in quantities many times greater than current human needs, and if practically all of it is recycled, why do we face a growing water crisis?*

The answer is as complicated as the hydrologic cycle itself. In a later section of this book we will analyze cases demonstrating why water is scarce and growing scarcer. A quick reply to our own question at this point needs to emphasize the *quality of the water supply* and its *distribution in time and space.* A common perception of stream use by city residents is that

*Energy and the Earth: The Hydrologic Cycle*

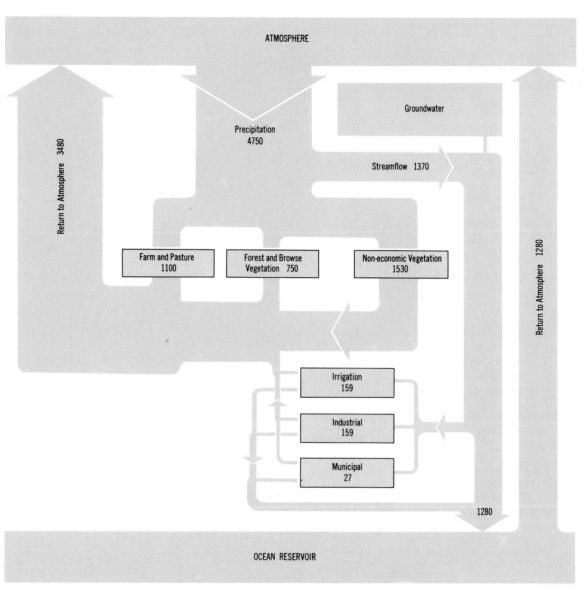

**Figure 5-17 Systems diagram of the hydrologic cycle of the United States** This diagram represents the same water flow as depicted in Figure 5-16 and Table 5-1. (Source of data: op. cit. Figure 5-16)

upstream water flowing to the city should be pure and drinkable. On the other side of the city, rivers are seen as convenient sewers for the citizens and industry. The difficulties, of course, are that usually more than one city utilizes a given river; that population is not always lo-

cated where water supplies are most abundant; and that a serious lack of fit often occurs between the spatial organization of human systems and the inherent spatial structuring of natural ones.

## Breakdowns in the Hydrologic Cycle

When asked on an examination to define the "hydrologic cycle," a student once replied that it is "a kind of vehicle used to cross swamps." While that answer is long on imagination and short on fact, it contains a basic perception that has an element of truth. The hydrologic cycle is in many ways like a complex and fragile machine. It can easily be put out of adjustment by human tinkering, and should be interferred with as little as possible.

Because we invoke the concept of environmental unity wherever possible and emphasize a synthesis of several approaches into our description of human/environmental relationships, we have already mentioned the problem of city water supplies in Chapter 3 and will look at other water-related issues again in Chapters 9 through 13. Rather than trying to pour all references to water into the present discussion, let us close this chapter with a brief look at two ways in which the hydrologic cycle can "get out of whack" through human intervention.

### Urbanization and runoff

Line-area relationships come immediately to mind. Earth surfaces covered with natural vegetation absorb precipitation in large quantities and slow the rate of runoff. It takes a long time for the water poured on an area by a storm to filter leaves, the litter on a forest floor, and through a deep soil mantle into the streams which drain an area. The overall effect is gradual and the runoff is spread over a time interval which allows existing channels to carry rainfall off without excessive flooding. Quite the opposite is true in urbanized areas, where impervious surfaces such as streets and parking lots prevent the percolation of water into the soil. The growth of paved space is illustrated in Figure 5-18. Where the automobile dominates, nature suffers. Man-made channels and gutters are usually closely spaced and thus shorten the time between the moment of precipitation and the eventual accumulation of runoff in larger channels. Furthermore, natural stream channels with their meanders or curves are often

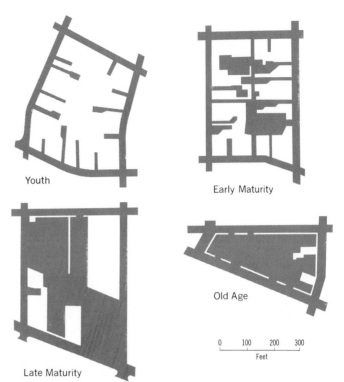

Youth

Early Maturity

Old Age

0    100    200    300
Feet

Late Maturity

**Figure 5-18  Stages in the expansion of automobile territory**  Examples are from four blocks in East Lansing, Michigan, and represent surfaces hardened for automobile use. (Ronald J. Horvath, "Machine Space," *Geographical Review,* vol. 64, April 1974, p. 175, copyrighted by the American Geographical Society of New York)

straightened and paved when planners and engineers set out to improve the system. As a result, the velocity of the runoff is slowed much less than under normal stream conditions and no water can be absorbed by permeable channel beds. Finally, people have very little awareness of the danger of flooding and build their settlements on vulnerable flood plains. (This topic is discussed again in Chapter 11.) Given all the above human-induced conditions, it is not surprising that the potential for flooding, that is, increased magnitude and frequency, increases in urbanized areas. Studies by S. D. James and Luna B. Leopold indicate that the discharge from a catchment area after it has undergone urbanization with all its attendant paving and destruction of vegetation can increase from 1.5

*Energy and the Earth: The Hydrologic Cycle*

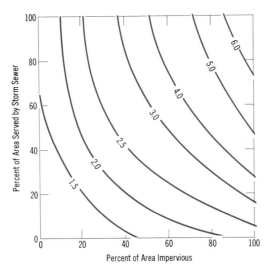

**Figure 5-19 Increase in runoff as a function of impervious surfaces and availability of storm sewers** Values on curves are ratios of discharge after urbanization to discharge before urbanization. (After Leopold, *Hydrology for Urban Land Planning: A Guidebook on the Hydrologic Effects of Urban Land Use,* U.S. Geological Survey Circular #554, U.S. Government Printing Office, Washington, D.C., 1968)

to 6 times the preurbanized amount. Figure 5-19 indicates this relationship.

An urbanized and essentially paved area must be served by adequate storm sewers. If the sewer distribution (a line-area relationship) is inadequate, flooding results. For example, in the diagram, if 80 percent of the area is impervious and only 20 percent of it is served by storm sewers, then discharge will increase by 250 to 300 percent. Nor does the problem end there. Even if sewers can handle increased runoff, the usual pattern is that storm waters are directed into sanitary sewers designed to ordinarily handle only the discharge from toilets and similar sources. The result is that sewage treatment plants are overwhelmed during rainy periods and must either have storage basins to hold the increased but unsanitary load until it can be slowly purified, or, as is often the case, must simply let it bypass the plant and empty into the nearest river, lake, or bay. New York, Boston, Detroit, Philadelphia, and Washington, D.C.,

were among the 1,943 communities in the United States serviced by such *combined sewers* in 1964. Conditions had not improved by 1974. The resulting pollution of natural water bodies is one of the major sources of water contamination in the nation.

*Saltwater intrusion*

We have already mentioned the interface between salt water and fresh water along the coastlines of the continents. This contact zone is kept in spatial equilibrium (that is, it remains more or less in the same place) by the hydraulic pressure of fresh groundwater moving downslope from the land. Salt water pressing inland from the sea wedges under the fresh groundwater because of its greater density (Figure 5-14). This relationship creates a situation easily disturbed by human activity. As water consumption increases in urbanized areas, surface streams cannot meet the demands placed upon them and more and more water is pumped from the groundwater supply beneath the cities. The result is that the surface of the water table is lowered and hydraulic pressure decreased. When this happens in coastal areas, the saltwater wedge moves inland and wells become brackish or salty.

There are many examples of this type of human-induced saltwater intrusion. Aquifers underlying parts of New York City and nearby Nassau County which once were fresh are now filled with salt water. As early as 1930 Brooklyn was forced to stop using groundwater because of saltwater intrusions into the area. Subsequent studies of the problem have revealed that if 25 percent of the water withdrawn from the aquifers underlying Long Island were replaced, the invasion of salt water could be halted (Figure 5-20). The rate of advance in this case is relatively slow because the geologic formations through which the salt water must penetrate are nearly impermeable clays and glacial materials. As a result, according to simulation of the problem conducted by the Massachusetts Institute of Technology, the intrusion advances at less than half a mile per century. On the other hand, this scarcely perceptible invasion may mean that public awareness of the impending difficulty may come too late. Also, once the damage is done, corrective measures, such as recharging the aquifers with fresh water, will take equally long to reverse the situation.

In other areas the rate of intrusion can be disastrously rapid. Urban development along the shores of Biscayne Bay in Florida, with the accompanying pumping of groundwater supplies and the digging of drainage and access canals, has allowed major saltwater intrusions to occur in less than six decades. In Chapter 13 similar intrusions are mentioned in the discussion of Los Angeles as a human ecosystem. This situation has also become an unpleasant fact on many Pacific islands and elsewhere around the world. However, it is not our purpose to catalog all such events. Rather we hope to show how the interaction of mankind with the physical environment results in critical situations which a knowledge of geography can help clarify and correct.

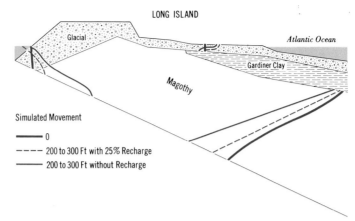

**Figure 5-20 Simulated saltwater intrusion, Long Island, New York** A model simulating saltwater intrusion was used to estimate the behavior of the saltwater body under different water-management schemes. The diagram has extreme vertical exaggeration. (After M. A. Collins and L. W. Gelhar, *Groundwater/Hydrology of the Long Island Aquifer System,* M.I.T. Hydrodynamics Laboratory Report 121, 1970, Cambridge, Mass.)

*Energy and the Earth: The Hydrologic Cycle*

# 6 | EARTH AS THE ECOLOGIC HABITAT FOR LIFE: CLIMATE AND EVAPOTRANSPIRATION

Life exists in a thin film on the surface of the earth, spread like an oil slick on water. And like an oil slick reflecting sunlight, it is colorful, complex, and always changing. We are part of that film and depend upon it for our existence. Life-forms constantly influence one another by changing each other's environments, sometimes benefiting, sometimes destroying, a particular species. In recent centuries man has become the dominant ecologic element in the total life environment. Many of the changes we are causing have ominous consequences for our own survival. We are well advised to appreciate and protect other life-forms such as vegetation and the organisms which live in the soil. To do this we must first understand the processes of which they are parts.

The possibilities for life vary greatly over the surface of the earth. Life-forms are configurations of giant molecules that use energy to maintain internal organization and local environmental conditions, to pass material through their systems, and most remarkably, to create organic structures which are capable of reproduction. This process requires a physical and chemical environment which cannot vary beyond certain limited ranges. The planet earth seems to be a rather special place where such environmental ranges can occur.

If we seek beyond our own world for other life-forms, we may find it lonely in our corner of the galaxy. Conditions suitable for sustaining life as we know it are seldom found on the planets. It is most unlikely that we have living neighbors in the solar system. Mercury shows one burning face always to the sun; the other is forever frigid and dark. Venus is too dry and too hot and is choked in carbon dioxide. Mars may have simple forms of life, but its atmosphere is unsuitable for humans and its temperatures are too low for comfort. The outer planets—Jupiter, Saturn, Uranus, and Neptune—are relative giants with dense, freezing atmospheres of hydrogen and helium. Pluto is so far from the sun that it remains frozen and dim. It is most unlikely that there are any planets suitable for humans circling nearby stars. Since we are far removed from inventing and perfecting an intersteller spaceship, we had better learn to live with and within the thin skin of habitable earth environments we have inherited.

## Earth Climates as Earth Environments

Energy—predominantly solar energy—acts upon the atmosphere, hydrosphere, and lithosphere of the earth. Materials at and near the surface are constantly reworked and rearranged into forms and distributions which, in various combinations, provide homes for many types of life. We have already discussed the distribution in space and time of insolation upon the surface

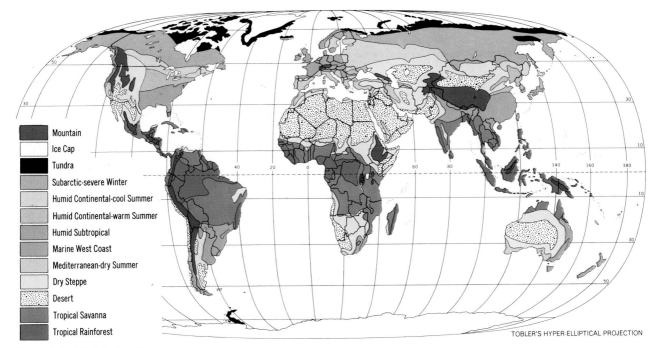

TOBLER'S HYPER-ELLIPTICAL PROJECTION

**Figure 6-1  World climates**

of the earth. We have also given some indication of the movement of air and water from place to place, and of the consequent redistribution of available energy and moisture.

A broad view is necessary in order to understand the environments created on the earth's surface by the above processes. In the words of Barry Commoner, "Everything depends on everything else." This attitude forms the basis for the geographic concept of *environmental unity*, and provides a suitable mode for summing up much of the subject matter presented in the preceding chapters. Such a summary can for convenience be phrased in the familiar terms of climatic distributions. Climate is an in-place characterization of the combined effects of energy and moisture availability. Weather can be considered a moving phenonmenon. A squall line marches across the landscape; great pinwheels of high- and low-pressure cells spin majestically from coast to coast. Conversely, we always think of climate's being associated with a particular spot. Thus, the combined effect of

many different kinds of weather makes up the climate of a given area. Weather and climate, in turn, have much to do with sculpturing the earth's surface where particular combinations of energy and moisture occur. Given the same underlying materials, an arid landscape is completely different from one formed under humid conditions. Thus, climate is a major control not only of vegetation and soils, but also of the shape of the land itself. It is logical then to begin with an understanding of climate before considering total earth environments.

*Plants as indicators of climatic characteristics*

Climate is an abstraction. Conditions of moisture, temperature, wind, and many other variables change from moment to moment at any given point. If we attempt to summarize the elements of climate at some geographic location, we will be faced with two problems. We must choose a time span of suitable duration

*Earth as the Ecologic Habitat for Life: Climate and Evapotranspiration*

**115**

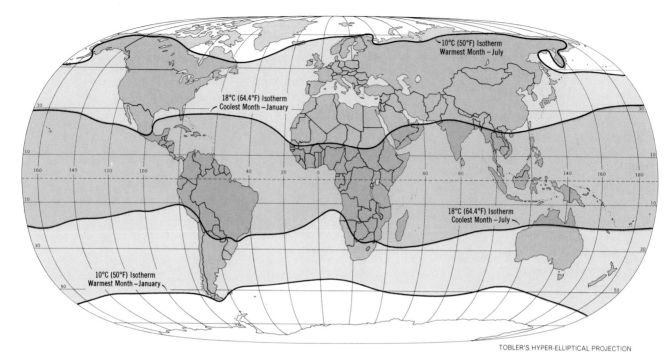

**Figure 6-2   10°C (50°F) isotherms of warmest month—18°C (64.4°F) Isotherms of coolest month**   An isotherm is a line of equal temperature. The 10°C isotherms for the warmest month (July in the Northern Hemisphere mark the most poleward extension of trees—the arctic tree line. The 18°C isotherms for the coolest month define a band around the equator where true palm trees are able to survive.

and we must specify the limits of the area. In essence, the dividing of the world into climatic zones is an exercise in region building. To make such regions, we must generalize the data available to us. While this can be done in a variety of ways, one of the quickest is to let nature help out. Vegetation of all kinds is a sensitive indicator of the natural conditions generalized as climate. Every plant has a particular set of physical needs for energy and moisture, and such needs must be met throughout the entire year in a fixed sequence of events. Thus, the timing of climatic events is as important to a plant as are absolute annual accumulations of water and solar energy. If vegetation is allowed to develop relatively undisturbed by human activities and for a long enough period of time, the geographic distribution of plants will reflect in large part the geographic distribution of

climatic types. There are many exceptions to this rule, particularly where man is concerned, and it must be used with caution. Nevertheless, most climatic distribution maps such as that shown in Figure 6-1 are in the final analysis based on the distribution of vegetative types.

Two examples should illustrate this point. The location of an isoline marking the distribution of points having an average July temperature of 50°F, or 10°C, (Figure 6-2) roughly corresponds to the poleward limit of normal tree growth. Thus, the July 50°F isotherm in the Northern Hemisphere is sometimes called the *arctic tree line* and marks the transition between the summerless climates of the high latitudes and the mid-latitude climates where summer alternates with winter. Conversely, true palms survive only equatorward of the isoline marking approximately 18°C (64.4°F) during the coolest month of the year (Figure 6-2). In each of these cases, special microenvironments may encourage the growth of these plants beyond their stated geographic limits, but such occurrences are relatively rare. In the case of climatic maps, as in the case of city maps, our perception of the edge of a particular region depends upon the scale and detail with which the maps are drawn. It also is important to remember that the neat lines on maps showing the break between climatic regions are in reality zones of transition between sets of conditions which themselves may contain considerable variance. For example, the arctic tree line viewed at a larger scale (Figure 6-3) actually interfingers with the tundra to its north. Finally, the oceans have climates just as much as does the land, although many climatic maps neatly terminate the conditions they depict at the water's edge.

Nevertheless, the relationship between vegetation and climate is such that we need to discuss it at some length. The response time of plants to climatic conditions provides us with a convenient time scale for looking at environmental events, since humans also operate in nature within approximately the same temporal dimensions. However, before going on to the vegetative and soil systems which support mankind, let us attempt a quick synthesis of the variables which account for the world distribution of climatic types.

## Climatic controls: A recapitulation

If the world were a completely homogeneous sphere without land and water or variations in topography, there would be an orderly temperature gradient from the equator to the poles. Climates would be arranged as a series of latitudinal bands girdling the earth with symmetrical occurrences on either side of the equator. However, many other climatic controls serve to scramble this theoretical symmetry. The list of controls that follows should not be considered complete, although it does include the major elements which help to shape the climate regions of the earth. We, in passing, will mention each control and its effect on climate. Since some of these elements have already been

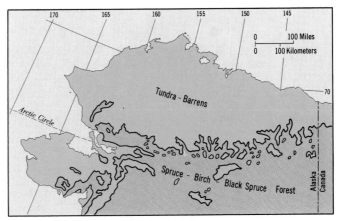

**Figure 6-3  Arctic tree line**  The arctic tree line viewed at a neighborhood scale reveals interfingering and isolated patches of both tundra and trees.

treated at length, only a few will receive special attention at this time.

Latitude-temperature

Pressure belts and winds

Land-water relationships

Ocean currents

Cyclonic systems (pressure cells)

Mountain barriers and highlands

**Latitude-temperature**  We have already mentioned the temperature gradient from the equator to the poles. It is interesting to note that just as ancient Greek geographers spoke of the "torrid, temperate, and frigid" zones, it is convenient to broadly specify three climatic types: (1) the equatorial-tropical group, which has no true winters; (2) the middle-latitude group, where winter alternates with summer; and (3) the arctic-polar group, which has no true summers. We can use this observation to summarize many of the others we already have made on temperature distributions. The three temperature zones are shown in Figure 6-2.

**Pressure belts and winds**  So much has already been said on these twin subjects that we summarize them here in the briefest possible

*Earth as the Ecologic Habitat for Life: Climate and Evapotranspiration*

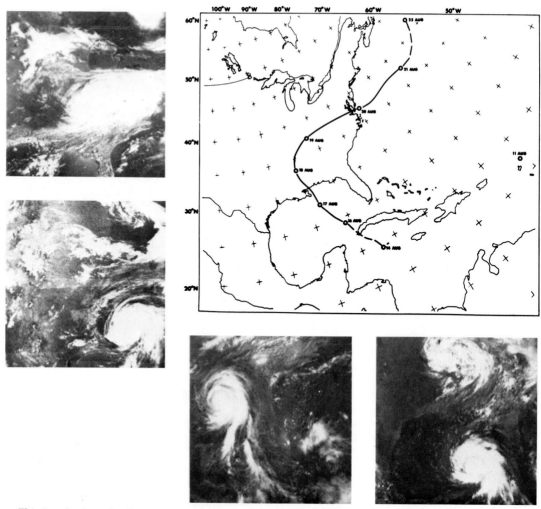

This is a daytime, day-by-day photographic history of hurricane Camille, one of the deadliest hurricanes in recent history. These pictures were taken by the National Aeronautics and Space Administration's Nimbus III meteorological satellite as the storm developed into a tropical storm, slammed into the coast of the United States on the Gulf of Mexico, and traveled northeast out into the Atlantic. These pictures show that the storm continued to pack a wallop even as it was moving across Virginia toward the ocean. (NASA)

way. While recognizing that the latitudinal distribution of high and low pressure rarely takes the form of neat "belts," it is still possible to identify bandlike pressure regions alternating from the equator to the poles. The equatorial low-pressure area is associated with convectional rainfall; subtropical high-pressure areas are typified by general aridity; the subpolar low-pressure belts are less well defined but correspond with relatively greater precipitation; finally, the poles are noted for their piled-up masses of frigid, dry air and scant precipitation. Only the lack of solar energy, and consequent low polar and subpolar temperatures, allows the accumulation of surface water in large quantities at high latitudes.

**Land-water relationships** Climates can also be divided into two classes based on their association with bodies of either land or water. *Marine climates* are relatively mild with warm but not hot summers and cool but not cold winters. This is the result of the insulating effect of water as compared with land. These climates are also noted for winter precipitation maxima associated with storms originating at sea. *Continental climates*, on the other hand, tend to be much more extreme, with hot summers and cold winters. Again, the poor insulating quality of the land accounts for this. For the same reason, isotherms bend equatorward over the centers of continents in the wintertime and poleward in the summer. The annual range between highest and lowest temperatures is far greater for continental climates than for marine ones. Precipitation is usually greatest in the summertime and derives from convectional showers over the heated land. Also, in the interiors of the continents, far from the oceanic sources of atmospheric moisture, aridity can be a problem. At the same time, high summer temperatures over the land can produce low-pressure areas which, in turn, allow the onshore movement of marine tropical air masses. The mechanism for such transfers is incompletely understood, but the famous Indian monsoon winds with their accompanying rains and the autumn hurricane season along the southeastern coast of the United States are undoubtedly associated with this. The result is that the southeastern shores of the continents in the Northern Hemisphere are often better watered than might be expected.

**Ocean currents** We have already described the movement of warm and cold currents along the continental shores. Such currents tend to bend isotherms poleward along the west coasts of the continents, while shores at comparable latitudes on the eastern coasts are comparably much colder. This effect is shown in Figure 4-10. Since cold currents cool the air above them just as warm currents somewhat heat the atmosphere, there is a distinct correlation between cold currents and dry coasts and warm currents and moister shores. Warm, moist air

moving inland is lifted and cooled, thus often giving up precipitation. Cold, dry air moving onto the shore is warmed and takes up additional moisture. The former condition contributes to the humid character of eastern coasts, while coastal deserts such as the Namib in South-West Africa and the Atacama in Peru owe their existence to the presence of the cold Benguela and Humboldt Currents, respectively.

**Cyclonic systems (pressure cells)** Alternating masses of dense, cold air and lighter, warmer air move eastward around the globe in a stately procession at mid-latitudes. Originating over the major bodies of land and water, these pressure systems are most evident in the Northern Hemisphere in the winter months when their tracks swing southward. The contact between warm and cold air results in the frontal precipitation relied on by so many of the world's people.

**Mountains barriers and highlands** Topographic barriers are important elements in the production of orographic rainfall like that of the northwest Pacific Coast and the Khasi Hills of India. Mountains also serve as major barriers limiting certain kinds of weather and climate. Typical of this is the north-south orientation of climatic regions in the western United States as opposed to the east-west trend of similar climates in Europe. Both owe their alignment to major mountain chains. There are other cases where a mountain barrier has profound effects. The Tsing Ling Mountains of central China effectively halt the northwestward movement of the monsoonlike winds that blow onto the China coast from the South China Sea. The southern Andes deflect the eastward movement of cyclonic storms and create the Patagonia Desert in Argentina. Even small barriers like the hills of San Francisco, California, can hold back masses of air (photo, p. 120). Sausalito, on the eastern side of the Marin County Peninsula, will oftentimes be sunny and in the clear while the oceanside is shrouded in fog. The same thing is true to the south, where Palo Alto frequently has open skies while fog fingers poke

*Earth as the Ecologic Habitat for Life: Climate and Evapotranspiration*

Fog over San Francisco Bay   The low mountains and hills on the ocean side of San Francisco Bay are sufficient to hold back the ocean fog, which dissipates as it descends on their landward flanks. The fog is able to enter San Francisco Bay through the gap of the Golden Gate, and it is along this route that the fog for which the city is famous travels into the city. The landward suburbs are usually clear. (Courtesy of A. Miller, Meteorology, California State University, San Jose.)

unsuccessfully through the Santa Cruz Mountains to the west.

Where plateaus and major uplands stretch for miles, the effect is the creation of new climatic regions. The wet-and-dry, semiarid uplands of East Africa are a case in point. The highlands of Kenya counteract that nation's equatorial location insofar as a wet-and-dry, semiarid climate prevails there instead of the year-round convectional rainfall normally expected at the equator. The Tibetan Plateau and the Altiplano of Bolivia also represent large upland areas with colder, drier climates than their latitudinal locations alone would dictate.

### World Climate Regions

A number of scientists have attempted to devise logical regional classifications of world climates. The most famous is that of the Austrian Wladi-mir Köppen, whose work illustrates the idea that the distribution of vegetation is a clue to the distribution of climate types. A somewhat more recent classification by C. W. Thornthwaite, an American, also concerns itself with the relationship of vegetation to moisture and solar energy but takes these three elements into consideration in a more systematic manner.

Köppen's method of classification is operational at three scales: global, continental, and subcontinental. At each scale letters of the alphabet are assigned to indicate different climatic elements. At the global scale he distinguished five major climatic types which can be areally designated. These are the (A) tropical rainy, (B) arid, (C) mid-latitude humid cool winter, (D) mid-latitude humid cold winter, and (E) polar climates. The polar climates are separated from the mid-latitude climates thermally at the 10°C (50°F) isotherm which we have already noted approximately parallels the arctic tree line. Tropical climates fall between the 18°C (64.4°F) isotherms north and south of the equator. The two mid-latitude climates are further subdivided thermally and spatially by the −3°C (26.6°F) isotherm, while a somewhat more complicated formula plotting the excess of evaporation over precipitation defines the boundary between wet and dry regions.

At a slightly larger scale Köppen distinguished three types of tropical climates: rainy, wet-and-dry, and monsoon. Arid climates were subdivided into steppe and desert. Mid-latitude climates fall into five categories (shown in Table 6-1), and polar climates become tundra and ice cap. The effect of this further subdivision is the creation of twelve types of climate regions in place of the original five. He continued his exercise in regionalization by adding eight possible sub-subdivisions to the tropical climates; six sub-subdivisions to the arid climates; thirteen to the mid-latitude humid climates; but no finer subdivisions to the polar category. The end result is a system of spatial classification with at least twenty-seven possible subdivisions. This proved so complicated that geographer Glenn T. Trewartha simplified the scheme and used only seventeen categories. The system of climatic classification proposed in 1955 by Thornthwaite and Mather involves the computation of a moisture index ($I_m$) which takes into account the seasonality of water availability and water need. Almost every place in the world has some variation in the amount of insolation and precipitation it experiences. Very often need and availability of water do not coincide and there may be deficits and surpluses in the same year. Their moisture index recognizes this and is expressed

$$I_m = \frac{S - D}{n} \times 100$$

That is, the moisture characteristics of a given place ($I_m$) are equal to the water surplus ($S$) or runoff minus the water deficiency ($D$), divided by the biotic and atmospheric demand for water ($n$). (This latter demand is called "potential

**Table 6–1  Köppen's System of Climate Classification**

A  Tropical rainy climates

    Af    rainy
    Aw  wet and dry
    Am  monsoon
         +8 sub-subdivisions

B  Arid climates
    BS  steppe
    BW  desert
         +6 sub-subdivisions

C  Mid-latitude humid cool winter climates

    Cf   Temperate, moist year round
    Cs   Temperate, dry summer
    Cw  Temperate, dry winter

D  Mid-latitude humid cold winter climates

    Df   Cold, moist year round
    Dw  Cold, dry winter
         +13 sub-subdivisions for C and D combined

E  Polar climates

    ET  Tundra
    EF  ice cap
         no sub-subdivisions

Note: The upper- and lower-case letters preceeding the climatic descriptions were used by Köppen, Geiger, and others as a shorthand or code for convenience. As such, they do not concern us here. Any standard work on climatology discusses the code in detail.

evapotranspiration" and is discussed in more detail shortly.)

While the values of the humidity index can range along a continuum from −100 where there is zero precipitation to more than +100 where rain falls in large quantities, Thornthwaite and Mather divided the values into five principal moisture categories with three variations (Table 6-2).

A moisture index of zero indicates that neither surplus nor deficit occur. When these values are plotted on a map of North America (Figure 6-4), the zero line is of considerable

*Earth as the Ecologic Habitat for Life: Climate and Evapotranspiration*

**Table 6-2 Principal Categories of Thornthwaite and Mather's Moisture Index***

| Moisture Index | Symbol | Name |
|---|---|---|
| 100 and above | A | Perihumid |
| 80 to 100<br>60 to 80<br>40 to 60<br>20 to 40 | B | Humid |
| 0 to 20 | $C_2$ | Moist Subhumid+ |
| −20 to 0 | $C_1$ | Dry Subhumid |
| −40 to −20 | D | Semiarid |
| Below −40 | E | Arid |

*D' (Subpolar) and E' (Polar) are special cases of aridity, the conditions of which are described in the pages that follow.
+This category is subdivided for reasons of particular interest which are explained in the text.

importance in that it separates the dry west from the humid east. We will have more to say about this when we discuss soil systems and classifications in Chapter 9.

The essential difference between the Köppen and Thornthwaite classifications is one of approach. Köppen's method is inductive, using vegetative patterns to establish lines of temperature and precipitation as separators. Thornthwaite's method is deductive, deriving indices directly from the climatic processes, primarily temperature and precipitation.

We have already mentioned some of the major weaknesses of schemes such as these. Boundaries are shown as clean lines of demarcation rather than as more realistic transition zones. Moreover, vegetation distributions and climatic conditions do not clearly parallel each other everywhere on earth. It is also important to note that when eight or seventeen or twenty-seven regional types are applied across the heterogeneous earth with subsequent repetition of the same type in several spots, the maps that result can be difficult to interpret. In this sense, the above methods of describing and locating climate regions of the world become ends in themselves. That is, the resulting maps become the reasons for the effort it takes to produce them.

If our intentions are more practical and are oriented toward the definition and subsequent solution of geographic problems, we must ask ourselves the reasons why we want to define and locate types of climates. Our questions are threefold: What kind of environments have resulted from interactions between the energy and moisture systems we have already described? What kind of human interactions take place with different types of environments? And what are the spatial relationships between different subsets of the total human-environment system? Therefore, our knowledge of climate should give us some method of understanding the dynamic elements underlying a given distribution of climate types, as well as a simple way of classifying energy-moisture environmental relationships. Having these in hand when we turn to the real world, either Köppen's or Thornthwaite's maps of climate distributions will serve our purposes, while the number of climate subtypes we use should reflect our own judgment of the number we need.

### Climatic controls—a simple synthesis

The list of climatic controls briefly reviewed in the last few pages offers us a quick and relatively simple method of understanding climatic

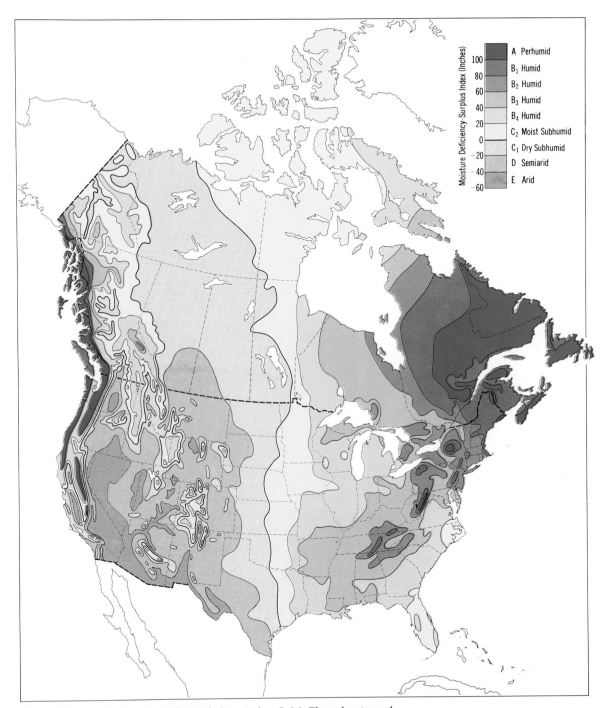

**Figure 6-4  North America moisture index**  (After C. W. Thornthwaite and
J. R. Mather, *The Water Balance,* Publications in Climatology, vol. 8, no. 1,
Laboratory of Climatology, Drexel Institute of Technology, Centerton,
New Jersey, 1955; and M. Sanderson, "The Climates of Canada According
to the New Thornthwaite System," *Scientific Agriculture,* vol. 28, no. 11,
November 1948, pp. 501-517)

Moisture Deficiency-Surplus Index (Inches)

| | |
|---|---|
| A | Perhumid |
| B₁ | Humid |
| B₂ | Humid |
| B₃ | Humid |
| B₄ | Humid |
| C₂ | Moist Subhumid |
| C₁ | Dry Subhumid |
| D | Semiarid |
| E | Arid |

distributions. If we create a hypothetical continent like the shield-shaped one shown in Figure 6-5, we can endow that continent with a few basic characteristics drawn from our list. The shield shape roughly approximates the shape of at least three world land masses. At the same time, we can move our continent of Hypothetica to any latitudinal position we wish.

We have chosen in Figure 6-5 to place Hypothetica at the same latitude as North America. We thus expose it to various conditions of aridity and moisture based on the world pressure belts shown in Figure 6-5A. Next we add a slightly drier interior portion of the country as well as a better-watered southeast coast (6-5B), in recognition of warm offshore currents and the monsoon effect. The result is a continent having the basic moisture regions shown in Figure 6-5C. We next impose temperature zones on this configuration. Such zones are the simple

**Figure 6-5 Hypothetical continent** The effects of the energy and moisture conditions create climatic regions postulated for a hypothetical continent. *A*. Pressure—precipitation. *B*. Moisture conditions: continentality and east coast monsoon. *C*. Moisture regions. *D*. Temperature zones. *E*. Climatic regions.

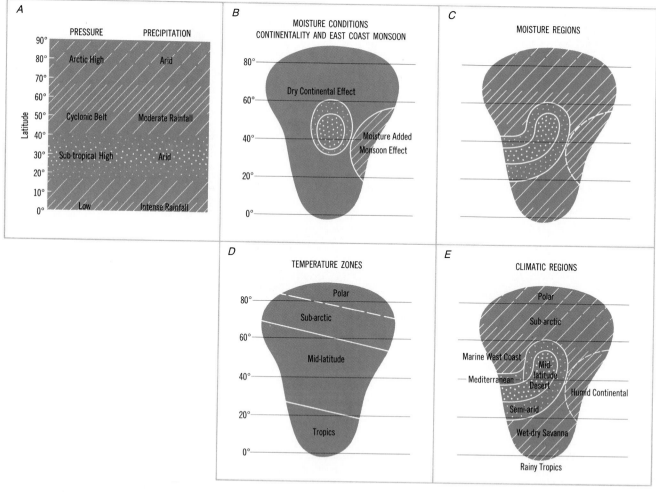

three-way division discussed earlier. The only embellishment to this triple scheme is a slight north-south skewing of the bands as the result of ocean currents (Figure 6-5D). The result shown in Figure 6-5E comes close to the actual pattern of climates in North America. It would be even more similar if we were to add a major topographic barrier along the west coast. However, we leave this to your imagination.

Beginning in the extreme south near the equator, let us quickly comment on the eight climate regions we have specified. A small bit of the rainy tropics comes first. This would correspond to portions of Central America. Directly to the north is an area where a wet season alternates with a dry one. This area is still within range of the vertical rays of the sun during the June solstice and receives convectional rainfall for part of the year. When the sun's vertical ray is in the Southern Hemisphere, this same area experiences a winter dry season. As we move farther north up the west coast, we encounter increasing aridity reflecting the subtropical high-pressure belt as well as more localized aridity resulting from cold offshore currents. Gradually semiarid conditions give way to a true desert regime. This region of little water is extended inland and to the north by the effects of the dry continental interior far from sources of moisture. North of the west coast deserts the Mediterranean and marine west coast climates show seasonal variations paralleling in time the movement of the vertical ray of the sun farther to the south. The Mediterranean climate region receives winter rainfall as the westerly storm tracks shift southward in the winter. But the summer is long and dry as the wet winds and cyclonic systems shift northward. The marine west coast type of climate is much the same but being farther north receives more rain in winter from westerly storms as well as some precipitation in the summer. Energy demands on moisture in the marine west coast climate region are considerably less because of the lower angle of the sun's rays.

Along the east coast the monsoon effect in large part offsets the aridity of the subtropical high, although not completely. At the same time, the warm, moist offshore air masses make this southeastern coast mild and moist. These are the humid continental climates of the eastern parts of the continents. Finally, all across the north, increasing aridity related to the polar high-pressure area is matched by increasing cold. What moisture is available seldom escapes from its frozen state on the surface. These arctic-polar climates are relatively undifferentiated across the roof of the world. In all cases, we have created a reasonably accurate pattern with references to only a few variables. Our continent of Hypothetica cannot match the complexities of a real continent, but with a little thought you should be able to derive most of the major climatic regions of the world by moving Hypothetica into new positions and considering the same variables that we have. A final word of advice might be to pay somewhat greater attention to the effect of mountain barriers and highland areas when looking at South America, Europe, and Asia.

### Energy-moisture environments— a simple classification

We do not need an elaborate classification of energy-moisture relationships in order to discuss man-environment systems at a general level. A simple 3 × 3 matrix with very few modifications provides us with a simple classification which matches equator-to-pole energy-moisture environments with reasonable accuracy. Figure 6-6 has three rows or divisions along the ordinate corresponding to equatorial-tropical, mid-latitude, and arctic-polar climates. The three columns represent arid, semiarid, and humid moisture conditions. The nine cells formed by the intersecting rows and columns need only slight changes to match the climates described for the continent of Hypothetica. In this preliminary diagram the box representing arctic-polar humid conditions has been crossed out because nowhere in high latitudes does the polar high-pressure cell permit heavy precipitation. Other changes relate to the availability of energy and moisture as expressed by the vegetation which flourishes under various conditions. In order to modify the matrix, let us consider how vegetation relates to energy and moisture.

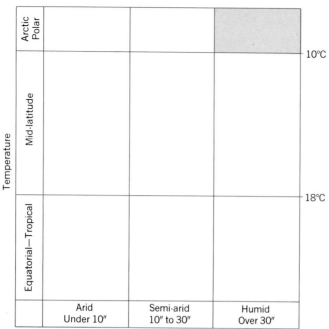

Figure 6-6 **Matrix of climates** A classification of climates by combinations of energy and moisture conditions creates a matrix which may be used to describe world climates.

## Evapotranspiration and the Water Balance

As the energy available in the atmosphere increases, plants maintain proper body temperatures by increased transpiration from their stomata. Evaporation from leaf surfaces also helps to cool plant tissue. *Evapotranspiration* indicates the combined water needs of vegetation. *Potential evapotranspiration* is the water that would be used if unlimited amounts were available, while *actual evapotranspiration* is the amount of water actually passed through the earth-plant-atmosphere system. The negative difference between these two amounts is a measure of the resulting water shortage or deficit referred to in chapter. C. W. Thornthwaite has devised a technique for computing the balance between water need and water availability.

The Mediterranean climate offers a good example of the march of the seasons in terms of water and energy availability. Figure 6-7 shows the water balance for Antalya, Turkey, a small town on the south coast. Typical of Mediterranean regimes, rainfall occurs almost entirely in the winter or low-sun months. This is indicated by line *p—p*, while line *0—0* shows the increasing energy, that is, the potential for evapotranspiration, within the environment as the summer advances. In the winter months, available water exceeds the water needed to meet the combined demands of plant transpiration and evaporation. By April, rainfall has dropped off to practically nothing, while the sun at noon rides higher and higher each day in the southern sky. Potential evapotranspiration increases steadily, and at first, the negative difference between potential and actual evapotranspiration, the water deficit, is replaced by groundwater drawn from the soil reservoir by the roots of the plants. However, as the summer progresses, the ground gives up at a slower and slower rate what little water it still holds. The water shortage increases, and by summer's end and the coming of the first fall rains a significant deficit exists. If this deficit cannot be satisfied, growing plants will experience severe drought. By September increasing rainfall has filled the lesser demands placed upon the system by the sun as it retreats into the Southern Hemisphere. November and December rains actually exceed potential evapotranspiration, and in those months excess water refills the dry soil reservoir. The soil becomes full as the new year approaches, and additional rains in January, February, and March form runoff on the land surface. Thereafter, as spring turns into summer, potential evapotranspiration once more increases; the rains diminish; and the cycle repeats itself.

The amount of water available to plants depends not only upon precipitation, but also upon the water storage capacity of the soil involved and the depth to which plant roots can reach. The water deficit is shown in Figure 6-7 by the solid red pattern between the lines indicating potential and actual evapotranspiration. Loamy soils combine storage capacity with permeability, which makes an ideal combination not only for storing water but also for retrieving it with minimum difficulty. Thus, loamy soils give up water to plants during a drought better than would either clays or sands. Clays might contain a large amount of moisture but would lack permeability, while sands are too permeable and lose their contents either to evaporation or to underground runoff. The increasing water deficit resulting from different types of soils is shown in the diagram by each added increment. It should also be noted that this diagram shows conditions relating to deep-rooted crops such as alfalfa, pasture grasses, and grapevines. Crops with more shallow root systems (beets, carrots, peas, etc.) or moderately deep roots (corn and cotton) would have less groundwater available to them and would wilt sooner. Orchards and mature forest with longer roots could survive the onslaught of drought somewhat longer.

Other climate regimes are shown in Figure 6-8. Seabrook, New Jersey, an east coast location with year-round humid climate has a relatively insignificant water deficit during the peak period of potential evapotranspiration. This situation is interesting, for even with these humid conditions, carefully controlled irrigation applied with the use of water balance analysis yields surprising increases in crop yields. Soil type also makes a difference. Just a few miles south of Seabrook in the Pine Barrens of New Jersey, loosely packed, sandy soils allow rapid escape of precipitation. The result is a region of dry soils and large water deficits suitable only for drought-resistant pines.

The next water balance (Figure 6-8) is from Manhattan, in the Flint Hills of eastern Kansas. There, native grasses can reach a height of 6 feet but few trees grow. In that area the winter and spring surplus of water is just about equal to the summer deficit. The annual net water balance, and therefore the moisture index, equals zero. Thornthwaite has observed that grasses such as these often make up the native vegetation in regions where the net water balance shows neither a large surplus nor a deficit. In this particular case, the potential evapotranspiration peaks almost at the same time as does precipitation, although a significant water deficit does occur in late summer.

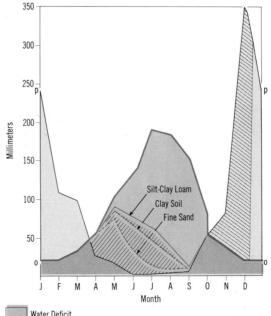

Water Deficit

Water Withdrawn from Soil

Soil Moisture Recharge

Water Surplus

——— Precipitation
══─══ Actual Evapotranspiration (3 Soil Types)
——— Potential Evapotranspiration
——— Water Deficit for Silt and Clay Loam Soils (Storage Capacity: 250 mm/Meter)
— — Water Deficit for Clay Soils (Storage Capacity: 200 mm/Meter)
------ Water Deficit for Fine Sands (Storage Capacity: 100 mm/Meter)

**Figure 6-7 Water balance diagram of Antalya, Turkey** The diagram is a monthly summary of the energy and moisture conditions affecting vegetation growth. The potential evapotranspiration is the amount of water, measured in inches, which would be drawn into the atmosphere if an unlimited supply of ground moisture were present. The actual evapotranspiration is the transfer of available water to the atmosphere. If potential exceeds actual, a water deficit condition exists. This depends, in turn, upon the amount of incoming precipitation and the water storage capacity of the soil.

The ranchers of the Flint Hills recognize a relationship between the growth cycle of the native grasses and their own activities as cattlemen. Native grasses provide the most nutritious forage during the months of April, May,

*Earth as
the Ecologic Habitat
for Life:
Climate and
Evapotranspiration*

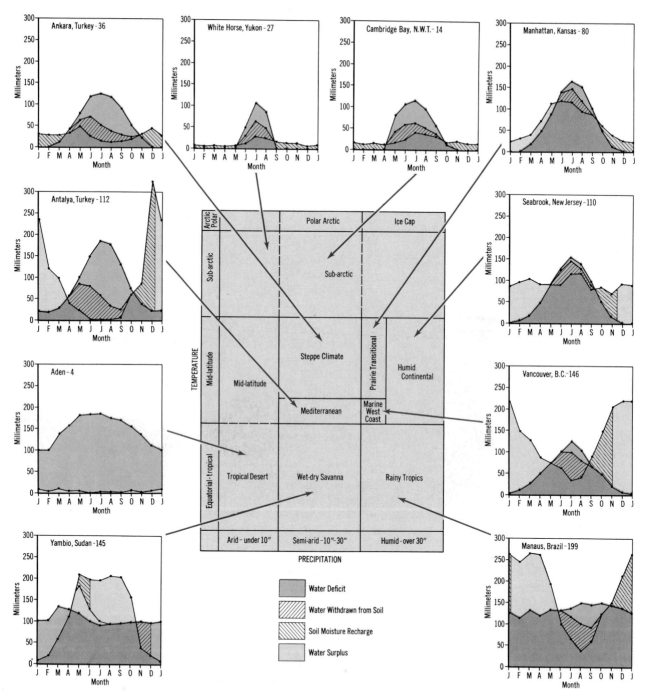

**Figure 6-8  Revised matrix of climates with sample water balance diagrams**  Numbers after place names represents average annual precipitation in centimeters.

and June. July marks the beginning of the summer drought, and the grasses no longer use quantities of plant nutrients for stem and leaf formation. Instead, nutrients are translocated into the root systems of the plants and stored in order to provide energy for the next year's growing season. Another portion of the nutrients is utilized for seed formation. The result of this natural adjustment to the onslaught of drought is that after the first week of July the grass has roughly "the same nutritional value as baling wire."

The stockmen recognize this and also know that their cattle gain at least 80 percent of their summer weight before the July water deficit begins. Supplemental feeding is expensive; and a good hay crop cannot be cut from the pastures, because the split-second timing necessary to catch the grass at its maximum growth but before it loses its nutrients is next to impossible. As a result, few local ranchers attempt year-round operations, and instead, most buy feeder cattle in the spring, fatten them on the early grass, and sell them as late summer comes on. All of this illustrates an important aspect of energy-moisture relationships. Timing is extremely important. The annual rainfall in the Flint Hills of eastern Kansas is between 31 and 38 inches, which places the region well within the humid set of climates. The late summer drought, however, influences the natural vegetation, the cattle that eat it, and the humans who utilize the cattle. Thus, time and space, energy and moisture, man and nature are all important in defining the environments upon which we depend.

## Climate as Evapotranspiration Types

The example of the Flint Hills can be expanded to other environments, other climates. Our view is that climate, vegetation, and soils are completely interconnected. More important, we consider their definition and their potential use as resources to be defined by human perception and human activities. We have already described three different water balances; we now offer seven more. The ten diagrams shown in Figure 6-8 form a simple and useful introduction

to world climates with the exception of the polar ice caps. Using these, we will discuss the development of landforms in Chapter 8, and in Chapter 9 some climatic sequences in terms of the vegetation, soil, and human systems that help define them.

The table in Figure 6-8 is a revision of the simple 3 × 3 matrix of energy and moisture in Figure 6-6 and reflects the ten water balances shown. Remember, this is not a map, although some of the suggested climatic juxtapositions can and do occur in nature. Mid-latitude and tropical deserts are treated together. There are differences between the two, but the overriding consideration is the continuous water deficit. This is shown by the extreme case of Aden on the southwest corner of the Arabian peninsula. Next, a distinction must be made between two types of tropical and mid-latitude semiarid climates, those which have a distinct dry season followed by a distinct rainy season and those which have some but not much precipitation throughout the year. The former, savanna-type climates, are most often associated with periods of high and low sun; the latter are often on the margins of true deserts or behind mountain barriers which block rain-bearing winds and are associated with steppe regions. The Mediterranean climate is also a wet-and-dry climate and is distinguished in the diagram from the semiarid steppe climates of the middle latitudes. The humid continental climates can be divided into cool and cold winter types, but we consider this unnecessary for our purposes. However, it is useful to distinguish the climate of both the long-grass prairies and the marine west coast type. The former receives more than 30 inches of rain but shows distinct semiarid characteristics, while the latter has seasonal periodicity resembling wet-and-dry climates but nevertheless has an annual total precipitation far in excess of 30 inches. We next recognize a poleward portion of the mid-latitudes which because of lower temperatures can retain precipitation, however scant, long enough for tree growth to occur. This is the subarctic climate that overrides all three moisture types. Finally, the arctic-polar climates form another group, again cutting across the three categories of moisture.

Within this latter set it is possible to distinguish *arctic climates* where some vegetation can grow and *ice cap climates* which are sterile and locked in ice and cold. Each water balance chart gives a reasonable description of the moisture-energy relationships prevailing under the condition it represents. Plant life has made adjustments to each of these environments. In fact, we often name the climate type after the vegetation that grows in association with it—for example, the savanna climates. Now, having set the scene, in the next two chapters we will consider the solid stuff of which the earth's surface is composed, how such materials are endlessly moved through a cycle of their own, and how under different climatic conditions they come to form distinctive landscapes.

# 7 | ENERGY AND THE EARTH'S CRUST: THE LITHOLOGIC CYCLE

Moving water is the plane which smooths the earth. Rivers are the major transporting agents for weathered materials. Thus, the *lithologic cycle,* which incorporates the wearing down and rebuilding of the continents, is in a very real way tied to the hydrologic cycle. The geologic processes involved in continental growth and destruction operate at longer time scales than do hydrologic ones, but we must still consider them in order to round out our picture of the dynamic planet on which we live. To do this, we will start with a small portion of exposed rock at the earth's surface, and follow a few grains of its material as they are weathered free, transported, incorporated in new rocks, depressed beneath the earth's surface, altered, and finally raised again and reexposed through erosion to another round of the *lithologic,* or *rock, cycle.* The path (or paths) followed by such material is shown in Figure 7-1. While we are fairly certain about the surface processes the diagram depicts, our ideas concerning subsurface processes within the earth are being radically revised in this decade. Therefore, we will do our best to tell you something about both the old and the new theories of the earth's crust and its formation.

## Elevation, Relief, and the Conversion of Energy

The unequal distribution of energy across the surface of the earth results in its flow from one place to another. Winds and currents are the principal components of such systems; the hydrologic cycle is another such principal component. Whenever an energy flow, in any form, encounters some obstruction, turbulence occurs and work is done. One form of work is the movement from place to place of part of the materials which make up the earth's crust. High peaks, craggy shorelines, and steep slopes of all kinds make excellent barriers to energy flow. As a result, turbulence, work, and erosion are often concentrated at such places. This is a function of both *elevation* and *relief.* By *elevation* we mean distance above sea level. For example, the Tibetan Plateau has an average elevation of more than 10,000 feet although its surface is relatively smooth and horizontal. The Tibetan Plateau, therefore, has high elevation and relatively low relief. A sea cliff which reaches only a few hundred feet above the surf has low elevation, but its relief is great. These relationships

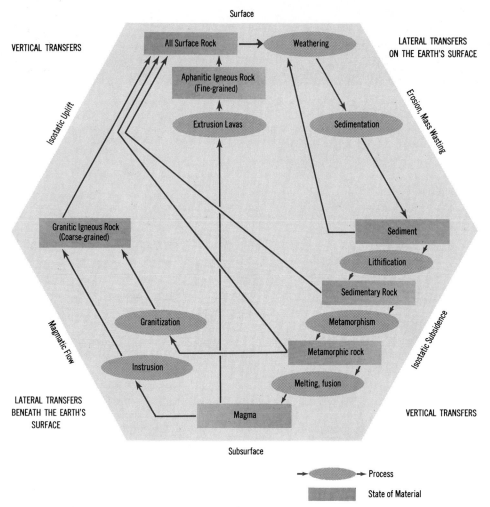

**Figure 7-1 The lithologic cycle** Materials of the earth's crust are in continual movement from subterranean depths to the tops of mountains and back again. This diagram shows several of the paths that a particle of rock might follow during this process. Movement is clockwise in the diagram. The longest cycle would go from surface rock to sediment, to sedimentary rock, to metamorphic rock, to magma, to coarse-grained igneous rock, and back to surface rock. The processes would involve weathering, sedimentation, lithification, metamorphism, melting, intrusion, and uplift. Other, shorter cycles can also be traced.

are shown in Figure 7-2. Water raised to high elevations has significant amounts of potential energy capable of performing large amounts of work. Steep slopes—that is, high relief—allow the ready conversion of potential energy into kinetic forms. For this reason weathering and erosion are intense near the summits of steep mountains, and that is where we will go to choose the rocky material we will trace throughout the lithologic cycle.

## Types of weathering

The play of atmospheric energy upon the rock surface beneath results in two basic types of rock alteration and disintegration. Stony materials can be literally broken up into smaller and smaller grains under the influence of *mechanical weathering.* Frost is the predominant agent in this process. Moisture in small amounts penetrates cracks and crevices within the rock and upon freezing expands with great force, thereby wedging off particles from the parent mass. Abrasion of surfaces by particles borne by wind and water may also be considered mechanical in character, but it is difficult to draw the line between the breaking up of rock (weathering) and its being transported elsewhere (erosion). Another form of weathering occurs when gases and water from the atmosphere combine with rock-forming minerals to cause chemical decomposition of the stone. *Chemical weathering* of this sort includes *solution, hydration,* and *oxidation.* Mechanical weathering is most common at high elevations and high latitudes, where freezing and thawing regularly take place. Chemical weathering predominates in warm, moist climates, particularly in the lower-latitude tropics.

If we begin our rock odyssey with a granite outcrop high on a mid-latitude peak, it is likely that a combination of mechanical and chemical weathering will help free the particles we seek. The rock there can be penetrated by microscopic portions of water, which, working inward from the surface by means of a series of ionic exchanges, can cause changes in the chemistry of the surface layer and bring about its subsequent expansion. Free water might also freeze and thaw in tiny cracks caused by the swelling of the altered rock-forming minerals. All this results in a spalling off of layer after thin layer of rock, much as the dry outer skin of an onion wears away. This process is called *exfoliation* and results in rounded boulders such as those Robinson Jeffers describes as "blunt bear's teeth."

In other climes, at other places, rock might be dissolved by a weak *solution* of carbonic acid often formed by a mixture of water and decay-

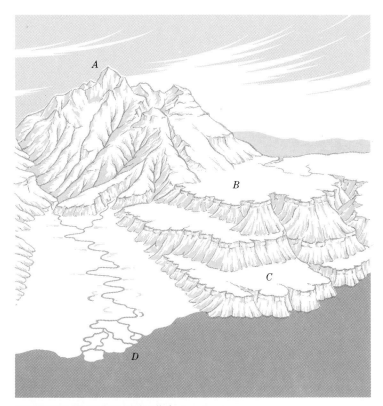

**Figure 7-2 Elevation and relief** By *elevation,* we mean the altitude or distance above sea level; *relief* means whether or not steep slopes exist. Thus, plateaus can have high elevation and yet be flat and with little relief. The various combinations of these two elements are shown in the diagram. *A.* Mountains—high elevation, high relief. *B.* Plateaus—high elevation, low relief. *C.* Sea cliffs—low elevation, high relief. *D.* Delta—low elevation, low relief.

ing organic matter. Limestone is particularly susceptible to this kind of destruction. Organic colloids resulting from the breakdown of plant materials into various compounds also can help create acidic solutions or *humic acids,* which are effective chemical agents. Sometimes the dissolving action of water is more direct if the rocks contain soluble salts. When chemical bonding between minerals and water takes place (as with our exfoliating granite), the process is termed *hydration.* If oxygen from the air and/or water combines with other elements, *oxidation* may

An approaching dust storm, Colorado, 1938. Wind erosion, aided by man's loosening of the topsoil and his destruction of its natural grass cover. (U.S. Dept. of Agriculture photo)

Gully erosion resulting from water running across improperly prepared and contoured fields. The farm has been abandoned. (U.S. Dept. of Agriculture photo)

result. Oxides are often soft and easily worn away, though sometimes concentrations of less soluble oxides can result when silica ($SiO_2$) is removed by alkaline solutions.

### Erosion by wind and water

The grains and pieces of granite removed by weathering fall from the parent surface and come to rest nearby or a distance away, depending upon the steepness of the mountain slope. Thereafter, their journey might continue in one of at least four ways until the grains are caught up by some stream. *Wind* might transport smaller particles for a distance. If rain were to fall at such a height, the runoff moving across the surface as a thin sheet might carry materials along by *sheet flooding*. If conditions were right, the material might fall upon either a snow and ice field or a glacier surface and begin a slow downhill movement by *glacial transport*. Finally, under somewhat different conditions, *mass wasting* might serve as the transporting agent. Mass wasting requires little or no water, and then only as a lubricating medium for the downhill movement of materials. Sometimes this form of movement will be abrupt, as when a landslide releases millions of tons of materials in some spectacular plunge. At other times the process will be imperceptibly slow. For example, a sand grain, a pebble, or even a boulder resting on a slight slope may be lifted at right angles to the slope by the expansion of water freezing in the surface layer of soil. When the frozen ground melts, gravity will bring the particle vertically back to the slope in its thawed configuration. Figure 7-3 shows the millimeter-by-millimeter progress of a pebble down a shallow slope by this method of freezing and thawing.

At some point, however, it is likely that the material we are following will fall or be washed into a running stream. Depending upon the geologic history of the area and the structure of the underlying rock strata, streams represent various patterns of line—area relationships. Where the underlying material is of relatively uniform hardness and texture, the runoff from the surface can take a general treelike configuration, or *dendritic* pattern (Figure 7-4). Perhaps 90 percent of all stream patterns are of this sort. If the water flows from the slopes of an isolated peak, the stream pattern can be *radial,* but if the underlying rock consists of alternating layers of hard and soft material, the result can be either *trellis* or *annular* drainage patterns. In every case, the immediate effect of the streams will be to carry away weathered materials. This removal can be either of solid materials or of dissolved materials in solution.

Where rainfall is relatively scarce, as in the basin of the Colorado River, the amount of solid material removed from a unit of land by a stream will be much greater than the material carried off in solution. Conversely, in humid regions less solid stuff will be removed from the land by running water, but the stream load of dissolved materials will go up proportionally. Table 7-1 gives some regional erosion rates in the United States.

Material carried to the oceans by rivers moves in one of three ways: *solution* (dissolved load), *suspended load,* and *bed load.* Material in bed load is moved in traction by rolling, sliding, and saltation (skips and jumps). Generally speaking, these methods of locomotion along a

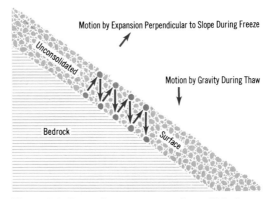

**Figure 7-3 Downslope progress of a pebble in freeze and thaw cycle** The process of freezing lifts the pebble at right angles to the surface, but when thawing occurs, the force of gravity moves the pebble vertically downward. This results in a gradual progression downslope of the surface mantle. Other processes serving to move materials downslope would be slump, slippage, and transport by running water.

stream channel are listed in order of increasing particle size and increasing speed of the water. The smaller a particle, the less energy will be required to lift it from the bed of the stream and to carry it freely in suspension. Extremely fine

**Figure 7-4 Stream patterns formed by surface runoff** Runoff on unconsolidated or unstructured materials often results in stream patterns which are treelike (dendritic), as shown in *A.* If a central high point exists, such as a volcano, the runoff may form a radial pattern (radial), as in *B.* Where parallel layers of rock crop out on the surface, thereby giving considerable structure or grain to the land—as in the Appalachian Mountains—the stream patterns draining the area may resemble a trellis (*C*), with short stretches crossing the ridges of more resistant rock at right angles to longer stretches in valleys formed on softer materials.

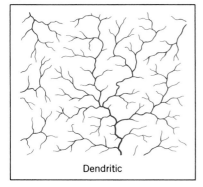

Dendritic

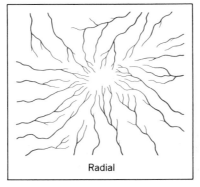

Radial

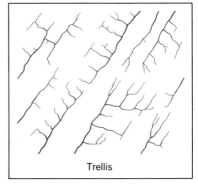

Trellis

**Table 7-1  Rates of Regional Erosion in the United States**

| Drainage Region | Drainage* Area Km² (thousands) | Runoff m³/sec | Load tons Km²/yr Dissolved | Load tons Km²/yr Solid | Load tons Km²/yr Total | Erosion cm/1000 yr | % Area Sampled | Average Years of Record |
|---|---|---|---|---|---|---|---|---|
| Colorado | 629 | 0.6 | 23 | 417 | 440 | 17 | 56 | 32 |
| Pacific slopes, California | 303 | 2.3 | 36 | 209 | 245 | 9 | 44 | 4 |
| Western Gulf | 829 | 1.6 | 41 | 101 | 142 | 5 | 9 | 9 |
| Mississippi | 3238 | 17.5 | 39 | 94 | 133 | 5 | 99 | 12 |
| South Atlantic and Eastern Gulf | 736 | 9.2 | 61 | 48 | 109 | 4 | 19 | 7 |
| North Atlantic | 383 | 5.9 | 57 | 69 | 126 | 5 | 10 | 5 |
| Columbia | 679 | 9.8 | 57 | 44 | 101 | 4 | 39 | <2 |
| Totals | 6797 | 46.9 | 43 | 119 | 162 | 6 | | |

*Great Basin, St. Lawrence; Hudson Bay drainage not considered.
Source: Sheldon Judson, "Erosion of the Land, or What's Happening to Our Continents," *American Scientist*, vol. 56, no. 4, 1968, table 1, p. 363.

materials like clays and silts will cling to the bottom longer than sand grains, which have more surface exposed to the force of the water, but once in suspension fine particles will travel farther than large, heavy ones.

Just as water molecules make many trips between heaven and earth during the hydrologic cycle, so do eroded materials pause frequently in their trip to the sea. Our original particles would come to rest after each period of sheet flooding had passed, or when freezing and thawing were inactive. Once in a stream, they subsequently might be deposited temporarily as *sediment* along the banks of some winding, or *meandering,* river. They might also come to rest temporarily in a lake or reservoir, where the energy of the flowing water is dissipated into the larger body of water. Shifting of the river meanders could eat away the deposited sediments and bring them back into motion. Or again, through continued downcutting, the lake

This photo and the next two show a river system at different places along its length. In this picture, water in a small stream high in the mountains begins its trip to the sea near Ketchum, Idaho, during the spring thaw. (U.S. Dept. of Agriculture photo)

might eventually disappear or the silted reservoir be scoured clean. All this deposition and removal depends upon the dynamic equilibrium of the stream. Near its headwaters, every river system has a zone of active channel erosion where constant downward cutting takes place. Farther downstream a transitional zone is found where sedimentation alternates with further erosion. Still nearer the mouth of the river is another zone of active deposition where water-borne sediments, or *alluvium*, at last come to rest (Figure 7-5). The steady extension of sedimentary deposits into the relatively still water of the sea results in the formation of deltas such as that of the Nile River shown in Chapter 8. The velocity of stream water entering a larger body of water diminishes directly with distance from the river mouth. As a result, heaviest materials are deposited nearest the shore, smaller grains are dropped beyond them, and fine particles extend outward still farther. Dissolved substances travel farthest of all and are precipitated under special conditions in the centers of basins.

All of this relates to the concept of the *base level*, or lowest point, toward which the streams are flowing. At the lowest point, which ultimately is mean sea level, the potential energy of the water is completely exhausted. The ability of the water to perform work is ended, and all transport ceases. Lakes along the course of a stream, or interior basins with no outlets to the sea, may create *local base levels* which may seem quite permanent to us, but in a geologic sense they exist for only one wink of eternity's eye. If the streambed is viewed in its entirety from headwaters to mouth and delta foot, we can talk about the *stream profile*. When the energy of the flowing water is evenly distributed along the length of the profile, we say the stream is in *equilibrium*. The profile of a stream in equilibrium would look something like that in Figure 7-5.

The Snake River near Route 26 in the Salmon National Forest. The river is larger now and is a major erosive agent. (U.S. Dept. of Agriculture photo)

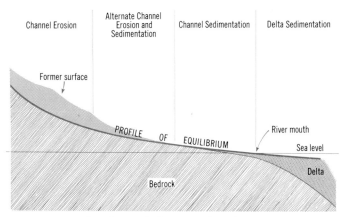

**Figure 7-5 Profile of a graded stream at equilibrium** A stream is probably seldom in a state of perfect equilibrium. However, in a perfectly stable stream system the volume of the materials eroded away from the headwaters of the stream would be deposited in its lower reaches so as to form a smooth, equilibrium profile.

The Columbia River downstream from its confluence with the Snake River. A giant river providing quantities of hydroelectric power as well as being a major transporter of sedimentary materials. The Dalles Dam, Oregon. (Photo by J. Nystuen)

### *An orderly view of stream processes*

Rivers have many forms, the variety of which might seem to defy analysis. Rushing like arrows, twining like snakes, mixed and multichanneled as a snarled braid, they nevertheless can be explained in systematic terms. We have already described the profile of a river in equilibrium, but there is more to stream mechanics and fluvial processes than can be properly treated in this book or even in a single volume devoted entirely to the subject. The discussion which follows is a very brief review of some of the more important aspects of systematic stream analysis. Our point is not to be exhaustive but to indicate that it is possible to understand complex natural processes in relatively simple, abstract terms.

### Stream flow as moving energy and mass

Water and eroded materials move ceaselessly from the raised parts to the hollow places on the earth's surface. In doing so, energy is carried with them as well as being dissipated along the way as heat by various fluvial processes. It is the relationship between this moving mass of liquid and solids and the energy it contains which accounts for the shapes of rivers and the landscapes which they help to sculpt or build. In the next chapter we will discuss landforms, including those resulting from the work of rivers. Here, we restrict the discussion to rivers themselves, though of course, in reality, the two cannot be separated.

The flow of a stream past a given point can be fast or slow; the water may be clear or murky with the materials it carries. The volume of water (*discharge*, measured in cubic feet or meters per second), and the materials moved along the streambed (*bed load*) or in suspension (*suspended load*) or in solution (*chemical* or

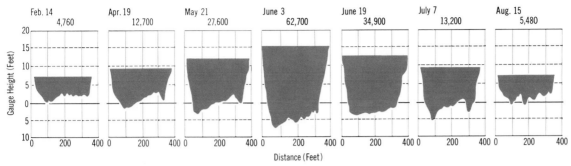

Figure 7-6 Water levels, scour and subsequent fill during flood passage, Colorado River at Lees Ferry, 1956 (After Luna B. Leopold, M. Gordon Wolman, and John P. Miller, *Fluvial Processes in Geomorphology*, Figure 7-13, p. 228, W. H. Freeman and Company, San Francisco, 1964.)

*dissolved load*) depend upon the velocity of the stream and the area of the cross section of its channel. If a stream neither gains nor loses water, its discharge must be the same everywhere along its length. Therefore, in this theoretically possible but unlikely situation, the stream must move rapidly where its channel has a small cross section and slowly where the channel is large. This means not only that the slope, or *gradient*, of the stream will affect its velocity, but also that the shape of its channel is important. Similarly, when additional water enters a river as the result of rain, snow melt, or the removal of vegetation from its watershed, additional energy will also enter the system. The ability of the stream to do work is increased, and not only will its waters overflow the banks in flood, they will also carry more load and scour its channel deeper. Thus, discharge and erosion capability can vary from one time to another for a single stream (Figure 7-6). In the same manner, as discharge diminishes, so will a stream's velocity and its kinetic energy. The ability of the water to lift and carry materials will also decrease, depending upon the particle size of the load. The bed load may slow and stop, the suspended load sink to the bottom, and the channel become more shallow. The

relationship between these elements is shown in Figure 7-7.

Figure 7-7 Curves of erosion and deposition for uniform material Erosion velocity shown as a band. (After Marie Morisawa, *Streams, Their Dynamics and Morphology*, Figure 4.4, p. 50, McGraw-Hill Book Company, New York 1963)

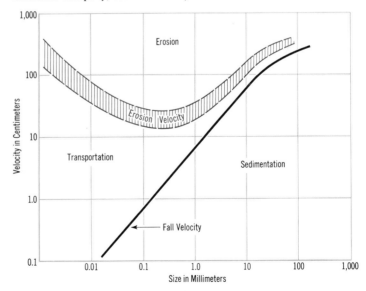

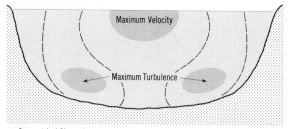

*A.* Symmetrical Channel

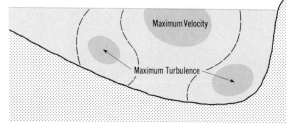

*B.* Asymmetrical Channel

**Figure 7-8 Zones of maximum velocity and turbulence in (A) symmetrical and (B) asymmetrical stream channels** (After Marie Morisawa, *Streams, Their Dynamics and Morphology*, Figure 3.6, p. 36, McGraw-Hill Book Company, New York, 1963.)

**Turbulence** Very little of the energy in flowing water erodes or carries materials. Most of it (as much as 95 percent) is dissipated as heat given up in turbulent movement or through friction of the water with the channel walls. Figure 7-8 shows the location of maximum turbulence and maximum velocity in the cross section of a stream. The smooth movement of water near the center of the stream is described as *laminar flow*. However, currents of gas or liquid all display turbulence.

*. . . when a stream gets stalled against its boundaries or against another stream moving in the opposite direction. The stalled stream breaks into pieces that roll over on themselves. Right at the boundary, the flow of the stream has zero velocity—which is why little particles of dust can ride on the blade of a fan without being blown off, and why you cannot blow fine pieces of dust from the surface of a table—only large pieces that stick up into the breeze. At increasing distances from the*

*boundary, the flow moves with higher velocities, and the difference in rates of flow causes the stream to trip over itself, to curl around on itself, just as a wave curls when it stubs its toe rushing up the beach.*[1]

Perhaps one of the most interesting forms in nature results when turbulent flow forms and dissipates in a periodic pattern. As water flows past large rocks or possibly the abutment of a bridge, eddies will peel off and move downstream. First one side of the obstacle will produce a swirl which will grow into a vortex, and then the other. As each vortex grows, it will reach out, in turn, and draw material from the one opposite. A flow of such opposite turning eddies, diminishing downstream, is called a vortex street and beautifully illustrates L.F. Richardson's commentary on the movement of energy everywhere in nature:

> *Big whirls have little whirls,*
> *That feed on their velocity;*
> *And little whirls have lesser whirls,*
> *And so on to viscosity.*

**The geometry of streams** Just as most ecological systems can be thought of in terms of spatial hierarchies and diagrammed as trees (Figures 3-19 and 3-20), the flow of water off the land arranges itself in a hierarchical pattern. Rivulets receive the runoff from small areas; streams with many tributary rivulets drain larger areas; and rivers with many side streams receive the combined waters of all such smaller flows. These *drainage basins* and the *drainage nets* of streams which empty them can be considered at many scales—from local ones to those nearly continental in size. But the general rules of river geometry by which they operate remain the same. Arthur N. Strahler has devised a scheme for ordering such hierarchies of streams. A *first-order stream* is usually tiny and has no tributaries. *Second-order streams* are formed by the joining of two or more of the first order. A *third-order* is formed by the joining of

[1]Peter S. Stevens, *Patterns in Nature*, An Atlantic Monthly Press Book, Little, Brown and Co. (1974), p. 54.

Alternating eddies created downstream from an obstacle in a flow of gaseous or fluid material is called a vortex street. The vortex street shown was created in an air flow filled with oil droplets from an aerosol spray. The photograph was taken by Owen Griffin and Steven Ramberg of the U.S. Naval Research Laboratory in Washington, D.C., and is reprinted with their permission.

two second-order streams, and so on. This principle is illustrated in Figure 7-9.

Parallel to the ordering of streams from small to large is their increasing length and the increase in the area drained. The frequency of streams in each order, however, is inversely proportional to their size. That is, the smaller the order number, the more numerous the streams (Table 7-2).

**Braiding and meanders** As we have mentioned above, an ideal profile of a stream reflects

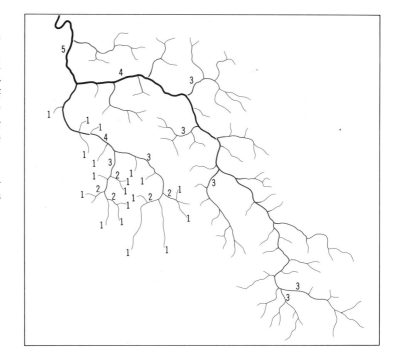

**Figure 7-9 Stream orders** The branching system of a stream may be divided into a hierarchy of segments. The smallest fingertip channels are first-order segments. A second-order segment is defined where two first-order segments join. Two second-order segments join to form third-order segments, and so forth. However, where a lower-order segment joins a higher-order segment, no change in order occurs.

**Table 7-2 Number and Length of River Channels of Various Sizes in the United States (excluding tributaries of smaller order)**

| Order* | Number | Average Length (miles) | Total Length (miles) | Mean Drainage Area, Including Tributaries (sq. miles) | River Representative of Each Size |
|---|---|---|---|---|---|
| 1 | 1,570,000 | 1 | 1,570,000 | 1 | |
| 2 | 350,000 | 2.3 | 810,000 | 4.7 | |
| 3 | 80,000 | 5.3 | 420,000 | 23 | |
| 4 | 18,000 | 12 | 220,000 | 109 | |
| 5 | 4,200 | 28 | 116,000 | 518 | |
| 6 | 950 | 64 | 61,000 | 2,460 | |
| 7 | 200 | 147 | 30,000 | 11,700 | Allegheny |
| 8 | 41 | 338 | 14,000 | 55,600 | Gila |
| 9 | 8 | 777 | 6,200 | 264,000 | Columbia |
| 10 | 1 | 1,800 | 1,800 | 1,250,000 | Mississippi |
| Total | | | 3,250,000 (approx.) | | |

*The definition is that of Strahler: Order 1 is a channel without tributaries; order 2 is a channel with only order 1 tributaries, and includes only the segment between the upstream junction of order 1 channels and the junction downstream with another order 2 channel.

a condition of equilibrium between available energy, volume, and load. Time is an important variable in these considerations. Almost every stream has periods of high water, when erosion and scouring are increased, and periods of low water, when sedimentation occurs. There may also be variation in these processes from the headwaters to the stream's mouth, depending on the location of high water along the profile as a flood crest moves oceanward. This is in addition to similar variations dependent upon slope and elevation, mentioned earlier. If a sufficient time span is considered:

*A graded stream is one in which, over a period of years, slope is delicately adjusted to provide, with available discharge and with prevailing channel characteristics, just the velocity required for the transportation of the load provided . . . The graded stream is a system in equillibrium . . .* [2]

[2] J. Hoover Mackin, "The Concept of the Graded River," *Bulletin* of the Geologic Society of America, vol. 59, p. 471 (1948).

Mackin then goes on to say that whenever some change occurs in the factors controlling the shape of the stream, there will be further displacement which will tend to counter or offset such changes.

While no two experts agree completely on the causes of *meanders* and *braided* (multiple) channels in streams, there is an undeniable relationship between such features and stream equillibrium. It has been suggested that braided channels occur when the velocity or energy of a stream is too small to move the materials in its bed load. This condition might result from the presence of very coarse materials which are too large to be lifted. It may also happen when a stream with highly variable discharge flows through easily eroded materials providing a large amount of load during high water which must be dropped when the stream's volume diminishes. In the latter case, as bars and riffles build up in the river, the channel is narrowed and the velocity of the remaining water is increased, thus facilitating its onward flow.

(The photograph on page 192 in Chapter 8 contains an example of a braided stream.)

The causes of meanders are even more complicated and apparently interrelated. Local variations in turbulence may permit local erosion and deposition, with parcels of the bed load moving downstream from one side to the other of the river in short trips. The resulting barriers and excavations might gradually deflect the energy of the stream more and more until meanders form. Another theory is that the slope of a stream is so completely adjusted to the material it carries that if the slope is steeper—and therefore the required energy greater—than that needed to do the work of transportation, the river will increase its length by meandering until the proper gradient is reached. For example, if a stream falling 5 meters in 15 kilometers produces excess energy, then it may readjust its course by meandering an extra 3 to 5 kilometers, thus lessening its slope. Still another theory suggests that meanders occur in streams with suspended loads of fine clays and silts, rather than in those carrying coarser materials. The real cause of meanders is undoubtedly a combination of these and other factors.

In any event, regardless of the size and length of a stream, the shape and spacing of meanders is remarkably similar. Sedimentation occurs on the inner or convex banks of curves, while cutting and scouring are found along the outer, concave edges (Figure 7-10). The main channel of the stream, with the greatest water velocity and depth (called the *thalweg*), moves back and forth across the bed of the meandering stream. The meanders thus formed by progressive cutting and filling move downstream along a sinuous path, the length of a given meander being about seven to ten times the stream width at that point. Sometimes a meander may increase its curvature until it cuts through its own neck

**Figure 7-10 Meandering stream** Sedimentation occurs on the inner or convex banks of the stream, cutting and scouring at the outer or concave edge. Cutoff meanders create ox-bow lakes. A portion of the Red River near the village of Campti in northwest Louisiana (USGS topographic sheet).

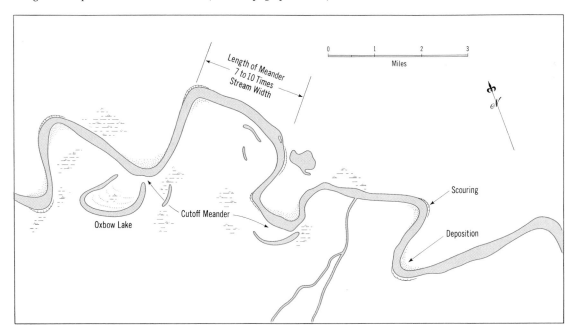

and straightens the course of the stream. The channel path that is left behind is called a *cutoff meander* and may form an *ox bow lake* (Figure 7-10). Meanders, ox bows, and other such forms will be discussed in the next chapter in connection with human occupance of the land. Meanwhile, the technical description of fluvial processes outlined above, although barely a drop in the bucket, hopefully provides us with some appreciation of the interaction between the hydrologic and lithologic cycles.

*Erosion and the role of man*

The above discussion may seem very far removed from our earlier discussion of city models and the works of man. In reality, erosion and deposition of earth materials concern us all and are affected significantly by human actions. We have already commented on how man's activity has increased world erosion rates about 2.5 times. Such an increase is geographically unequally distributed in terms of types of land utilization. A recent study of the Washington, D.C., area by M. G. Wolman has estimated the rates at which the surface of the land has been lowered by erosive forces. In precolonial times the forest floor was lowered by about 0.2 centimeters every 1,000 years. As the colonists cut the trees and plowed the ground, the rate increased to nearly 10 centimeters for the same unit of time. Then, with the expansion of the city and the abandonment of farmland on its edges, second-growth woodlots and temporary pasture reduced erosion to about 5 centimeters per 1,000 years. When modern construction began, rates as high as 10 meters per 1,000 years were reached for short periods of time at active sites. Only when the completely paved city took over did erosion drop to a negligible amount. While Wolman discusses erosion in a historical perspective, it is possible for us to view the city as a kind of time machine, with the urban center representing the most recent types of human occupance and the spatial sequence of land use leading away from the center paralleling the city's historic sequence of growth. Figure 7-11 illustrates the city as the focal point for various rates of erosion based on different land uses.

It would be a bit too fanciful to change scales and talk about the urban areas of the world being surrounded by concentric zones of erosion rates (as we speak in Chapter 10 of the world's being organized around an urban core), but it is quite reasonable to predict that as more and more humans harness and use more and more energy for construction purposes and for increased agricultural production, significant changes in global geology may well occur. The only thing that makes this prediction conjectural is the short time man has been an effective agent on earth. Urbanization of any note has taken place only in the last 2,000 years; city growth covering significant areas is less than 100 years old. Man, in one form or another, has occupied the earth for something in the neighborhood of 2 million years, a mere tick of the geologic clock. It looks as if our chances of lasting as long as the giant dinosaurs did, some 100 million years, are none too certain. But if we do survive, the earth will be a different place geologically as well as geographically because of us. If we survive. . . .

## Subsurface Processes

Our particles have by now found their way into the ocean waters surrounding the continent of which they were a part. From here on, their subsequent travels and transformations can be learned only by indirect observation. As a result, we must present at least two theories of continental formation. The first, and until recently accepted, one is based on the concept of stable continental masses being renewed by essentially vertical convection currents in the earth's interior. This can be called the theory of *isostatic equilibrium.* The second theory, which is currently being explored and tested, is based on a more dynamic model of the earth where large lateral movements occur over vast periods of time. This is referred to as the *plate tectonic theory.* Both of these theories consider operations at continental and global scales. Before discussing their details, let us look quickly at the

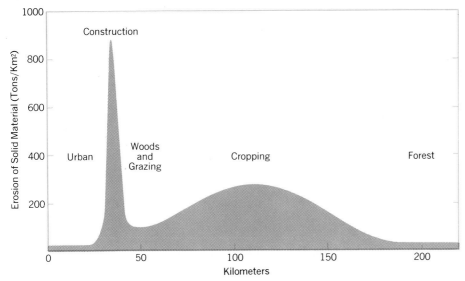

**Figure 7-11 The city as a focal point of erosion**  Little or no erosion can take place within the paved limits of the city. On the suburban periphery, where construction is in progress and the protective layer of vegetation has been removed, erosion rates are greatest. Abandoned farmland just beyond has a slow rate of erosion which rises slightly in still more remote areas of active cropping. Untouched forest experiences rates of erosion almost as low as paved areas, but what a difference! It is interesting to note that the original figure suggested by Wolman used time rather than distance as the measure along the abscissa. Thus, he considered a single location once in a forest but now at the urban core, and what happened to it over the century or more during which its use shifted from one activity to another. In this case, time and space have an interesting substitutability. (Suggested by Wolman, *American Scientist,* vol. 56, 1968, figure 9, p. 363)

changes in quality and composition that rock-forming materials can undergo when caught up in such giant processes.

*Rocks and rock formation*

As an offshore area receives more and more waterborne sediments, layer on layer of unconsolidated materials pile up. Near the shore they will consist of sands. Farther out, silts and clays will be deposited. Finally, in some areas calcareous, or limy, deposits will be formed from either the shells of sea animals or the precipitation of materials carried in solution. As the weight of the overlying sediments increases, materials in the lower layers will become compacted and eventually cemented together. Sands will form sandstones; silts and clays will become shales; and calcareous materials will change to limestone. All such rocks which are formed from sediments are called *sedimentary rocks* and are characterized by their occurrence in layers, or strata, and by their relative softness.

With increasing depth beneath the pile of sediments, pressure and temperature increase. Sedimentary rocks can be subjected to great lateral pressure near the surface as well as intruded by molten materials from below. Such rocks can also be depressed to great depths beneath the crustal zone. In any of these events, the pressed, baked, and distorted rocks assume new characteristics. These processes of change are called *metamorphism,* and the rocks which result are known as *metamorphic rocks.*

*Energy and the Earth's Crust: The Lithologic Cycle*

Sandstones become quartzites; shales transform to slate; and limestone turns to marble in the earlier stages of metamorphism. Further melting and fusing or perhaps chemically induced changes at lower temperatures create *gneiss* and other metamorphic rocks.

When rocks have been so transformed and melted that they become molten, or behave at great depth as viscous substances, we refer to them as *magma.* Magmas can move vertically or horizontally and often intrude into the solid rock surrounding them. After their intrusion they cool, and if the rate of cooling is slow, new rock will form, composed of crystals large enough to be seen by the unaided eye. These include the light-colored *granites* and their darker cousins the *gabbros.* If magma breaks through to the surface and extrudes as *lava,* the subsequent cooling will be quick and the rocks will be fine-grained. *Basalts* are typical of these. In either case, rocks with magmatic origins are given the general term *igneous.*

Though we have described this transition from one rock type to another as an unbroken sequence, the lithologic cycle has many loops of different dimensions within it. Thus, sediments, sedimentary rock, metamorphic rock, and igneous rock may all be exposed to the atmosphere and subsequently weathered and eroded. The multiple paths in Figure 7-1 are indicative of this.

### Isostasy and geosynclinorial theory

With few exceptions the view held by earth scientists of the internal structure of the earth and of its crust was until recently a static one. It was, in brief, something as follows. The earth is layered like an onion, with a dense core and thin outer skin. The *innermost core,* with a specific gravity of from 14.5 to 18 times that of an equal volume of water, has a radius of approximately 780 miles. It exhibits the properties of a solid body, but this is conjectural because of the difficulties of observation. The next 1,380 miles of radius behave as a viscous *liquid outer core.* Beyond that for another 1,800

miles is the *mantle* of the earth. The mantle, while dense and rigid, flows under the immense pressure of overlying rock. One might think of it as something like tar or silicone putty (though much more dense) which will break if struck a sharp blow but which will slowly spread if given enough time. Floating on the mantle and constituting the *crust,* or outermost zone, of the earth "onion" is a layer of crystalline rocks. This crust is from 10 to 25 miles thick and is separated from the mantle by a clear-cut discontinuity of rock type. Called the *Mohorovicic discontinuity,* or simply the *Moho,* after its discoverer, a Yugoslavian seismologist, this discontinuity is revealed by the sharp change in velocity that earthquake waves make when passing across it.

The crust, being outermost and being the home of man, attracts the most attention. It is composed of a lower, continuous layer of dark, heavy, basaltic rock with a specific gravity of about 3.2. Composed of *si*lica and ferro*ma*gnesium minerals, it is called the *sima.* Floating on the sima just as a coin can float on a pool of mercury are discontinuous "islands" of lighter, granitic rock with a specific gravity of 2.6. These islands which form the continents consist of *al*uminum-, potassium-, sodium-, and calcium rich minerals which also have *si*lica as a major constituent. They are called the *sial.* A kind of equilibrium exists between the sialic blocks, their thickness, and the manner in which they ride—like an iceberg in the ocean—in the sima. Near the center of the continents where the sial is thick, a corresponding root of granitic material rides like a deep keel beneath the mountains on the surface. On the margins of the continents where the blocks taper out, the continental edges, which are submerged by overlapping sea water, are underlain by relatively little sial. Beyond the edges of the continents are the open ocean basins floored with basaltic sima. The submerged margins of the continental blocks, the *continental shelf,* extend with surprising regularity to a depth of about 100 fathoms (600 feet). At that depth they are abruptly terminated in many places by a steep slope which plunges to the extreme depths of the open sea. These relationships are shown in Figure 7-12.

In accordance with the above ideas, it was thought that as erosion lowers the interiors of the continents, the sediments deposited on the continental shelf or in the near-shore depths stack up layer on layer of strata until immense prisms of sedimentary rock are formed. The weight of the overlying materials presses those beneath deeper and deeper into the lower layers of the crust until their relative buoyancy prevails and brings them back to the surface. The process of downwarping and accumulation occurring in *geosynclines* is followed by another in which the geosynclinal prism is raised during periods of *isostatic uplift.* These latter periods can be thought of as times of *orogeny,* or mountain building. If the pressure is great enough, the lower portions of the geosyncline may fuse or be invaded by magma from beneath. This accounts for some lateral transfer of materials deep within the earth. Given enough time, such massive movements of the earth, or *tectonic* activity, will return the particles we are tracing to the tops of new mountains, where they will again be in position to begin another trip through the lithologic cycle.

## Continental drift and plate tectonic theory

Finding a suitable explanation for the subterranean lateral transfer of materials presented some difficulty in the theory of isostatic adjustment. Also, if geosynclinical accumulations of immense size occurred in the past, they should have contemporary counterparts, but no neat corollaries can be found today. This lack of contemporary examples runs counter to the concept of *uniformitarianism,* basic to all geologic thought. First suggested by James Hutton, the theory of uniformitarianism states that the geologic processes going on at present are the same as those processes which in the past created the earth features we now observe. This and other contradictions made scientists look for a more comprehensive theory to account for continental construction and crustal tectonics.

Four major and many minor clues led to the

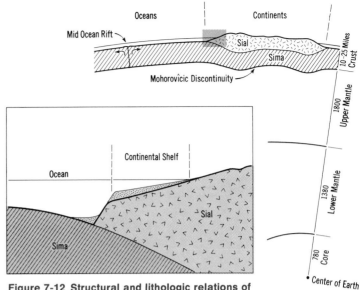

**Figure 7-12 Structural and lithologic relations of the earth's crust** From a geographic point of view, the thin crustal layer of the earth is most important. It consists of two types of materials, sima and sial. Sima is a dark rock which underlies the continents and covers the floors of the ocean basins and is heavier than sial, which constitutes the continents. The two materials are separated from the upper mantle of the earth by a discontinuity which shows up in the different rates at which earthquake waves travel above and below it. Sial floats upon the sima much as a coin might float on a puddle of mercury.

currently developing notions about the crust of the earth. (1) In 1912 A. Wegener pointed out that by removing the Atlantic Ocean a remarkably close fit existed between the continents of the Eastern and Western hemispheres (Figure 7-13). He suggested that these continental blocks had split apart and drifted away from each other across the underlying sima. However, many people found it difficult to imagine "ships made of thick continental crust plowing through a passive sea of oceanic crust." (2) In the years that followed, other scientists observed that evidence of ancient glaciation as well as the spatial distribution of many fossil life-forms occurring on the continents of the Southern Hemisphere could best be explained if

*Energy and the Earth's Crust: The Lithologic Cycle*

**Figure 7-13 The matching of continental forms according to Wegener's hypothesis** As early as 1912, W. Wegener had observed that if the oceans of the world were removed, it would be possible to fit the continents together into one or two super-continental blocks. His theory suggested that the continents had "floated" apart over millions of years. Subsequent developments in plate tectonic theory indicate that Wegener was largely correct in his hypothesis, except that the continents did not float apart; they were carried along as the sea floor spread.

those land masses had at one time been continuous. (3) Little was done to actively investigate Wegener's *hypothesis of continental drift* until after World War II, when oceanographic explorations in the mid-Atlantic discovered a major north-south trending ridge, or mountain chain, midway between Europe and America. Subsequent investigations of this ridge and the ocean floor on either side of it revealed facts of major importance. The ridge itself has along its center a gigantic, steep-sided valley much like the rift valleys of East Africa. At the same time, dredging and coring of the ocean floor showed that the sediments near the ridge are thin and young, while those progressively farther from it are thicker and contain older sediments. The implication is that the sea floor of the mid-Atlantic is spreading outward away from the ridge with new material welling up from great depths. In other words, *new area is being created in mid-ocean* and by its appearance is forcing the continents farther apart. (4) The fourth major clue came when the epicenters, or exact location of greatest intensity, of all earthquakes that occurred during the period 1961–1967 were plotted on a world map. The plotted points (Figure 7-14) revealed major zones of tectonic activity and earth movement on a hitherto unappreciated global scale. The plotted earthquake locations also outlined huge plate-like areas of the earth's crust. These plates included not only whole continents but also portions of the major ocean basins. In other words, according to the *theory of seafloor spreading*, the continents are not drifting away from each other across the underlying sima so much as that the crustal plates described above are growing on one side. But what is happening on their opposite edges?

It had long been recognized that the Pacific Ocean, in particular, is surrounded by a series of arc-shaped island archipelagoes, famous for their active volcanoes, such as the islands of Japan. Similarly, the American shores of the Pacific have inland chains of active or recently active volcanoes paralleling them. Furthermore, just seaward of the Asian island arcs are deep sea trenches of enormous depth. It was along these trenches and islands that the plotted earthquakes indicated significant tectonic activity. The epicenters of the earthquakes showed that they occurred near the surface of the earth's crust in the ocean trenches and at progressively greater depths within the crust back under the island arcs and toward the mainland of the continents. The outcome of all these

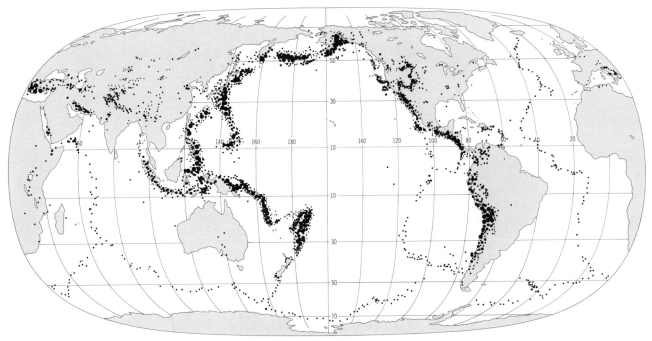

**Figure 7-14 Earthquake epicenters** The epicenters of major earthquakes are plotted on this map. The pattern reveals the mid-ocean ridges important in plate tectonic theory. It also shows the subduction zones along the
island arcs of the Pacific. Volcanic activity is also closely associated with the
pattern of epicenters. Thus the "Rim of Fire" title given the volcanoes bordering the Pacific Ocean gains new significance.

observations was the discovery that the crustal
plates which are forming along the Atlantic
Ridge are in all likelihood disappearing or
being *subducted* again into the earth along the
island arcs. Movements along the descending
plate account for the earthquakes at increasing depths away from the oceanic trenches
which mark the line of underthrusting (Figure
7-15). *Area is literally being created in one
place and consumed in another.*

Mountain building, too, can now be given a
new interpretation. The accumulation of sediments beneath sea level on the continental
margins eventually is brought into contact with
other continental portions of the moving crustal
plates (Figure 7-16). The subsequent pressures
crumple the geosynclinal masses much as a
specially designed car bumper forms accordion

pleats in order to take up the shock of a collision. In most cases a continental block with its
shield of sedimentary deposits will strike
against an outlying island arc. However, in at
least one case, where the subcontinent of India
appears to be being subducted under the
Tibetan or greater Asian continental mass, two
continental blocks have come into contact. Figure 7-17 gives some indication of the crumpled
topography where they meet. The result is the
highest mountains in the world, the Himalayas.
Though our knowledge of plate tectonics is as
yet incomplete, our view of the world as a
dynamic place is enormously enhanced by these
new ideas. In the same way, the particles of
sediment which we have followed from moun-
taintop to ocean basin can now be raised upward again through the crumpling of the con-

*Energy and
the Earth's Crust:
The Lithologic
Cycle*

**149**

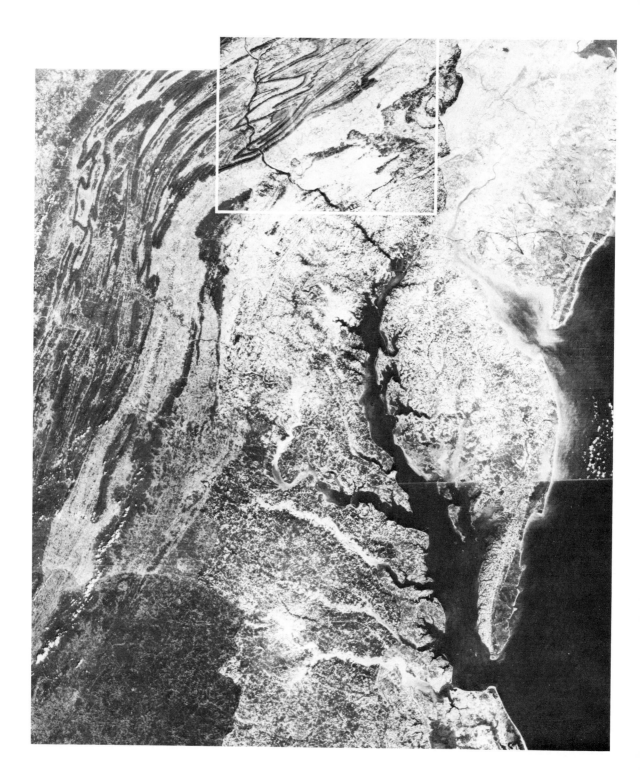

An Earth Resources Technology Satellite I photo (taken from 914 km) of the Allegheny Mountains, which run diagonally from lower left to upper right. The Susquehanna River cuts across them as a strong black line. The ridges and valleys, which form a zigzag pattern, are typical of the pressure ridges along the contact zones of tectonic plates (see Figure 7-16). (NASA photo)

tinental margins. Then, too, the enormous forces involved in these processes allow us to imagine rocks at great depths being metamorphosed, magmatized, and injected to become crystalline granites or extruded as fine-grained basalts. Thus, for the time being, we can consider as complete the round trip of the rock particles we have been following.

This photograph is a mosaic made from a number of individual satellite photo images. Chesapeake Bay and the Atlantic Ocean are on the right; the folded Appalachian Mountains are in the northwest. The area outlined in white is shown in the enlarged picture on the opposite page. (Mosaic courtesy of Georgraphic Applications Program, ERTS, U.S. Geological Survey)

*Energy and
the Earth's Crust:
The Lithologic
Cycle*

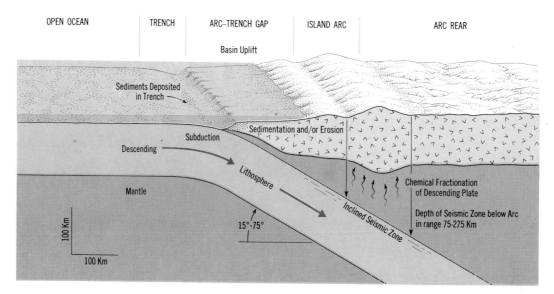

OPEN OCEAN | TRENCH | ARC–TRENCH GAP | ISLAND ARC | ARC REAR

Basin Uplift

Sediments Deposited
in Trench

Sedimentation and/or Erosion

Descending · Subduction

Lithosphere

Chemical Fractionation
of Descending Plate

Mantle

Inclined Seismic Zone

Depth of Seismic Zone below Arc
in range 75-275 Km

100 Km

15°–75°

100 Km

**Figure 7-15 Generalized section of an arc-trench system** According to plate tectonic theory, surface is continually produced along mid-ocean rifts and continually consumed along the edges of the island arcs. This diagram shows the zone of subduction where materials return to the depths of the crustal zone. The oceanic trenches found in front of the island arcs mark the zone of subduction.

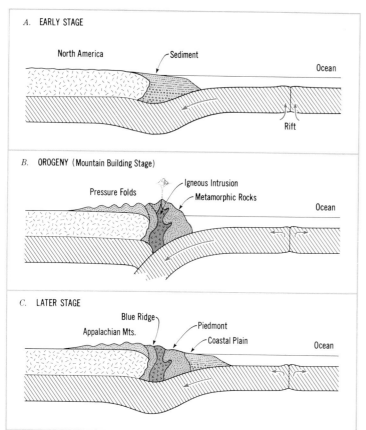

A. EARLY STAGE

North America — Sediment

Ocean

Rift

B. OROGENY (Mountain Building Stage)

Pressure Folds

Igneous Intrusion
Metamorphic Rocks

Ocean

C. LATER STAGE

Blue Ridge
Appalachian Mts. — Piedmont
— Coastal Plain

Ocean

## Human implications of crustal formation

It may seem overbold to consider man and his activities in terms of the immensities of time and space involved in crustal dynamics. And yet the deck of spaceship earth upon which we tread can toss like that of a ship on a stormy sea. People living in earthquake zones, those tectonically active areas of the earth's crust, must take special precautions or pay the price. The price can be a high one, too. On January 24, 1556,

**Figure 7-16 Mountain-building sequence** According to plate tectonic theory, orogeny, or mountain building, takes place when materials are carried up against the continental blocks by moving plates of the lithosphere. The resulting collision, although taking place over immense stretches of geologic time, nevertheless crumples the materials at the point of contact. The subsequent pressure folds and upthrusting of materials produce mountain systems parallel to the zone of plate contact.

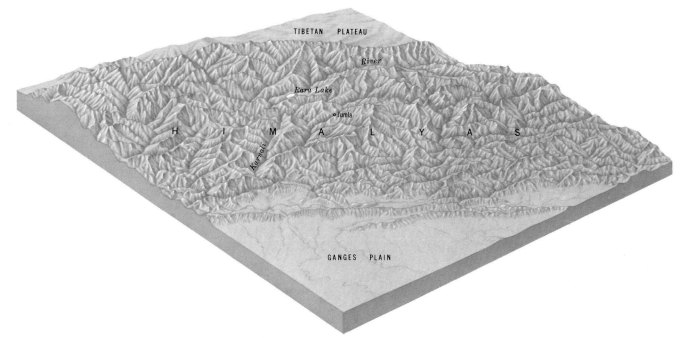

**Figure 7-17 Mugu Karnali region of Nepal** This diagram shows the incredibly rugged terrain along the Himalayan front. It is theorized that this marks the line of contact between two colliding blocks of continental materials. The Gangetic Plain in the foreground and the Tibetan Plateau beyond the mountains represent those blocks. The Himalayas represent the crumpling resulting from the ongoing orogenic process.

approximately 830,000 people died as the result of an earthquake in Shensi Province, China. On December 16, 1920, another 180,000 died in the neighboring Kansu Province. The famous Tokyo quake of September 1, 1923, killed 143,000, while on August 19, 1966, another 2,529 people perished at the village of Varto in eastern Turkey. In fact, a very conservative estimate of the number of people killed by earthquakes since 1500 exceeds 2,250,000.

We cannot go into a lengthy discussion of the exact causes of earthquakes beyond those we have already implied. However, there are a variety of reasons why so many people die when earthquakes strike. In China, Shensi and Kansu Provinces are essentially treeless, and many of their inhabitants dwell in unsupported caves dug into thick layers of wind-deposited soil called *loess*. The slightest earth tremor can collapse these caves upon their helpless inhabitants. The Tokyo quake took its toll because the city at that time lacked earthquake-proof buildings on firm foundations. In many parts of the world, including the Mediterranean region and Turkey, buildings either are poorly built or consist of stacked stone walls held together by simple mortising. These too collapse quickly and with little warning when earthquakes strike. While we do not want to sound like doomsayers, San Francisco and other coastal California towns are located literally on top of the San Andreas *fault zone* along which crustal plates move horizontally in opposite directions relative to one another. Such *transform faults,* as they are known in the parlance of plate tectonic theory, can produce earthquakes as dangerous to man as those occurring along subduction zones, like the one that wrecked

*Energy and the Earth's Crust: The Lithologic Cycle*

**153**

Side-looking radar image of the terrain near San Francisco, Calif. In this image the Golden Gate and Oakland Bay bridges can be seen clearly. (NASA photo)

This picture is taken from a slightly different angle from the one above. San Francisco Airport is marked by the X of crossed runways in the top center. The San Andreas Fault is clearly marked by the strong linear topographic elements across the center of the picture. The diagonal white slash above the fault in the right half of the picture is the mile-long track of Stanford University's experimental linear accelerator. This seems a strange site for such a delicate instrument. Place tectonic movements take place along the fault line, causing linear displacement of the areas on either side of it. (NASA photo)

Tokyo. Proper construction techniques, building codes permitting only low structures with proper reinforcement, and avoidance of steep hillsides and actual fault areas can all reduce the loss of life from earthquakes. As yet we have no real way of predicting when an earthquake will take place. Research may help us in the near future to foretell such events, but an ounce of prevention is still worth many pounds of cure. The San Francisco earthquake of April 18, 1906, took 700 lives, and that was at a time when the relatively few people who lived in the area occupied, for the most part, flexible wood frame buildings. The possibility of another major quake in California is very real; the results could be calamitous. There is talk of reducing the threat of big earthquakes by triggering a series of smaller ones along major fault zones by the injection of lubricating liquids into deep wells or even by exploding small atom bombs beneath the surface. But who will be able to take the responsibility of such an experiment, or who would dare? It seems that earthquakes will remain a major problem for a long time to come.

On the positive side, some imaginative scientists view subduction zones as a possible means of disposing of our dangerous waste materials. The reasoning behind this idea is that holes could be drilled into those portions of the crustal plates which are in the process of descending beneath the overiding crust (Figure 7-18). However, before this can be put into practice we must be certain that the sedimentary layer containing radioactive and toxic wastes would not simply be scraped off and returned in a relatively short time to the surface. Much more research is needed before this disposal plan could become a reality, although it has fascinating possibilities as the solution to this very urgent problem.

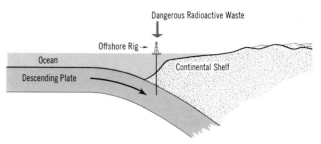

**Figure 7-18 Proposal for long-term disposal of radioactive wastes** The proposal to dispose of dangerous radioactive wastes takes advantage of the subduction zone in the plate tectonics process. Wastes could be implanted in deep wells penetrating the zones. Thereafter, plate movement would carry the materials deep into the earth's crust. They would remain there for millions of years before being cycled back to the surface.

## The orders of relief

A complete discussion of man's occupance of the earth's crust could fill a volume by itself. Instead of continuing, let us end this chapter from a slightly different point of view. We have already referred to the difference between relief and elevation. Relief can be thought of as the steepness of the slope separating the highest and lowest portions of the earth's surface at any given scale. For convenience, we distinguish three major orders of relief. The first is at a global scale and considers only the relative elevations of the ocean basins and the continental blocks. The second order of relief views the surface of the continental blocks in particular—although we could also include the floors of ocean basins as individual entities for this purpose. When a single continent is our frame of reference, the second-order variations in the relief of its surface that would be most prominent are mountains, plateaus, and plains. At an even larger scale, a mountain, or a plateau, or a plain will have a multitude of smaller features

upon its surface. These valleys, hills, swales, closed depressions, and numerous other minor topographic elements make up the third order of relief.

The highest point on earth is Mount Everest, 8,848 meters (29,028 feet) above sea level. The lowest point is the bottom of the Mariana Trench just east of Guam in the Pacific Ocean, 11,093 meters (36,198 feet) beneath the surface. The distance between these highest and lowest points is just about 19.9 kilometers, or 12 miles. Figure 7-19, below, shows a small arc segment of a circle 3.26 meters (10.7 feet) in diameter. If the earth were reduced in size to a sphere just that large, the nearly 20 kilometers of first-order relief described above would be contained in the .5-centimeter-thick band shown. Embedded in this band is a thin line only .04 centimeter in width. This thin thread represents that portion of the earth, between sea level and 1.6 kilometers (1 mile) high, where most of the world's population is found. Such are the dimensions of our abode.

**Figure 7-19 Reduced arc segment of the earth**

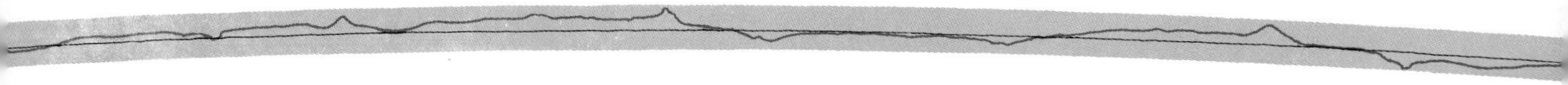

# 8 | ENERGY AND THE EARTH'S SURFACE: LANDFORMS AND MAN

Very much like the hero or heroine in a continued adventure story, our previous chapter left humankind occupying a thin and fragile film on the earth's surface. If we were shrunk in the same proportion as was the globe in that example, a six-foot-tall person (1.8 meters), would be reduced to .5 micron (1 micron = .000,001 meter). That is about one-tenth the diameter of the largest bacterial cell. Though seldom aware of it, we have such a size relationship to an imposing variety of landforms which dwarf us by their magnitude. Sometimes we are caught up in natural disasters and events with which we can scarcely cope, while at others we are as troublesome and perhaps as deadly to the earth as any bacterial swarm that might infect a human. If we are to survive and prosper, Lilliputians as we are on this big apple, we must learn to appreciate the morphology of the earth's surface, how its many shapes came into being, and how humankind affects and is affected by the processes which lift the continents, shape the mountains and the plains, and sculpt the hills and valleys on which we live.

## The Fundamental Elements of Land Formation

The questions suggested above can be dealt with systematically. Landforms result from the interaction of constructive and destructive processes: those that build and rebuild the earth's surface, and those that wear it away. We have already examined such processes in their broadest outlines, including descriptions of the rock and unconsolidated materials of which the surface is made. Now let us consider some additional factors of importance in the sculpting of the land. These include the relationship between energy and moisture available at various locations throughout the biosphere, and the shapes or profiles of the land which result from those relationships.

Chapter 6 has shown how an understanding of climatic processes helps us conceptualize energy-moisture relationships. Chapter 7 introduced the orders of relief and some notions of how differences in elevation come about at a global scale. It also dealt with some basic elements of weathering and erosion. We now must combine these things into a scheme which can be used to describe and understand the land around us. We will devote much of this chapter to discussing individual *landforms*, the particular shapes taken by the rocks of the earth's crust or by the *regolith*, and the unconsolidated mantle of rock fragments, weathered rock, and soil resting upon it. Sometimes, though, it will be more convenient for us to talk of particular *landscapes*, totalities including not only characteristic landforms but also soils, vegetation, bodies of water, and, wherever pertinent, the works of man.

## The building blocks and how they get there

Processes which shape the earth's surface can be divided into three basic types: *diastrophic*, *erosional*, and *depositional*. Diastrophism includes all massive earth movements except those involved in vulcanism. However, we will consider the latter along with diastrophism because of the dramatic contributions to the landscape made by molten rock in its many forms and processes. Among the diastrophic processes are folding, faulting, and warping of the earth's crust. By vulcanism we refer to all things associated with the movement of molten, fragmented, and gaseous materials from the earth's interior to its surface. Having already defined erosion and weathering in the previous chapter, let us continue our description of Mother Earth's fascinating face.

One day in early summer, one of us had the pleasure of bouncing across part of central Idaho in a jeep with Hoover Mackin, a geologist for whom landforms held a lifelong fascination. As we broke from the pines onto the open slopes far above the Payette Lakes, the author exclaimed with surprise, "How did those lakes get there?"

"The real question," said Mackin, demonstrating his ability to get directly to the point, "is how did the hole they're in get there?"

His question can be rephrased but not improved. How do differences in surface elevation and slope occur? Diastrophism at different scales makes a good place to begin looking for an answer.

## Landforms at a global scale

Major differences in elevation at a global scale occur between high-riding, sialic continental blocks and the floors of ocean basins. In accordance with plate tectonic theory, it is thought that these blocks once formed a single world island called Gondwanaland, which broke up and began drifting apart in Mesozoic times.

The two tiny figures in the foreground are dwarfed by the West Ridge and summit of Mount Everest several thousand feet above them. The photograph was taken during the 1962 American Expedition, which successfully climbed the world's highest mountain. In that effort, four Americans and two Nepalese Sherpas reached the summit. (Photograph by Barry C. Bishop, © National Geographic Society)

Subsequently, these stable masses of crystalline rock, called *shields*, became the cores around which continental formation has continued. Although the material of these shields is tough,

*Energy and the Earth's Surface: Landforms and Man*

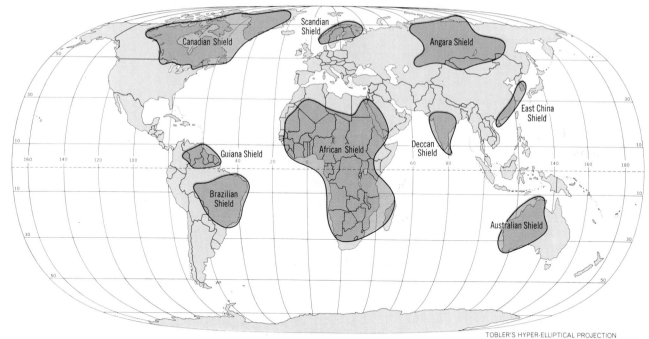

TOBLER'S HYPER-ELLIPTICAL PROJECTION

**Figure 8-1 Continental shield area** Shield areas are stable masses of crystalline rock which form the core of continents and which have remained above sea level throughout most of the geologic eras. They sometimes contain veins of metallic ores such as uranium, nickel, copper, and gold.

resilient, and nearly impervious to folding and other distortions resulting from lateral and vertical pressures in the earth's crust, because of their immense age their surfaces have been eroded nearly to sea level. Figure 8-1 shows the location of shield areas throughout the world.

Erosion of these shield blocks provides materials for deposition as layered sediments in the basins surrounding them. The sediments, in turn, are carried up against the rigid shields by plate tectonic movements. The resultant folding and buckling of the layered rock form mountain rims around the shields. In other places, large portions of the earth's crust sag as the result of compensating tensions or downward currents in the mantle beneath. These sags become depositional basins, which from time to time may be raised above sea level as the result

of further crustal adjustments. The newer highlands also contribute to these geosynclines and provide materials for still younger chains of folded mountains.

This highly simplified description provides us with a basic model of landforms at a continental scale. The familiar continent of Hypothetica has been adorned with these elements in Figure 8-2. Somewhere on each continent is an ancient, low-lying shield area. On one or more of its edges are folded mountains of varying age. Elsewhere, sedimentary basins collect alluvial materials brought down to the sea by rivers. However, in many cases, the gently sloping edges of these basins will be above sea level and form the structural platforms for vast plains. Thus, crystalline shields, folded mountains, and sedimentary platforms are apparent in this broad panorama.

*Physical Geography: Environment and Man*

## Landforms at a regional scale

But the above picture provides us with few details. Let us abandon our satellite view of the earth and discuss what diastrophic and volcanic landforms might be seen from a glass-bottomed 727 flying a few thousand feet above the surface.

**Folds and domes**  The scene becomes more complicated. The rimming mountains display a series of convex upward folds, *anticlines*, alternating with concave downward ones, *synclines*. Softer layers near the summits of anticlines may sometimes be breached and their interiors eroded away. The result is long, narrow valleys enclosed by outward-sloping ridges. Conversely, sometimes hard strata will preserve synclinal structures near the tops of hills. The long axes of these fold structures often *pitch* beneath the surface, and anticlinal valleys may be closed at one end while synclinal ones open outward. The resulting parallel ridges and amphitheaterlike closed valleys are distinctive additions to the landscape (Figure 8-3). The satellite picture of the Appalachian Mountains on page 151 gives some indication of these shapes.

In the same way, sedimentary beds can be pushed upward by the intrusion of molten materials from below or by a combination of lateral forces in the crust. The result is a dome structure or, if it is eroded, a closed ring of mountains, their gentle slopes facing outward where the surfaces of the layered rock are exposed by erosion; their inward-facing slopes are steep where erosion slices the strata like a layered cake cut by a knife (Figure 8-4). Very often the materials exposed in the interior of domes will be crystalline granites or some similar rock. These rocks are part of the molten intrusion which subsequently slowly cooled under a mantle of sedimentary beds before they were removed by erosion.

**Faults and fault blocks**  Sometimes tectonic forces exceed the flexibility of the rocks involved. Then ruptures or breaks occur along planes of movement throughout the crust. Such breaks are called faults; the plane in which

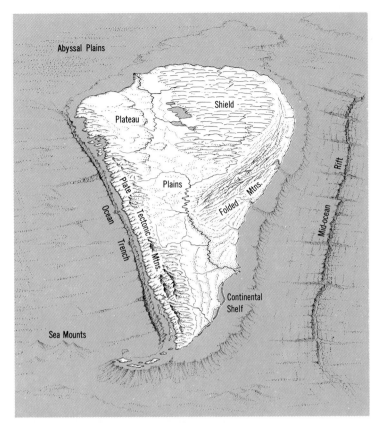

**Figure 8-2 Landforms of the continent Hypothetica**  Most of the major landform types found on the real continents are shown here, along with an indication of the plate tectonic movements of the adjacent ocean floor which contribute to the pattern. Look for the stable continental core, folded mountains, depositional plains, shallow seas, plateaus, and mountain ranges. Consider the tectonic and erosional processes and how they would combine in various places on the continent to form the general landscape.

movement of the blocks on either side occurs, relative to each other, is called a *fault plane* or *fault zone*. If the crustal forces are esentially tensional—that is, pulling apart—a center block may drop relative to the two on either side (Figure 8-5). These have been given the name *graben* (Ger. grave) and form spectacular *rift valleys*, such as those extending from East Afri-

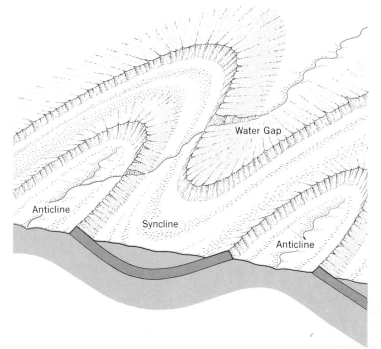

**Figure 8-3 Folded ridge and valley landscape**
In regions with layers of sedimentary rocks, upward-folded anticlines and downward-folded synclines erode into characteristic canoe-shaped or zig-zag ridges and valleys. Harder rocks form ridges, and softer rocks form valleys. Water gaps are created by streams which were in place before the folding occurred.

ca to the Dead Sea. If compressional forces squeeze a fault block upward, it is called a *horst*. The steep faces, or *fault scarps*, of these blocks form some of the most dramatic elements in the landscape. In other cases, movement of one block is horizontal relative to that of its neighbors. These planes of movement are referred to as *strike-slip faults*. Needless to say, faults and folds can occur at many scales, including some small enough for a child to step across. But large or small, the energy represented by such crustal displacements is enormous, and when it is released the resulting earthquakes can be dangerous.

## Vulcanism and associated landforms

There remains another set of basic landforms, sometimes deformational and sometimes depositional. Vulcanism thus makes a good transition to the discussion of erosion and deposition still to come.

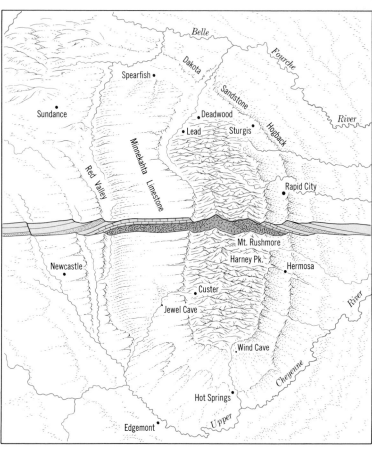

**Figure 8-4 The Black Hills, South Dakota** The Black Hills are a dome mountain structure which has thrust upward into the northern Great Plains region. Layers of sedimentary rock have been eroded away, revealing the crystalline rock center. The faces carved on Mount Rushmore (see photograph on page 88) are cut into granite. Wind Cave and Jewel Cave, both part of the National Park system, are found in the Minnekahta limestone deposits, the inner edge of which creates an inward-facing cuesta. Caves are normally found in limestone, a sedimentary rock. The second cuesta ringing the feature is a deep red sandstone named Dakota sandstone. Both of these rock layers were somewhat harder than the other sedimentary layers and stand out in relief as the entire area slowly erodes.

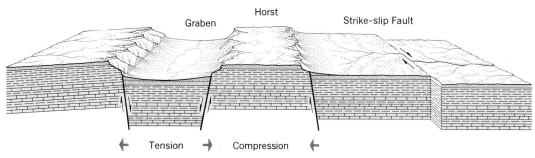

Horst

Graben

Strike-slip Fault

← Tension → Compression ←

**Figure 8-5 Faults and fault blocks**  Faults are sharp breaks in rock structure. Blocks of earth may move along the fault planes relative to one another. Blocks that have dropped downward relative to their surroundings are called *grabens.* Death Valley in California is a graben, the surface of which has dropped below sea level. Blocks that are thrust upward due to compressional forces are called *horsts.* Sometimes the blocks slip horizontally relative to one another. These create *strike-slip* fault zones. The infamous San Andreas Fault passing through the San Francisco Peninsula is a strike-slip fault. See the photograph on page 154.

**Vulcanism at the surface**  The most dramatic and obvious forms of vulcanism are volcanoes (photo on page 2). We began this book with a description of the destruction of Pompeii by Mount Vesuvius, which belongs to a class of volcanoes noted for their explosive qualities. These high, relatively steep-sided mountains are formed by the alternating deposition of layers of ash and lava. Acidic in chemical composition, they tend to vent gases explosively from their craters. Their lavas, poured forth and cooled on the surface, are light-colored and often the fine-grained equivalent of granite. Sometimes, when the central vents become plugged by solidified lava from former eruptions, violent explosions may result. Clouds of glowing rock fragments (*lapilli*) and gases may burst unexpectedly from side vents with disastrous results. The destruction of the city of St. Pierre on the island of Martinique in 1902 is typical of such calamities. These *strato volcanoes* account for the sublimely tapered cone of Mount Fuji in Japan and for the more rugged peaks like Mount Shasta, rising above the lower slopes of the Cascade Mountains.

Volcanoes with chemically basic eruptions ooze lava in great sheets from central vents, like those in the Hawaiian Islands. Mountains of this type may be of enormous size and cover vast areas with their accumulations. Their rounded profiles suggest the name they have been given, *shield volcanoes.* Chemically basic lavas can also issue from enormous fissures, and entire plateaus can be formed by their deposits. The Columbia Plateau in the Pacific Northwest, the Deccan Plateau in India, and the highlands of Ethiopia are all souveniers of such eruptions. When these lavas cool, they form dark-colored *basalt.*

If the material underneath a volcano is erupted in very large quantities, the cone may collapse in upon itself. The remaining supercrater, or *caldera,* can be dry, like Ngorongoro in East Africa, or filled with collected precipitation, like Crater Lake in Oregon. On the other hand, the molten materials in volcanic vents may sometimes solidify and survive long after subsequent erosion removes less durable ash and rock from around them. The resulting *volcanic necks* or *plugs* can be startling landscape features, such as Ship Rock, New Mexico (Figure 8-6), and Devil's Tower, Wyoming.

**Vulcanism below the surface**  Molten rock can either extrude onto the surface, where it cools rapidly, or it can intrude into existing strata without being exposed to the atmosphere. In the latter case, slow cooling under insulating layers of rock allows larger, easily visible crystals to form. Of course, these coarse-grained

*Energy and the Earth's Surface: Landforms and Man*

Mount Shasta in northern California is a dormant volcano which has been deeply eroded by the action of ice and water. A secondary cone of more recent origin developed on its flank.

*igneous rocks* have to be exposed by erosion or remain solely for the gaze of lonely miners.

Erosion reveals a variety of intrusive igneous forms. When narrow sheets of molten rock force their way at an angle across layered sedimentary strata, and when subsequently the softer sediments are eroded away more rapidly than are the igneous materials, a wall-like barrier, or *dike*, may be exposed. When intrusions force their way between layered rocks like jam in a sandwich, the resulting solidified forms are called *sills*. Sometimes what starts out to be a sill bulges the overlying rock upward, although the layers below remain relatively undisturbed. This type of mushroom-shaped intrusion is called a *lacolith* and appears on the surface when exposed by erosion, much the same as a dome structure. The difference between igneous domes and lacoliths is that the intrusive rock of the former structure extends downward to great depths, while the latter have essentially

Shiprock in the northeastern corner of New Mexico is a volcanic *neck* or *plug*, which consists of erosion-resistant basalt that was once the molten core of a volcano. The less durable ash and rock, layed down by the eruption, have long been removed by erosion.

flat-layered sedimentary floors. When igneous intrusions reach regional size and implace themselves by melting and eating, or *stopping*, their way into the rocks which they intrude, they are called *batholiths*. These form the crystalline cores of many complex mountain re-

**Figure 8-6 Volcanic landforms** Features of active vulcanism are shown to the right of the diagram. These include volcanos, lava flows, and cinder cones. Lava flows enter valleys in the existing landscape. To the left in the diagram are examples of remnants of igneous rock intrusions. The batholith consists of igneous rock which cooled and crystallized slowly while still well below the surface. It is shown here as being exposed by subsequent removal of overlying material through a long erosion process. The region of the exposed batholith is complex geologically because metamorphic rock, created through pressure and heat, is present at the margins of the batholith. An ancient, deeply eroded volcano, and table mountains and mesas of old lava flows, stand higher than their surroundings because they are more resistant to erosion than the original surface rock. A volcanic plug and accompany radial dikes, shown in the center of the diagram, are remnants of a previous intrusive stock, or core, of an ancient volcano.

Multiple layers of basalt are exposed at Dry Falls on the Columbia Plateau near the Grand Coulee Dam. They are part of a series of lava flows which make up the Columbia Plateau and extend over thousands of square kilometers in Idaho and eastern Oregon and Washington.

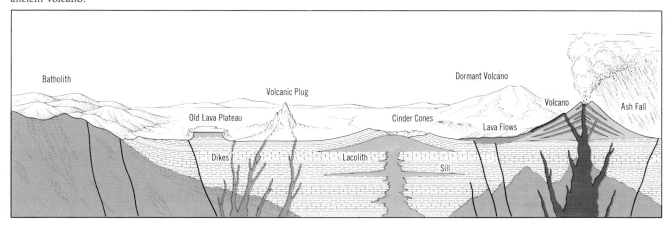

Molten lava sometimes forces its way into fissures in the earth, where it cools and solidifies. Subsequent erosion of the surrounding material reveals this hard, igneous rock in the form of a wall-like barrier called a *dike*.

gions, such as the Idaho batholith of the northern Rocky Mountains. These and other forms appear in Figure 8-6.

*Erosion and deposition*

The essence of geography inevitably asserts itself. That is, the unity of the environment is always revealed by the geographic point of view. No matter how hard we tried in the last section to talk only of the landscape's building blocks and of how they are put in place, erosion found its way into the discussion. Now that we have some basic shapes to consider, let us face up to the inevitability of erosion and see how landforms are sculpted and modified. We also will need to consider what becomes of the materials taken from those modified surfaces. Deposition is also an indivisible element in the landscape.

Folds, domes, fault blocks, and volcanic forms seldom retain their original shapes. Energy and moisture constantly work upon them. It is when climate passes its cool, moist or feverish, dry hands across the earth's face that its features take the shapes we know. Because of this and in order to be consistent with our previous discussions, we will return to our basic climatic classification and show how crustal materials and structures are modified according to their location on the earth. The presence or absence of vegetation also is important in this discussion. Deciding where to introduce plant geography into a general discussion of physical geography is always a problem. In this text we have reserved the next chapter for a complete

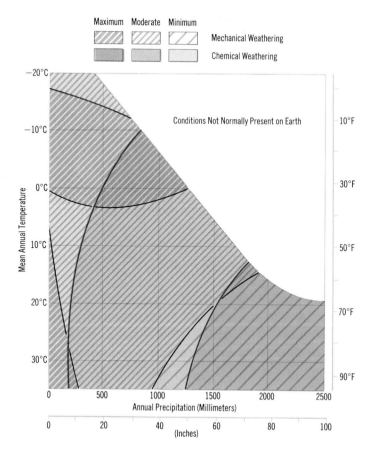

**Figure 8-7 Intensity of weathering in relation to moisture and energy availability** Mechanical weathering is greatest where vegetation is sparse, where freezing and thawing occur, where winds are common, and where a moderate amount of water is present. Chemical weathering is promoted in regions of higher temperatures and maximum water. All these conditions vary continuously from place to place. (After Van Riper)

description of evapotranspiration, soils, and vegetation. But as we talk now of landforms and landscapes, we will not lose sight of plant cover as a critical cosmetic on the face of the earth.

**Energy, moisture, and landforms**  The availability of energy and water largely determines the kinds of weathering and erosion that the lithosphere undergoes. It is possible to locate different intensities of mechanical and chemical weathering in terms of temperature and precipitation. Figure 8-7 does this, while Figure 8-8 places the same information on the 3 × 3 chart of worldwide energy and moisture conditions introduced in Chapter 6. It is necessary to remember that the relationships, as shown, have been greatly simplified, although some attempt to depict transition zones has been made.

Materials once loosened by weathering are subject to removal by wind, water, and mass wasting. The distribution of these agents and their effectiveness are illustrated in Figures 8-9 and 8-10. While the type of rock involved has a definite effect on the resulting topography, it is still possible to equate landforms and the processes which create them to the climatic environments suggested in these diagrams. However, before discussing a possible climatic classification of landforms, it is important to understand how slope profiles are formed and what is meant by the general cycle of erosion.

**Slope processes**  When a landform is first exposed to climatic forces, its surfaces begin to weather, and the resulting loose materials are carried away. William Morris Davis, an American geographer writing at the turn of the century, suggested that such processes produce slopes with convex upward profiles (Figure 8-11A). This shape, according to Davis, reflects rapid weathering on exposed upland surfaces and rapid movement downslope of the weathered debris. Materials pile up at the foot of the incline because of the inability of streams and other agents of erosion to remove them as fast as they arrive. It follows from this line of reasoning that slopes under these conditions become progressively less steep as time goes on.

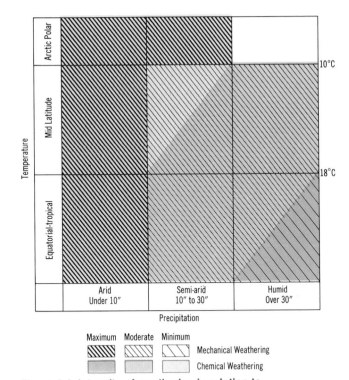

Maximum  Moderate  Minimum
Mechanical Weathering
Chemical Weathering

**Figure 8-8 Intensity of weathering in relation to climatic categories**  This figure corresponds to Figure 6-6 and abstracts information from Figure 8-7 showing the relative action of mechanical and chemical weathering processes by climatic categories.

Walther Penck, writing forty years later, suggested that slopes may develop and retain concave upward profiles throughout their histories. In Penck's scheme, slopes retreat, with their later profiles parallel to those formed at earlier times (Figure 8-11B). In this case, the removal of materials keeps pace with their arrival at the foot of the slope. The resulting profiles are concave upward, with solid rock a short distance beneath the surface. The surfaces thus created are called *pediments*.

Both types of slope profiles are found in nature and undoubtedly depend upon the balance between the production of weathered materials and the rate of their removal. The extreme result of Penck's slope-forming processes would be straight-sided canyons or even overhanging cliffs. This would occur if the energy concentrated by streams at the foot of the slope

*Energy and the Earth's Surface: Landforms and Man*

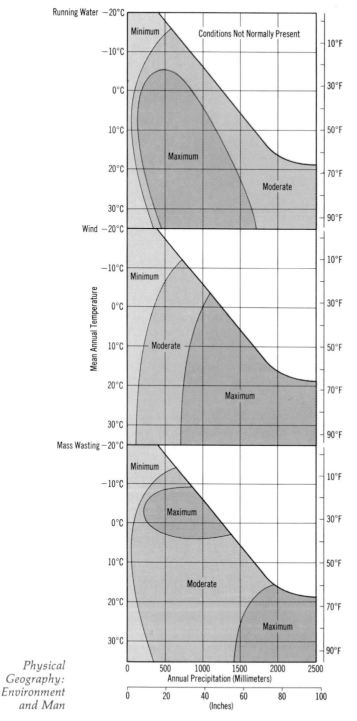

Running Water

Wind

Mass Wasting

Mean Annual Temperature

Annual Precipitation (Millimeters)

(Inches)

Minimum · Conditions Not Normally Present · Maximum · Moderate · Minimum · Moderate · Maximum · Minimum · Maximum · Moderate · Maximum

was more effective than the energy acting on the upper surfaces. These schemes imply a sequence of events in time which requires additional explanation.

**William Morris Davis** Davis worked in the Appalachian Mountains, where he observed that many of the summits, regardless of their geologic structure, were of the same height. These accordant peaks had gentle slopes near their tops and steeper sides. This suggested to him that they were the remnants of an older, flatter erosional surface. This ancient surface supposedly dated from the Cretaceous Period, a time during which erosion had worn down an earlier mountain range to a nearly featureless plain close to sea level. This ancient land surface, which he termed a *peneplain*, was thought to have subsequently been uplifted by renewed diastrophism and dissected by the streams which had originally meandered across it.

Davis reasoned that geologic history consists of periods of continental uplift followed by periods of erosion leading to *peneplanation*, that is, the reduction of the raised block to near sea level. He thus viewed every contemporary landscape surface as the result of the underlying *structure* of the materials eroded, the geologic *processes* involved, and the *stage* currently in progress during the *cycle of erosion*. Such cycles followed regular sequences of *youth*, *maturity*, and *old age* (Figure 8-12).

During their youth, uplifted continental blocks have flat surfaces or uplifted peneplains dating from the previous cycle. Streams from that cycle are rejuvenated by having their upstream portions lifted higher than their outlets. Such streams cut narrow, steep-sided canyons and have oversteepened gradients. Waterfalls are common. As this youthful stage progresses into maturity, the V-shaped valleys are broadened and their side slopes become more gentle. If rivers cut down into these surfaces at a

**Figure 8-9 Intensity of erosion by water, wind, and mass wasting in relation to moisture and energy availability** The effectiveness of removal of material from a site by various agents is related to moisture and energy conditions.

faster rate than uplift occurs, the result is a series of *incised meanders* reflecting the shifting meanders of those streams at an earlier time.

With rejuvenation, softer strata would erode more rapidly than resistant rock. New mountains carved by this erosion often have streams cutting across their crests rather than flowing parallel to them. Such a rejuvenated surface will have *water gaps* where downcutting has kept pace with the upthrust of the folded mountains. Some *wind gaps* are also found where uplift outpaced downcutting and diverted streams from partially formed transmontane alleys.

The period of old age described by Davis ended with peneplanation, meandering rivers

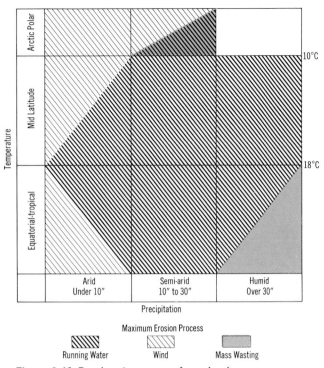

**Figure 8-10 Dominant process of erosion in relation to moisture and energy availability** This diagram is an abstraction of Figure 8-9. While all processes of erosion are at work in most places, the dominant one is shown for each domain.

**Figure 8-11A Davis's theory of slope development** According to William Morris Davis, the evolution of a slope proceeds through a succession of upward curves from steep to gently concave. This type of development appears to occur when material is carried down the slope faster than it can be carried away at the base of the slope by stream action. *B* **Penck's theory of slope development** According to Walther Penck, the evolution of a slope proceeds by a parallel retreat of the slope face. This type of slope development occurs when material deposited at the base of the slope is removed by stream action or other agents at the same rate as material moves down the slope face.

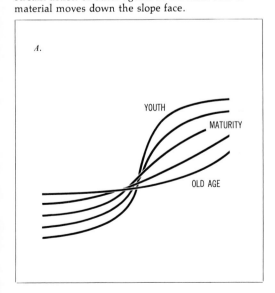

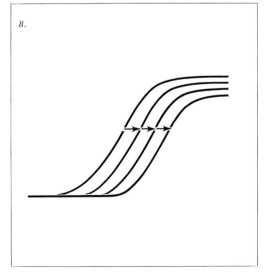

*Energy and the Earth's Surface: Landforms and Man*

**167**

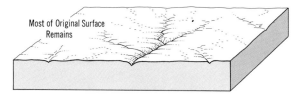

Most of Original Surface Remains

*A.* YOUTH – Vigorous Drainage, Headward Erosion

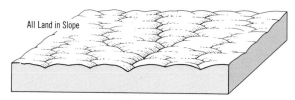

All Land in Slope

*B.* MATURITY – Well-established Drainage

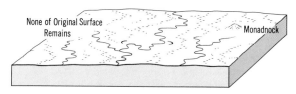

None of Original Surface Remains  Monadnock

*C.* OLD AGE – Sluggish Drainage, Meandering

**Figure 8-12 Davisian landform cycle** William Morris Davis thought of a region as passing through stages of erosion. Each stage was characterized by certain landforms. With the passage of time, the region reduces to a nearly flat sea-level plain. Subsequent uplift by renewed diastrophism starts the cycle over again.

with low gradients, and the remnants of the old upland surface surviving as a few low, isolated mountains, or *monadnocks*. These latter features were named after Mount Monadnock in Maine, which rises above part of the peneplain which Davis postulated extended along the Appalachian crests. Peneplains of this sort were supposedly the end result of an erosion cycle's having "run down." As such, they were thought to be removed from active erosion and in a stable or steady state condition.

This interpretation of landform processes,

while offering a useful way of thinking about the erosion of the earth, has several shortcomings. It now appears that continental uplift is a continuous process without long periods of stagnation. Moreover, uplift varies widely from place to place, and large areas seldom resemble peneplains. The variety of landforms and surface slopes is equally great. It would seem that there need to be too many cycles of erosion going on at once to categorize things neatly in Davis's terms, if indeed we should think at all about cycles of erosion.

**G. K. Gilbert** An alternate theory of landform development by G. K. Gilbert explains variation in slopes and surfaces without insisting upon continentwide uplift or peneplanation. Nor does his theory depend upon the starting up and running down of successive cycles of erosion. Put another way, his view of landforms is independent of historic or geologic time.

Gilbert's report on the Henry Mountains of Utah was published some time before Davis's essays, yet his ideas seem more in keeping with modern systems theory. His view of landscape formation saw erosion processes in a steady state where the import of energy and materials to any portion of the system is balanced by the export.

*Of the main conditions which determine the rate of erosion, namely the quantity of running water, vegetation, texture of the rock, and declivity, only the last is reciprocally determined by rate of erosion. Declivity[1] originates in upheaval or in the displacement of the earth's crust by which mountains and continents are formed: but it receives its distribution in detail in accordance with the laws of erosion. Wherever by reason of change in any of the conditions, erosive agents come to have locally exceptional power, that power is steadily diminished by the reaction of the rate of erosion upon declivity. Every slope is a member of a series receiving the water and waste of the slope above it, and discharging its own water and waste upon the slope below. If one member of the series is eroded with exceptional*

[1] By "declivity," Gilbert refers to "slope" as used in our discussion.

*rapidity, two things immediately result: first, the member above has its own level of discharge lowered and its rate of erosion is thereby increased; and second, the member below being clogged by an exceptional load of detritus has its rate of erosion diminished. The acceleration above and the retardation below diminish the declivity of the member in which the disturbance originated, and as the declivity is reduced the rate of erosion is likewise reduced.*

*But the effect does not stop here. The disturbance that has been transferred from one member of the series to the two which adjoin it is then transmitted to others and does not cease until it has reached the confines of the drainage basin. For in each basin all lines of drainage unite in a main line and a disturbance upon any line is communicated through it to the main line and thence to every tributary. And as a member of the system may influence all others, so each member is influenced by each other. There is an interdependence throughout the system.*[2]

A number of scientists, among them Leopold, Hack and Strahler, have recently established that equilibrium conditions exist almost continuously as surfaces are eroded. For example, the flatter upland surfaces at the Appalachian summits which Davis identified as peneplain remnants are in reality active erosion surfaces. Their shape is the result of conditions existing at present, and they are being actively lowered at the same time that they are being eaten away by the headward erosion of streams from a lower level.

The world we live in is a kinetic one. If the elevation of an area is steadily lowered by continuous erosion, different energy and moisture conditions will apply when it is elevated than when it is nearer sea level. In turn, the character of the erosion process may adjust to greater or lesser amounts of water and to different quantities of energy in the guise of air temperatures. But these changes, tied as they are to geologic processes, require very long periods of time. Within more restricted time

scales, climatic trends or seasonal variations may also place varying amounts of energy in the erosive system. However, changes in landforms are seldom easily perceived in the short run. On the other hand, geologic time scales may be too vast for us to appreciate. In deciding the proper time scale with which to view the shaping of the earth's surface, it seems reasonable to choose an intermediate span of time, such as those in which climates come to exist. Thus, the regional distribution of climatic types offers a reasonable time-space framework for looking at basic landforms and the processes which shape them.

## A Simple Landform Classification

The following classification is limited to five basic energy-moisture regimes: arid, arctic, semiarid, humid temperate, and humid tropical. They are presented in this order because of certain similarities which link them. For example, arid and arctic environments are both subject to lack of water, mechanical weathering, special forms of mass wasting, and wind erosion. Semiarid environments are transitional, with better-watered areas, and combine features of both arid and humid regions. The temperate and humid tropics allow chemical weathering, mass wasting, and the work of running water to a degree not found elsewhere.

A classification of this kind is subject to many oversimplifications and exceptions to whatever "rules" might be suggested. We hope, however, that the discussion that follows has a particular convenience which justifies the classification it presents. We say this because the observer will always find himself *in an environment* with special energy, moisture, vegetation, and surface conditions. What he sees will be a kind of unity resulting from a number of interacting processes. This is why we use a climatic approach rather than the more conventional descriptions which talk about *the work of running water, the work of wind,* etc., and lead the reader a merry chase from wet to dry and from hot to cold as though he could be everywhere at once.

[2] G. K. Gilbert, *Report on the Geology of the Henry Mountains,* U.S. Government Printing Office (2d ed., 1880), Washington, D.C., pp. 117–118.

*Energy and the Earth's Surface: Landforms and Man*

## Landforms in arid regions

Landform processes in arid regions are subject to the following conditions. Precipitation is scanty and irregular. Statistics giving average rainfall in the desert have little meaning as to either the amount or the time of its occurrence. For example, the settlement of Tamanrasset in the central Sahara had no rain from June 1933 to August 1939, and then in 1950 within three hours 3.6 centimeters (1.42 inches) fell in forty minutes. Vegetation is widely spaced or lacking and provides little resistance to immediate run-off. Because surface materials are often bone dry and unprotected by either plant leaves or root systems, wind erosion is significant. Surfaces have little or no soil as a result. Thunderstorms provide much of whatever rain falls. All these factors in combination result in sheet-flooding and the rapid accumulation of flash floods in valleys and canyons.

Streams are of three kinds: ephemeral ones, which flow only when precipitation is immediately available from unpredictable cloudbursts; seasonal streams, which flow only during the rainy season—a condition which may occur on the margins of the desert; and exotic streams, which acquire their flow in remote and better-watered regions and sometimes may not even reach the sea because of evaporation losses.

Lack of surface water inhibits chemical weathering. High daytime temperatures are followed by surprisingly low ones at night because of open skies and intense reradiation. This means that rocks may expand and contract ever so slightly, loosening materials on their surfaces. Windblown sand also has a scouring effect. Thus, deserts are subject to slight but steady mechanical weathering, the movement of windblown materials from place to place, and the irregular but violent impact of flash floods.

The diastrophic and volcanic landforms mentioned earlier occur as often in desert regions as elsewhere in the world. Arid conditions, therefore, have a wide range of ingredients upon which to work, and it is their secondary effects —including a special set of depositional landforms—that give deserts their distinctive appearance. The scarcity of soils and vegetation further enhances desert landscape features (Figure 8-13).

**Figure 8-13 Arid region landforms** The absence of a mantle of vegetation lays bare the surface geology in arid regions. Wind and water are erosion agents. Water does its work during infrequent, brief, local rainstorms. Material eroded from the mountain collects in *alluvial fans* or, as the fans coalesce, into a piedmont or *bahada* ringing the mountain remains with a gently sloping surface. Rocks layered in horizontal beds form mesas or table mountains, with the top level being more resistant rock. Isolated remnants form buttes, and old, worn-down igneous rock masses remain above the general basin level as *inselbergs*. In the foreground, *arroyos* and intermittent streams feed into a *playa* or flat, dry, lake bottom.

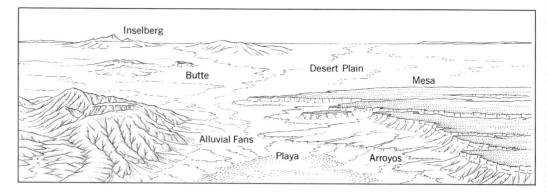

This walled village in Iran is in a desert landscape. In the foreground, very sparse vegetation is scattered over a stony desert floor and dry streambed. The barren mountain in the background shows evidence of wind and water erosion. Notice, along its base behind the village, an alluvial apron, or *bahada*. The action of water contributes to the formation of desert landscapes during infrequent rainstorms, which may be sudden and violent.

At a regional scale, closed basins with interior drainage having no outlet to the sea are frequent. There simply isn't enough water to fill the depressions; therefore, lakes are scarce. Where block faulting occurs, this results in a series of grabens partially filled with eroded materials and by uplifted horsts with steep slopes formed by fault scarps. The dramatic *basin-range topography* of Nevada and Utah is a good example of this. These basins, or *bolsons* as they are sometimes called, receive all the materials brought down from the surrounding mountains by ephemeral and seasonal streams. The alluvial materials collect in fan-shaped deposits at the mouths of canyons. Such *alluvial fans* often are so close together that they overlap, forming an unbroken slope, or *bahada*, at the foot of the mountains.

In other places, the mountain fronts may have retreated, leaving a gentle slope similar to that on the alluvial fill along the basin's flanks but cut in the underlying rock. These rock-cut *pediments* are often overlapped on their lower sides by alluvial deposits which fill the basin centers. If the erosion is taking place on crystalline rock, and if it has gone on for a sufficient time, there will be only mountain remnants, or *inselbergs*, similar to monadnocks rising above the surface. In the lowest areas of the basins shallow, ephemeral lakes may form when runoff is sufficient. But such lakes are short-lived and rapidly evaporate, leaving behind *playas*, glaring white salt deposits smooth as a billiard table. The Bonneville salt flats in Utah, where high-speed autos are raced, are such a feature.

Sometimes, particularly where permanent

In regions where horizontal bedding of sedimentary rock has occurred, erosion processes lower the general surface level, but in places more resistant layers tend to protect softer layers beneath them, and *mesas* and *buttes* are formed. An isolated tower of stone is usually called a butte. Tabletop mountains or uplands are called mesas. Adjacent mesas or buttes are often the same height because the same resistant rock layer forms the cap rock for the entire area.

streams flow out of or across a desert region, the surfaces may be stripped of their mantle of weathered debris. Wind can also be important in this removal. If the underlying strata are horizontal, and if hard and perhaps thick rock layers are included in the sequence, the landscape may include *mesas* and *buttes* rising above flat desert plains. Mesas, from the Spanish word meaning "table," are formed by the eroding away of softer beds underlying the harder ones on top. Steep cliffs develop, with shallow

caves worn in the softer underlying materials. At such places the overhanging rock breaks off, forming a *fall face*. The fallen material eventually pulverizes through further weathering and is removed by water and wind.

Steep-walled valleys (called *arroyos* in the Spanish-American Southwest and *wadis* throughout the Middle East) are a distinctive landscape element under these conditions. As the fall faces continue their retreat, arroyos become wider and the upland surfaces between streams smaller and smaller. Flat-topped mesas result and with further erosion become buttes, with little remaining of their upper surfaces. If the rock layers are tilted—for example by an igneous intrusion—steep-faced hills, or *cuestas*, with gentle back slopes may be found. If the beds are nearly vertical, *hogback hills* with extremely steep slopes will form.

At still more localized scales the desert is filled with a variety of special features. Rocky deserts from which all the finer materials have been winnowed by the wind are common. If these surfaces are bare, they are known in Arabic as *hamadas*; if they are covered with a layer of rocks and pebbles, they are called *regs*. Stones on reg deserts are often sculpted into faceted *venifacts* if the windblown sand which abrades them comes consistently from one direction. Stones can also have a varnishlike sheen or patina resulting from desert conditions.

But probably sand dunes most often catch the imagination of the desert traveler. These masses of unconsolidated sand are moved about by wind action. If there is relatively little sand moving across a flat surface, the dunes will be crescent-shaped *barchans*, with the horns of the crescent pointing in the direction of the prevailing wind. With more sand available and a strong wind from one direction, long whaleback shapes may form as *longitudinal dunes*. Where sand covers the entire surface as a true sandy desert (Arabic: *erg*), light to moderate winds will create transverse dunes at right angles to the flow of air. But if the winds blow from all directions, *star dunes* take shape and remain essentially in one location.

Much more could be said of arid landscapes and landforms, but we must resist the lure of the desert. Instead, we now need to consider some of the problems which such places present to human occupance. While thirst belongs more properly to the category of climatic problems, the search for water is never-ending. Particular problems of water procurement, however, are discussed at the beginning of Chapter 13 and will be bypassed here. Too much water can also be a problem in the desert. The configuration of drainage patterns, impervious surfaces, and

steep-walled arroyos combine to make flash floods a real danger. A group of French tourists recently were wiped out by such a flood while on their way to the ancient city of Petra in Jordan. The narrow canyon leading to the archaeological site was suddenly inundated and all were drowned. This incident belatedly prompted the installation of a flood-warning system at the entrance to the dangerous part of the route.

If wind instead of water is considered, drifting sands continually threaten to overwhelm oasis gardens. Even those farms not about to be engulfed by sand can have their crops sheered off by windblown sand as though by scissors. In many places mud walls are built around oases not only to keep out strangers but also to divert

Desert varnish is a black shiny stain of iron and manganese oxides, usually no thicker than a coat of varnish, deposited on rock surfaces. The rocks shown here are covered by such a varnish and also show surfaces flattened by abrasion from windblown sand.

wind-driven sand. And yet, the desert's surface is remarkably fragile from an ecological point of view. In some ways it resembles that of the moon, for wheel prints and campsites endure for long periods of time. Trash scattered on the surface decomposes slowly for lack of water and bacteria, and a moment's heedless damage or littering can endure for generations. This is particularly true in the American Southwest, where dune buggies, motorcycles, and multiterrain vehicles are raced for sport. Vast areas of the Mojave Desert have been brutalized in this way. The subsequent disruption of vegetation, slopes, and surfaces will last for years.

The relationship between the occurrence of deserts and human activities is therefore worthy of study. It has been suggested by Bryson that excessive cultivation and denudation by overgrazing result in a dust pall over semiarid regions and deserts. Such clouds tend to shade the ground beneath, with subsequent cooling of the surface. The cooler surface warms the atmosphere in contact with it less than that above warmer, exposed areas. The cool air is less likely to rise, and convectional rainfall is less likely to occur. Thus, a deviation-amplifying loop forms within the system. The cooler it is, the less the rain; the less the rain, the more intense the desert; the more intense the desert, the more the dust; the more dust, the greater the dust pall; the greater the dust pall the cooler it is.

Recent studies also suggest a similar mechanism relating to desert surfaces having high reflectivity, or *albedo*. If overgrazing bares the surface, greater reflection and reradiation occur. Less energy is available to heat the air above, and such cooling may produce less convectional air currents and precipitation. This seems to be the case along the 1948–1949 Egyptian-Israel armistice line on the east side of the Sinai Peninsula. A fence was erected there in 1969, and grazing was prevented to the east and north of it. Measurements of surface temperatures on either side of the dividing fence showed values from 3° to 5° C (6.3° to 10.6° F), cooler on the overgrazed and brighter side.

Again we see how environmental unity makes it difficult to discuss landforms without reference to vegetation and climate. But the processes of weathering and erosion prevalent in desert areas clearly contribute to the fragile character of desert plant cover and to subsequent differences in surface temperature and convectional precipitation. We might even theorize that if human occupance were to continue long enough, differences in rainfall could affect the basic erosional shapes in areas whose climate has been changed by mankind.

*Landforms in cold regions*

Landforms related to low-energy climate regimes and to the work of moving ice are found both in the arctic and antarctic, where such conditions prevail. They are also found in warmer areas which once were subjected to glaciation but are no longer. High mountains at lower latitudes are also sculpted in large part by cold climate processes. Rather than make an arbitrary division of these regions and their landforms, we will move from high latitudes and elevations to warmer, previously glaciated regions at lower latitudes and nearer sea level under the general heading of "cold regions."

Arctic, subarctic, and antarctic regions resemble arid ones in that liquid water is scarce and mechanical weathering predominates. In this case, weathering includes not only the spalling off of rock through freezing and thawing but also the scouring effects of moving ice in glacial form. We say that water is scarce because it is seldom available in liquid form, though ice and snow may abound. Thus chemical weathering is inhibited.

**Mountain glaciation** Glaciation can be of two kinds: continental and mountain. Mountain glaciers are still active in many areas, though many more have melted after helping shape the land (Figure 8-14). These glaciers may originate in icefields or begin their journeys downslope from smaller catchment areas or bowl-shaped *cirques* high on the sides of mountains. Cirques are formed when snow accumulates on the shaded side of a mountain or in some depression. There it crystallizes into icy *firn*, which in turn compacts into glacial ice. When the ac-

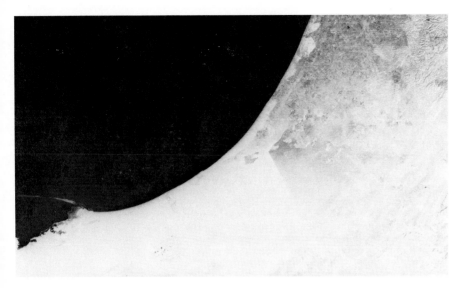

Human control of animal grazing affects the albedo (amount of reflected light) of desert surfaces. The effects are visible in this satellite photograph. A fence was erected along the 1948–1949 armistice line between Israel and Egypt in the Sinai-Negev region and had been in place five years when this ERTS-1 (Earth Resources Technology Satellite, no. 1) photo was taken on Oct. 22, 1972. During this interval, animal grazing was permitted southwest of the fence in the region now characterized by high albedo, but not in the currently darker region northeast of the fence. Removal of vegetation by grazing accounts for the difference. Energy conditions—and, therefore, moisture conditions—are now different on each side of the fence. (See Joseph Otterman, "Baring High-Albedo Soils by Overgrazing: A Hypothesized Desertification Mechanism," *Science*, vol. 187, pp. 531–533; NASA photo)

cumulation is great enough, the ice will begin moving downslope under its own weight. As it moves away from the original catchment area, it plucks frozen rock from the surrounding slopes. This plucking action gradually gives cirques their distinctive amphitheaterlike appearance. If a glacier melts, the scoured floor of its cirque may be lower than the rocky lip at its edge, and the depression may be filled with a *tarn* lake. The mountain peaks above the upper ends of valley glaciers are often sculpted into sharp *horns* or serrated ridges called *aretes* by the coalescing of glacial cirques. Sometimes two adjacent glaciers will carve away most of the mountainside between them, leaving a sharp *cleaver*, or ridge. Cold prevents chemical weath-

ering, but mass wasting and wind scour further sharpen the outlines of these landforms.

All the above processes take place at relatively high elevations in the *zone of accumulation*. Subsequent downhill movement may tumble the ice over cliffs or down very steep valley slopes with resulting *ice falls*. Eventually, higher air temperatures at lower elevations will melt or vaporize the snow faster than it can accumulate, and the glacier will gradually lose volume. This is the *zone of ablation*.

Moving glaciers scour materials from the sides and bottoms of the valleys they occupy. They also transport debris which falls onto them from higher slopes. Such materials form long stripes on the surfaces of glaciers. If these

*Energy and the Earth's Surface: Landforms and Man*

Figure 8-14 Mountain glacier landscape  Glaciers form in high snow fields and move plastically under their own weight down valleys, where they melt at lower elevations. The action of the ice cuts great amphitheaterlike *cirques* into the mountain. A mountain eroded on several sides by the action of glaciers will have a prominent *horn* at the summit and sharp-edged ridges or *aretes* separating the cirques. Valleys which have held glaciers are typically U-shaped, and as the larger glaciers have greater cutting action, side valleys are left hanging. They are often the site of waterfalls. The material removed from the heights is deposited in *lateral* and *terminal moraines* at lower elevations, where the glaciers have melted. Glaciated mountain landscapes provide us with some of the most spectacular mountain scenery found.

stripes are near the sides, they are called *lateral moraines*; if they are in the middle as a result of two glaciers coming together, they are called *medial moraines*. If the ice melts before reaching the sea, its load of materials is dropped and may build up at the snout of the glacier in the form of a *terminal moraine*. It is not uncommon to find water collected there in *moraine dammed lakes*.

During past periods of glaciation, valley glaciers descending from continental ice fields scoured U-shaped profiles in the former V-shaped river valleys which they occupied. When the ice melted, it left behind many distinctive features visible today. Mountain spurs are often truncated, and side valleys enter the main one at discordant levels high above the present valley floor. Such discordancies mark the height to which the glacier filled the main valley. These *hanging valleys* sometimes hold streams which plunge in spectacular falls to the newly revealed surfaces far below. Yosemite National Park, with its many waterfalls, is the most famous example of this type of landscape.

If mountain glaciers reach the ocean, scouring out their characteristic U-shaped valleys below sea level, when the ice melts the sea will fill those depressions. Thus long, narrow,

steep-sided inlets, or *fjords*, can be formed. Fjords penetrating coastal mountain ranges are found today in Norway, British Columbia and Alaska, and along the most southerly shores of western South America.

**Continental glaciation**  Antarctica, Greenland, and parts of Alaska and Canada still retain *continental glaciers* or *ice caps*. In those places the gradual accumulation of snow over the centuries has compacted into thousands of feet of ice, which stretch for miles with only an occasional rocky outcrop, or *nunatak*, punctuating the surface. These icy masses move slowly outward under the pressure of new accumulations or downslope with gravity to the sea. They may find their way through narrow valleys, as in Alaska, or advance along a broad front, as in parts of Antarctica. Though they are confined now to high latitudes, continental glaciers once covered much of North America and northern Europe. During their advance, they sculpted the land they moved across. When their ice melted, many deposits of glacier-borne materials were left behind, marking the land with their distinctive forms.

Continental glaciation represents one of the

This glacier, located in the Canadian Rocky Mountains, was larger and extended farther down the valley in the recent past. Notice the lateral moraine on the left-hand side of the valley and the lack of vegetation on the recently uncovered area in the foreground. The stream in the foreground issues from the leading edge or snout of the glacier and is opaque from the heavy load of finely ground rock material it contains, a product of the ice actions. The picture was taken in late summer, when the snow cover was minimum. There is permanent snow accumulation in the cirque visible in the background. Also the sharp horn and arete profile of the mountains are typical of glaciated mountain landscapes.

most impressive adjustments in the energy balance of the earth. During glacial periods, and for reasons largely unclear to science at the present, much of the available moisture is retained on the earth's surface as ice and snow. For example, glaciers at least 1.6 kilometers thick (1.0 mile) covered up to 30 percent of the earth's surface during the most recent, or Pleistocene, period of continental glaciation. Sea level during the Pleistocene was lowered by as much as 600 feet, and the earth's albedo was increased by the glaring white of those vast snow fields. Areas that are now deserts were apparently better watered because less energy was available for evaporation in subtropical regions. Crustal adjustments also occurred as the weight of the ice depressed the land beneath. Thereafter, when the ice melted, the land rebounded to its original level. Evidence of this can be found associated with major glacial features like Lake Michigan, which was in part scoured out by glaciation, depressed, and later filled with glacial melt waters. As the ice retreated from south to north, the southern shores and beaches rebounded first and the more northerly ones somewhat later. In fact, it has been suggested that when the final isostatic adjustments take place, the northern shores of Lake Michigan will rise and the lake will be tilted, so to speak, until it empties through the Illinois and Mississippi River system to the south. Adjustments such as these, along with the worldwide lowering of sea level and the accompanying changes in the base level of rivers everywhere, are landform responses to glaciation at a global scale.

There is evidence that continental glaciation also occurred in Cambrian and Permian times. But whether there is a cyclical nature to such events is unclear. Within the Pleistocene itself, there were several advances and retreats of the ice in Europe and North America. Table 8-1 shows the estimated times for these events. One matter of considerable interest is the length of the interglacial periods. Such periods have been as long as 50,000 years, and because the last ice retreated from the northern United States less than 20,000 years ago, it is impossible to tell if we are safely out of the ice age or whether another glacial advance is around time's corner. Fortunately, this is probably the least current of our environmental problems.

To the north in the shield areas, glaciation has just ended or still exists in places. There the landscape shows evidence of the scouring and stripping power of the glaciers. The surface has been cleaned of any soil that might have devel-oped in preglacial times, and is monotonously level, with lakes filling old scours everywhere. The present cold climate regime inhibits weathering and plant growth—something that will be dealt with in the next chapter—and the landscape to most people would seem inhospitable.

Farther south, the land has had sufficient time to recover from the overriding ice, and more solar energy is available to modify the effects of glaciation. Nevertheless, glacial sculpting and deposits are everywhere seen in modified, relic form. Most widespread of these and best seen at a regional scale are the terminal moraines of the major glacial advances. As the ice moved south, it carried an enormous load of materials scoured from the surfaces over which it moved. At some line of advance, the energy available in the environment was sufficient to melt the advancing ice edge, and the glacier stabilized in space. Sometimes ice would be abundant and the edge would assume a more southerly position; sometimes the sources diminished and the ice would retreat. That is, it would reach spatial equilibrium farther north. (The term "retreat," which sounds as though the ice pulled back like a tape measure into its case, is somewhat misleading.) Wherever such temporary equilibrium conditions were reached, rocks, boulders, and pulverized material piled up. These *recessional moraines* form festoons of low hills all across the Midwest (Figure 2-10A). Where the advancing ice plastered the surface with a mixture of glacial materials, this unsorted veneer forms *ground*

Table 8-1  Stages of Pleistocene Glacial and Interglacial Activity in North America and Europe

| Time Before Present | North America | Europe |
|---|---|---|
| 0–65,000 | Wisconsin-Iowan Glacial | Wurm Glacial |
| 65,000–125,000 | Sangamon Interglacial | Uzuach Interglacial |
| 125,000–180,000 | Illinoian Glacial | Riss I and II Glacials |
| 180,000–230,000 | Yarmouth Interglacial | Hotting Interglacial |
| 230,000–300,000 | Kansan Glacial | Mindel I and II Glacials |
| 300,000–330,000 | Aftonian Interglacial | — |
| 330,000–470,000 | Nebraskan Glacial | Gunz I and II Glacials |
| — | | |
| 538,000–600,000 | — | Donau I and II Glacials |

*moraines* or *till plains* (Figure 8-15). Sometimes piles of till might be streamlined by the glacial advance and form whale-backed hills called *drumlins*. Their characteristic shape is blunt-ended in the direction from which the ice came, and smooth and sloping in the direction of its advance.

When the continental glaciers disappeared, enormous quantities of melt water drowned the land in front of them with floods. *Outwash plains* are thus found in front of terminal moraines. Sometimes temporary lakes formed between the melting ice edge and more advanced moraines. The smooth surfaces of those former *lake beds* are also important postglacial features. Sometimes great chunks of ice buried in glacial deposits might take years to melt, and would leave closed depressions, or *kettles*, pockmarking the land. Where drainage streams flowed underneath the glaciers, sediments were deposited in their channels. Once the surrounding ice melted away, *esker* deposits shaped like winding snakes remained. If lakes formed between the ice margins and the adjacent upland, the alluvial deposits laid down in them today appear as *kame* terraces.

Our discussion of glacial landforms might go on and on. But since we have reached warmer climes, we must state a simple rule. The mode by which energy is applied to the earth's surface is in great part responsible for the number of different landforms. In the case of arid regions, mechanical weathering, wind, and sometimes rushing water predominate. In cold regions, grinding ice replaces flowing water in large part. Where more moisture is available in liquid form, chemical weathering and a steady flow of water dominate landforming processes. The surface is more smooth, softer in appearance, and concealed beneath a cover of soil and vegetation. For example, tilted sedimentary beds can be sharply sculpted cuestas or more smoothly contoured hills, depending on moisture and energy conditions. Since arid and cold regimes are typified by vivid landforms, we chose to begin with them. Now let us look at the landforms of warmer, moister regions.[3]

[3]Human problems associated with cold-climate landscapes are discussed in the next chapter because of the special nature of tundra soils and permafrost.

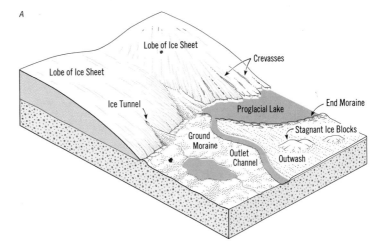

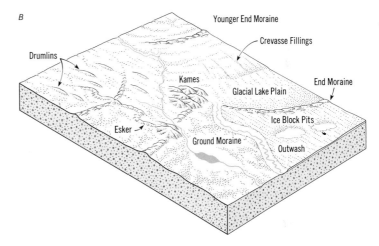

**Figure 8-15 Continental glacier landforms** (*A*) Landform formation during a period of continental glaciation. Notice ice and debris-dammed lakes, streams filled with meltwater, and streams depositing sediments in tunnels under the ice. (*B*) The same region shown in (*A*) after the ice has melted. An *esker* marks the position of the ice tunnel. Streams under fit their valleys. Drumlins mark the direction of ice movement; kames and moraines outline positions where the ice was stationary. Kettles are depressions left when ice blocks under the outwash plain finally melt.

## Landforms in semiarid regions

The characteristics of semiarid regions which contribute to their distinctive landforms include sufficient moisture to allow spatially continuous plant growth, often in the form of grasses. Sufficient chemical weathering to create a mantle of fertile soil is also important. The moisture regimes are either annual, with wet and dry seasons, or moisture is distributed more or less evenly throughout the year. In both situations, considerable variation can occur from year to year. In either case, dry spells allow wind erosion where plant cover becomes insufficient. Conversely, the greater moisture in these regions, when compared with the deserts they often neighbor, allows the fixing of wind-blown materials by vegetation. Thus, the smoothing of the land mentioned in the previous paragraph begins to be apparent under these conditions.

*Badland topography* is associated particularly with semiarid regimes. Soft, easily eroded materials, sometimes capped by more resistant beds, are susceptible to the intermittent rainfall in such regions. Unconsolidated clays in the Dakota badlands and volcanic ash and tufa in central Turkey's Cappadocian area have been eroded into fantastic shapes under these conditions. Resistant beds above softer ones in both cases provide protective caps which help all manner of cones, columns, and bare steep-sided hills to form.

Another distinctive set of landforms associated with semiarid conditions are *loess hills* composed of wind-transported soil. These deposits occur on the lee sides of deserts, where increased water vapor or actual precipitation tends

In the Cappadocian region of central Turkey, a strange landscape has been created through erosion of soft volcanic ash. Caves are easily dug into this material. Since before the Christian era, people have taken refuge in caves they had prepared in these formations.

to settle out desert dust from the air and trap it under an accompanying grass cover. Thus the lighter materials removed from reg deserts and hamadas find new shapes. The rolling Palouse Hills east of the desert scablands of central Washington State are loess deposits. They are also found in China east of the arid regions of Sinkiang Province and the Ordu Desert and in parts of western Siberia. In the American Midwest just east of the 100th meridian, where rainfall increases to about 20 inches (50 cm), are other loess deposits. (It should be noted that this is just east of Thornthwaite's "0 moisture index line," Figure 6-4.) In some cases the wind-blown materials originate not only on deserts but also where retreating glaciers have dumped loads of pulverized rock. Regardless of their sources, loess deposits have most often been brought into wetter zones by prevailing westerly winds. The character and problems of these soils will be discussed in the next chapter, but their distinctive smooth or gently rolling topography qualifies them as unusual landforms.

Semiarid conditions can also result from underlying rock if it is highly porous. Limestone is subject to erosion by solution, and underground flow through caves and expanded joint and fracture systems can rob the surface of most streams. Thin soils and sparse vegetation often result. The distinctive *karst* landscape formed by solutional erosion of limestones and dolomites is found in Yugoslavia, Yucatan, parts of Kentucky, and Florida. When solutional erosion begins, the surface topography may reflect less soluble overlying strata. But where water finds its way into a cave system, small enclosed basins then appear. These *sink holes, swallow holes,* or *dolines,* as they are called, eventually widen into larger pits. The famous sacrificial wells, or *cenotes,* of Mayan Yucatan are among these. Next, blind valleys may form, followed by major depressions, *poljes,* many miles in length. Parts of the Taurus Mountains in Turkey have spectacular examples of these latter forms. Finally, irregular jumbles of limestone blocks and in-place limestone remnants resting on underlying insoluble rock may be all that survive. When this happens, the desertlike character of the area may disappear and surface drainage again take over.

Generally speaking, the original building blocks continue to maintain their shapes in semiarid landscapes. Soils and vegetation play less of a modifying role than in wetter regions. Semiarid regions are thus transitional in character and find expression through their combination of grass cover but few trees, moderate to good soil formation, and a combination of landforms shaped by wind and water. Wind particularly can be a curse or a blessing under these conditions, for dust storms and rich loess mean very different things to farmers. More of this will be described in the next chapter and in the case study of the "Great American Dust Bowl" in Chapter 13.

## Landforms in humid regions

Increasing humidity means increasing chemical weathering, deeper soils, and abundant vegetation. Running water is at work everywhere, along with mass wasting, and in the tropics, leaching of minerals through solution processes. Floodplain deposits, with their special landforms, catch the eye. Here, if anywhere except under the grinding plane of ice, the original contours of the basic building blocks are subdued by erosion.

It is true that Hack and others have demonstrated that different rock types help determine the different elevations of erosion surfaces in the Appalachian Mountains and elsewhere. But it is, nevertheless, difficult not to emphasize the differences between arid and humid landscapes and landforms shaped from similar rocks. This returns us to the questions of system and process posed earlier in the chapter. There is apparently no simple answer.

*. . . the Davisian cycle of erosion is essentially a closed-system framework of reference wherein the initial conditions (uplift) determine the final product (peneplain). The open system, in contrast, embraces natural drainage basins wherein a dynamic equilibrium of form (hillslopes and channel profiles) and processes results. It is apparent that the classical or Davisian view tends to*

*exclude the equilibria which have been demonstrated in nature, whereas the open-system viewpoint is reconciled only with some difficulty with the facts of continuous denudation through time.*[4]

Rather than join in this ongoing tournament of ideas, let us simply describe a few of the landforms and processes encountered in humid regions. Because water is abundant, structural basins (e.g., grabens) are filled to overflowing. In other words, there are few empty, closed depressions, and lakes are the rule. By the same token, streams flow year round and are familiar features.

The constant flow of water removes weathered materials at a steady rate. Though soils are well developed, their thickness remains constant where removal and formation maintain a kind of dynamic equilibrium. This steady supply of alluvial materials, combined with those that may be transported downstream from other types of landscapes, are deposited in the lower courses of the rivers as *floodplains*. We have already mentioned the process by which rivers keep their profiles in dynamic equilibrium and how meanders and oxbow lakes are formed. Another common feature of floodplains are *natural levees* (Figure 7-9). These result from the overflow of the main channel during times of flood. As excess waters pond either side of the stream, their burden of alluvium is dropped in the slower-flowing water near the submerged banks. This has the effect of increasing bank heights as yearly flood follows yearly flood. Deposition also raises the channel bottom, and it is not uncommon for a stream's surface, even during normal periods of flow, to be higher than the surrounding floodplain because of confining levees. Of course, humans often assist this elevating process by topping natural levees with artificial ones. The overall effect is to make flooding more severe once the containing barriers are breached.

Poor drainage behind natural levees or because of very low stream gradients near sea level often results in swamps. In the tropics, where swamp waters can be particularly acidic, chemical weathering of adjacent rocks can be extreme. In fact, limestone cliffs adjoining tropical swamps frequently have a chemically weathered notch found along their base. Tropical karst often develops spectacular towers and "cockpits," where massive, thick bedded limestone is eroded in this manner.

Soils and decomposed rock become progressively deeper as one moves from the mid-latitudes toward the equator in humid areas. At the same time, a thick mat of roots and vegetation may allow very steep soil-covered slopes in the tropics, while more gentle profiles are common to the cooler regimes farther north or south. Soil creep accounts for the convex profiles in the latter regions, but slumping and landslides often scar and steepen tropical hillsides.

As the final base level of the sea is approached in the lower portions of rivers, their remaining load of alluvium is deposited in *deltas* located at their mouths. This landform was originally named for the triangular Greek letter "delta," for many such deposits take this shape (photo, page 183). Actually, several deltaic forms can occur as streams lose their velocity and subdivide into final *distributaries*.

Having reached the sea, we have also reached the end of our classification of landforms according to energy-moisture regimes. Our discussion has not been exhaustive, nor have we intended that it should be. Generally, the time frame in which landforms are created far exceeds that of human generations. Therefore, we are perhaps less concerned with them than with more ephemeral elements in our environments. There remains, however, another location where energy is concentrated to such a degree that changes in the shape of the land can happen there from one year to the next. We refer to the coastlines of the world.

## Coastal landforms

Everywhere from the tropics to the arctic the sea gnaws and the sea builds. Coastal features

[4]Lawrence K. Lustig, "Backgrounds of Modern Geomorphic Concepts," in *Deserts of the World*, W. G. McGinnies et al., eds. (The University of Arizona Press, 1968), pp. 99–106. The above quotation is a summary, in part, of Chorley's earlier comments.

formed by either deposition or erosion reflect conditions that override the climatic regimes discussed above. Only in a few places where glaciers may abrade the shore or buffer it from the attack of waves, or where coral or mangrove swamps may help build a shoreline, do we find exceptions to this rule.

The energy delivered against the shoreline by waves can be tremendous. One study estimates that in fair weather along a rocky coast the force delivered by wave action is 600 pounds per square foot. When larger waves batter the same shore, the force averages 2,000 pounds. When we learn that waves strike the shore about six times per minute, it is easy to see why erosion and mechanical weathering are major elements in shoreline processes. This happens in a variety of ways. The mass of water delivered to the shore has weight and a battering effect of its own. At the same time, wherever cracks and caves occur, the penetrating water delivers enormous hydraulic pressure. Waves also lift and hurl bottom materials against the shore as well as grinding or abrading the bottom. What comes in must go out, and the retreat of waters after the initial wave action can drag materials seaward in its rush. This erosive power of the sea was amply illustrated when a 2,600-ton block of concrete implaced on the shore near Wick, Scotland, was washed away during a storm! The depositional and transporting power of currents and waves along the shore is just as impressive. For example, Long Island, New York, is bordered on its southern shore by a string of sand bars paralleling the coast. These *offshore bars* are composed of water-deposited sands which are carried along the coast by prevailing currents. The flow of water moves 600,000 cubic yards of sand westward each year. Most of the sand is redeposited in quiet waters beyond the western tip of Fire Island at Democrat Point. This spit of land extends westward at an average rate of 212 feet per year (Figure 8-16). The original Fire Island lighthouse is now five miles back from the existing point, and the process of westward extension is continuing.

The general effect of the sea upon the land depends in large part on whether the shoreline is being actively uplifted or depressed by diastrophic movements. (Conversely, it can be

The two Egypts. Agriculture is possible in Egypt only where irrigation water is available. Topography restricts irrigation to the valley and delta of the Nile River. Beyond those areas, "boundless and bare, the lone and level sands stretch far away." The contrast between the two environments can be clearly seen in this satellite photograph. (NASA Gemini 5 photo)

argued that sea level can either rise or sink, creating the same effect.) It is therefore possible to distinguish *coastlines of emergence* and *coastlines of submergence* (Figure 8-17). In the former case, gradually sloping continental shelves which are the underwater extension of coastal plains may be exposed. The result is often a straight shoreline, with either very little slope or perhaps with *wave-cut terraces* paralleling the shore but some distance behind it. If the sea is shallow, offshore sand bars may result as depositional features. The general effect, though, is of a somewhat unvaried shore with few outstanding landforms.

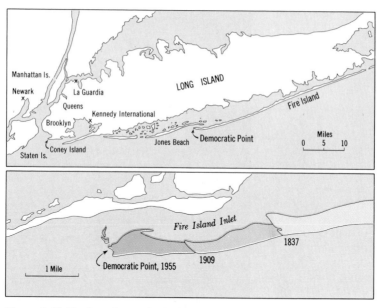

**Figure 8-16 Westward growth of Democrat Point** Offshore bar islands often change shape and location as shifting sands are carried along by surf and currents. The westward growth of Democrat Point on Fire Island over the last century is an example. The shifts are not necessarily slow and steady. Dramatic changes may occur during heavy storms such as hurricanes. Such events often cause severe damage to homes and facilities that have been built on the sand islands open to the force of the sea.

Submerging coastlines often present hillslopes with steeper profiles to the onslaught of the sea. In that case, the upper erosion surfaces controlled by atmospheric weathering and erosion may have their gentle profiles interrupted by wave-cut cliffs. The abrading power of the sea will steepen the shore slope, perhaps cutting a notch into the cliff face at water level, where the full power of the waves is concentrated. This creates a steep fall face and mass wasting on a large scale. Meanwhile, the erosive power of the waves is considerably less beneath the surface of the water. This allows the rapid retreat of cliff faces, but at the same time the creation of a *wave-cut bench* at low tide level. Where more resistant headlands thrust like battlements into the sea, *wave-cut notches* forming on opposite sides of the nose of land may meet. *Sea caves* will thus become *arches* which eventually collapse, leaving isolated columns of more resis-

tant rock called *sea stacks*. Along these coasts, hanging valleys are common and again reflect the different rates of land and marine erosion. In time, depositional as well as erosional landforms appear along submerged shores. Ridges are truncated by the sea; flooded valleys are closed at their mouths by sand bars or partially blocked by sand spits. Islands may also be connected to the mainland by similar bar deposits and form *tombolos* (Figure 8-17A).

As mentioned above, organic processes can also influence shoreline features. In tropical areas with shallow seas, brackish swamps and mangrove forests can prevent erosion by the massed twinings of roots. Eventually, such coasts may extend themselves seaward as land gradually fills in near the shore. The growth of *coral reefs* also creates coastal landforms, particularly along submerging coasts (Figure 8-18). These tropical marine organisms form colonies

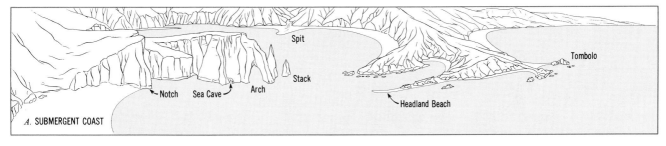

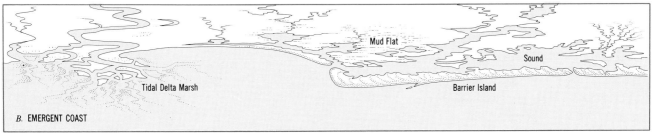

**Figure 8-17A Submergent coasts** The headlands along submergent coasts are the places to see wave-cut cliffs, sea stacks, arches, and tombolos (the last feature being a sea stack, or small island, connected to the mainland by a sand bar). Material cut from the headlands is deposited in sand bars and beaches in coves and protected areas. **B Emergent coasts** Emergent coasts are characterized by gently sloping coastal plains, straight shorelines, and barrier islands with lagoons and swampland behind them. These are slowly filling with material brought down by streams. Deltas form in places where large streams come to the sea.

which build protective calcium deposits in which to live. Such reef deposits also become the refuge of all manner of marine life. Coral polyps need sunlight in order to grow, and if the water in which they are found becomes either too cold or dark—both possible functions of increasing depth with changes in sea level—they will perish. When this happens, the more deeply submerged parts of the colony die, while new growth forms in the shallower surface waters. In this way coral reefs can grow upward, forming special islands called *atolls*. These ring-shaped islands, found especially in the Pacific Ocean, are composed entirely of coral and coral sands upon a rock basement, usually volcanic. It is thought that their unusual form derives from the slow submergence of isolated volcanic islands around which *fringing reefs* first form. Later, and with further subsidence, these are

transformed into *barrier reefs*. As mentioned above, the steady death and upward extension of the colony, with luck, keep pace with the rising waters until only a ring reef or atoll survives, enclosing the waters of a *lagoon*.

In all these examples, human occupance of the shore is a risky business. The time scale utilized by us in deciding to build along shores exposed to wave erosion is too brief. It should come as no surprise to anyone when the bad storm which comes once every fifty years washes away his summer home which stood so unprotected. For example, only slight damage occurred along the beaches at Fire Island from 1937 until 1962. But in March of the latter year 96 homes were destroyed by a spring storm. Other losses included damage to 195 houses and the washing out of 3,000 feet of roadway. All in all, the cost of the storm from New England to

*Energy and the Earth's Surface: Landforms and Man*

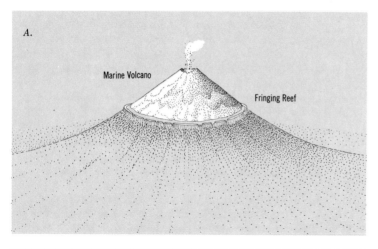

A.

Marine Volcano

Fringing Reef

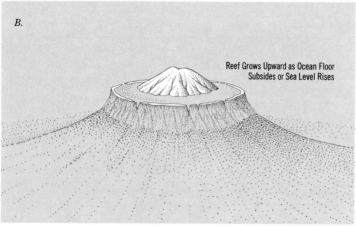

B.

Reef Grows Upward as Ocean Floor
Subsides or Sea Level Rises

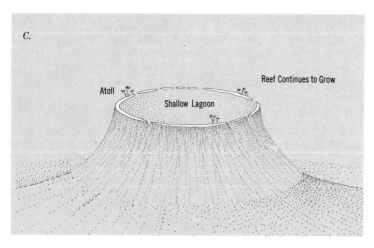

C.

Reef Continues to Grow

Atoll

Shallow Lagoon

Virginia was about 234 million dollars. Eight years later another storm wreaked 2 million dollars additional havoc. Similar events occur elsewh e in the United States and around the world. More significant are disasters like the high waves that swept into the Bay of Bengal on November 12, 1970. Pushed there by a particularly powerful cyclonic pressure system, 25-foot-high waves advanced toward the Ganges Delta at 100 miles per hour. The funnel shape of the bay, plus the driving winds, piled up water until it smashed onto the low shoreline. A breakdown in the early warning system left almost everyone living there totally unaware of their impending doom. This was despite the fact that the danger had been reported by weather satellites as much as a week before. Nearly 500,000 people drowned because of this combination of events. In every case, it is the concentration of energy along the shore interface between land and sea that shapes the coastlines of the world and makes them both beautiful and dangerous.

## Landforms and Man

The subject of the above title is endless. And yet, how often we overlook the deck of spaceship earth. Human activity must adjust to the forms the land takes. At the same time, mankind more and more frequently reshapes the surface of the earth, creating new landforms and new landscapes. Let us consider man as both a creator of and an adapter to landforms before closing this chapter.

**Figure 8-18 Coral atolls** Coral atolls consist of a string of coral islands and reefs forming a ring around a shallow lagoon and typically found in the South Pacific surrounded by very deep water. Charles Darwin, in 1842, suggested a theory for their formation. He postulated that an island such as a mid-ocean volcano slowly subsided at a rate which allowed the coral to build up the reef and remain near the surface. Eventually the island disappeared completely, leaving an atoll-ringed lagoon. Weathering of the mountain and a general rise in sea levels caused by the melting of the continental glaciers may have contributed to the process. The long white beaches of the tropics often consist of pulverized coral rock. The reefs provide an environment for a unique and rich flora and fauna.

Steep cliffs, sea stacks, and arches are characteristic of eroding headlands or capes along the Oregon and Washington coasts of the Pacific Ocean. Point of Arches, near the Pacific Coast area of Olympic National Park in Washington, is typical.

## Man-made minescapes

As the search for minerals becomes more intense (see Chapter 12), more and more of the land is disturbed by mining. Over 3.2 million acres have already been mauled by our machines in the United States. Table 8-2 shows the share of responsibility each type of mineral has in this destruction. Coal leads the list and promises even greater devastation in the future. This is because coal deposits underlie 50,000 square miles of the Appalachian Mountains while even larger areas are being opened to strip mining in the American West at places like Black Mesa, New Mexico.

Since strip mining of coal is so important and because it already is well developed in Appalachia, let us look at some of the man-made landscapes that result from this activity.

> Coal field! A term like tundra,
> Rain-forest, karst, savanna;
> A humanly created
> Topographical constant.
>> Donald Davie
>> "County Durham"

Appalachian coal was mined at first from tunnels and shafts. The resulting piles of mine waste or tailings are still a familiar sight in the

*Energy and the Earth's Surface: Landforms and Man*

**Table 8–2 Commodities Responsible for Surface Mining as a Percentage of Total Mining-Disturbed Land in the United States (as of Jan. 1, 1965)**

| Commodity | Percent |
|-----------|---------|
| Coal | 41 |
| Sand and gravel | 26 |
| Stone | 6 |
| Gold | 6 |
| Phosphate | 6 |
| Iron | 5 |
| Clay | 3 |
| All others | 7 |

Source: *Surface Mining and Our Environment*, U.S. Department of the Interior, Strip and Surface Mine Study Policy Committee (1967).

area from Pennsylvania to Alabama. While locally unpleasant, these disturbances of the natural terrain are point locations when viewed at a regional scale. However, with the increasing demand for energy since World War II and with the petroleum crisis of the seventies, new methods of mining have taken over. Operations have grown larger and larger, and small owner-operated mines are disappearing. Large corporations employing massive earth-moving equipment are now *strip mining* the region. This technique requires the unusable beds overlying the coal seams to be removed and dumped to one side, thus exposing the underlying coal so that it can be directly loaded onto giant earth movers by huge electric shovels.

The magnitude of these operations is difficult to imagine. The largest electric shovels can lift one-third of a million pounds of earth in their scoops. The gondola trucks which receive the burden of the scoops carry away 200,000 pounds of material each trip. Although such shovels can move only one-quarter mile per hour, their steady gnawing at the landscape consumes 65,000 acres each year, according to the National Coal Association. Strip mining has already

Candlestick Park is a baseball stadium located on land reclaimed from San Francisco Bay. The material for the land fill came from the adjacent hill, a sizable portion of which was used as fill. Although parts of San Francisco Bay are very deep, 70 percent of the original bay is shallow enough to permit reclamation, and land fills have occurred in many places. Major climatic changes, changes in aesthetic values, decline in wildlife, and other effects would occur if even a quarter of the potential reclamation took place. Local pressures have resulted in continual piecemeal reclamation, as in the case of Candlestick Park. Now there are also pressure groups whose purpose is to "Save the Bay." The Bay is a regional feature with regional value and should be subject to decisions arrived at through regionwide debate.

A strip mine at the outskirts of Potsville, Pa. Strip mining leaves severe scars on the landscape. It becomes economical to exploit coal seams by strip-mining techniques under a variety of local conditions, including thickness of material overburden relative to thickness of the coal seam. Available technology and current prices are also important. What is considered a resource or not may change with changes in technology and/or prices. Strip mining further depends upon legal, institutional, and social postures. Rather permissive rules exist in states where strip mining is common. Neighbors of the mines experience undesirable effects; no provision for reclaiming devastated lands exist or are enforced; and so on. Regulatory powers at various levels of government could be brought to bear on this issue. This would require political action on the part of all concerned. (Photograph courtesy of E. Bannister)

affected 1,500,000 acres in the Appalachians. Areas this large remain noticeable even at regional scales.

Where the underlying coal beds are on flat terrain, the shovels and earthmovers attack the surface from above. The land is rearranged into long parallel ridges with steep slopes and sharp crests. These *spoil banks* are separated by pools of water, which are often so acidic that no fish can live in them. The spoil banks themselves, being composed of mineral soil with little humus or weathered materials, are slow to reseed with vegetation. The overall effect is of a sterile accordian-pleated land unfit for life.

Alternatively, if the coal occurs in hilly country, the mining machines attack the seams from the side. *Contour stripping*, as this is called, grooves continuous notches around hills and on valley sides. The effect is somewhat like girdling a tree. Just as a continuous strip cut in the bark of a tree will kill it, so too do these stripped contours alter the water level of the land above them. They not only tend to drain it and dry it but they also increase erosion by changing local base levels and by exposing loose materials. Fifty-foot artificial cliffs along Appalachian hillsides are now a common man-made feature and have a total length of more than 20,000 miles! Their effect is increased by piles of useless overburden, which is bulldozed downslope and left lying.

The impact of strip mining does not end with an unaesthetic landscape. We have already mentioned the acidic waters in the spoil bank pools. Iron pyrites and marcasite found in the piles of rubble contribute iron sulfides to the

*Energy and the Earth's Surface: Landforms and Man*

water of the entire region. Some of this changes to sulfuric acid, which can kill fish, sometimes by the hundreds of thousands. Nor does chemical destruction represent the only problem associated with strip mining. The increased sediment load delivered to streams is impressive and deadly to aquatic life. Studies show that in Kentucky spoil banks produce nearly 8,000 tons of sediment per square mile per year, compared with about 6 tons from well-forested areas. Even the unpaved access roads yield as much as 3,750 tons of sediment per mile per year, which can find its way into local streams. Sedimentation like this has the overall effect of depleting fish stocks by removing their food supplies. A blanket of silt quickly covers the bottom, suffocating plants and small life-forms, such as insects. Denied their food, the fish either starve or move out of the area. If they remain, their eggs can be buried by the silt and destroyed.

There have been some efforts to either control strip mining or to restore the disturbed areas to their former state. Both have seen little success. At this writing, legislation is before Congress to limit strip mining and to make certain the mining companies must reshape the land they spoil as well as replant it. This represents an interesting problem in geographic scale. In nearly all cases where legislation has been introduced at the state level, the mining lobbies have been able to render it ineffective or to frustrate its being put into practice. This is because local legislators are often too open to political pressure, including that from voters who see the mines largely in terms of employment opportunities. At the national level lobbies still exist, but senators and representatives from nonmining states bring some objectivity to the law-making process. It should also be noted that as the public becomes more aware of ecological consequences, some local political campaigns have been run successfully on anti-strip-mining platforms.

When actual efforts are made to restore spoiled lands, many problems are encountered. Steep slopes on notched mountainsides or on spoil banks can be successfully planted only

about 20 percent of the time. Locust trees and pines are among the few large plants that can survive on the disturbed soil, and love grass, vetch, and some leguminous species provide the only cover in between. This is a far cry from the hardwood forests with their rich variety of growth that once covered the same areas. In mountainous regions, mining should be limited to slopes of less than 14 percent; in all cases spoil banks should receive much more attention than they do.

But who's to blame? While writing this book here in the Midwest, our room has been lighted with electricity generated fifty miles away by burning coal mined in Kentucky and hauled to Michigan. Moreover, though we have selected coal as the villain of our piece, many other mining operations vital to our life-style contribute to the disruption of the land. Which of the commodities shown in Table 8-2 would any of us be willing or able to do without? The organization of the world into complex economic and political systems is just as critical a part of environmental structuring as are the hills themselves. While laws controlling runaway mining are necessary, and while reclamation projects are vital, the problem of environmental deterioration will never be completely solved until it is viewed within the total context of the human ecosystem, and as a philosophical as well as an economic issue. We will discuss these aspects of resource use in Chapters 10 through 13.

### Human adjustment to the land: A political example

A fascinating example of the role of landforms in political and economic processes took place along the United States and Mexican border in the twin cities of El Paso and Juarez. In that area the Franklin Mountains extend southward almost to the Rio Grande River. The river curves southward around the mountains and has marked the boundary between the two countries since 1848 (Figure 8-19). The original border treaty stipulated that the main channel of the river should form the legal international boundary as long as any changes in the river were "effected by natural causes through the slow and gradual erosion and deposit of alluvium and not by the abandonment of an existing river bed and the opening of a new one."[5]

When the river shifted rapidly southward during the years until 1895, the Mexican government claimed that this constituted an unnatural movement. The United States claimed otherwise. At the same time, the mayors of each city in 1899 cooperated in having a meander or gooseneck bend of the Rio Grande cut through and shortened in order to reduce flood hazards in the twin towns. The subsequent cutoff territory was considered to be Mexican but remained in such an uncertain situation that no one was willing to build upon it (see accompanying photo). The disputed land to the south of the 1852 channel was known as El Chamizal (after a desert grass that grew there), while the cutoff meander was called Cordoba Island. The land in El Chamizal was used for industrial and other purposes incompatible with its location near the center of the urban area. That of Cordoba Island was totaly empty. At the same time, the bending of transportation routes around the Franklin Mountains, along with the foreign status of Cordoba Island, pinched El Paso nearly in half between the high ground and the Mexican border. The congestion which resulted created a serious problem for El Paso.

Subsequently, through a series of international negotiations, the two governments agreed to trade parcels of property in order to even out the boundary, thus easing congestion and making use of the land possible by ensuring title to it. The United States received the northern part of Cordoba Island while ceding to Mexico the southern portion of El Chamizal and another territory just to the east of Cordoba Island. The rectification of these territories has allowed the more rational development of the

[5]William M. Mallow, *Treaties, Conventions, International Acts, Protocols and Agreements between the United States of America and Other Powers 1776–1909* (2 vols.), 61st Congress, 2d Session, Senate Document No. 357, 1910, Vol. 1, pp. 1109–1110.

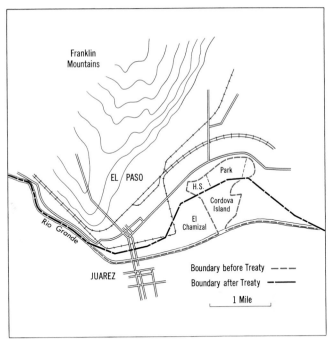

**Figure 8-19 Boundary treaty** The boundary between the United States and Mexico in the adjacent cities of El Paso and Juarez followed an old meander of the Rio Grande. This constrained transportation facilities and other urban land uses in El Paso to a narrow area between the Franklin Mountains and the Mexican border. Because the river had shifted channels, ownership of the El Chamizal district was in dispute, and this uncertainty also affected land use. A treaty was signed in 1964 straightening the boundary and ending the dispute. (See photograph on page 192)

two cities. This is clearly a case where the river and the mountain, though inanimate elements, helped shape the problems and the solutions rising from human occupance of the area. This example also shows how the shape of the land and the processes which continue to transform it must be dealt with in a systematic and knowledgeable manner. We could cite many more, but other important matters remain to be examined before returning to the subject of total human ecological systems in the concluding chapters.

*Energy and the Earth's Surface: Landforms and Man*

The El Chamizal area of El Paso, Texas, on the left and Juarez, Mexico, on the right. See Figure 8-19 for details. (Reprinted by permission from *Geographical Review,* © American Geographical Society)

# 9 | EARTH AS THE HABITAT FOR LIFE: VEGETATION AND SOILS

Life is an edge dweller. It takes sustenance from air, water, and land. However, the bulk of material found in organisms is composed of elements which in their pure form at earth temperatures are gases. We consist, for example, largely of hydrogen, carbon, oxygen, and nitrogen. Of these, even carbon assumes a gaseous form as carbon dioxide ($CO_2$). More than 99.5 percent of the biomass of the earth—that is, the entire mass of all living earth organisms, ourselves included—is by weight composed of these materials.

Although life-forms are made up of materials which occur in the atmosphere in large quantities, water is the medium through which they enter the stream of life as chemical compounds dissolved in water. Land—that is, the solid material of the earth's crust—also provides many metals and other elements critical to life.

This chapter examines the manner in which air, water, and land interact to provide a suitable environment for all kinds of life. Of all life forms, we consider vegetation first, for plants are the basic converters of air, water, and land into compounds upon which animal life depends. Even more important, plants are the only means by which animals can convert solar energy into usable forms. We begin by discussing molecular processes at scales smaller than the individual. The reasons for this choice of scale should become apparent from the discussion that follows.

## Nutrients, Plant Life, and Earth Systems

The fact that plants can use elements only in the form of water-soluble compounds is sometimes a problem. In regions of heavy rainfall, plant nutrients—that is, the chemical compounds just mentioned—may be leached out of the soil and washed away. In other words, they are dissolved by water percolating down through the soil and carried off by groundwater. In addition chemical barriers in water sometimes prevent plants from receiving nutrients. By this we mean that the soil moisture available to plants may have an excess of hydrogen ions ($H^+$) and be too acidic, or a shortage of hydrogen and an excess of hydroxyl ions ($-OH$) and be too basic. The amount of hydrogen ions in solution is expressed by a numerical index. This number, termed the pH, is the reciprocal of the logarithm of the degree of dilution of the hydrogen ions and ranges from 0 to 14. The water content of a soil may give a reading from pH 3 (very acid) to pH 10 (very basic or alkaline). Most plant growth requires a soil pH that does not vary from the neutral value pH 7 by more than 2 points either way. Marine organisms live only in a more restricted pH range from about pH 7.5 to pH 9. The problem is that the water solubility of chemical compounds changes with the pH. Plant nutrients can either precipitate out of solution or be blocked from access to the plant's system if the pH of the soil shifts outside the tolerable range. Also, certain parts of the plant organism may actually dissolve if the safe pH range is exceeded.

In every case, certain key processes associated with vegetative growth make inert elements available to all forms of life. *Photosynthesis* is one of the most important of these. The photosynthetic process is one in which green plants possessing chlorophyll convert solar energy into chemical energy which is stored in the form of glucose. Carbon dioxide, water, and energy are thus converted into sugar, with oxygen molecules as a by-product. Glucose is a basic building block from which most of the other compounds needed by living organisms are made. Plant sugars are also the "fuel" by means of which animals, most of which cannot manufacture the basic organic compounds upon which they depend, maintain their cellular activities.

Another key process is *nitrogen fixation*, in which nitrogen gas ($N_2$) is combined with hydrogen to form amino ions ($+NH_2$), which are the raw materials for amino acids. Amino acids are used in the synthesis of proteins and enzymes which make up the incredibly complex structures and control mechanisms of lifeforms. Despite its importance, free nitrogen gas, which makes up 79 percent of the atmosphere, is unavailable for the great majority of life-forms, which cannot use it directly. It must first be "fixed" by a few organisms capable of combining it in compounds which can be incorporated into other life-forms. Most nitrogen is fixed by microorganisms which inhabit the soil in association with certain plants such as legumes. Among other nitrogen fixers are the blue-green algae, of particular interest because they are also photosynthesizers.

Man has learned to produce fixed nitrogen industrially. Fritz Haber and Karl Bosch developed a process in Germany in 1914 primarily to meet the demand for nitrates for explosives in World War I. These processes are in sharp contrast to those in nature. The nitrogen fixed by man requires temperatures of 500°C and several hundred atmospheric pressures. Bacteria accomplish the same thing with the help of special enzymes. Their process requires normal temperatures and one atmosphere of pressure. The human process uses nickel as a catalyst to bring about the necessary reactions, while bacteria need trace amounts of cobalt and molybdenum.

The need for heavy metals is an example of the third medium vital to plant growth. For example, legumes such as beans until recently could not be grown in many parts of Australia. After much research it was discovered that as little as 2 ounces of molybdenum per acre counteracted poor soil conditions and stimulated plant growth.

The many elements and minerals required from the land are made available by weathering of rocks from the surface of the continents. The root systems of plants take up elements in ionic form as they come into solution in the soil. Ocean life depends upon the presence of such ions in the discharge from rivers. Since the original occurrence of trace elements in the lithosphere is limited to a very few spatial locations, the redistribution and diffusion of crustal materials by the hydrologic-lithologic cycle are vital geographic aspects of worldwide plant nutrition.

One ominous fact is that while many elements are needed for life processes, certain heavy metals and other substances found in the earth are poisonous to life. Human activity in recent centuries has resulted in an increase in the amount of these substances available to various organisms. Mercury mined from the earth, concentrated, and subsequently dumped as waste in lakes and streams is one example. At the same time, we are adding new and poisonous compounds to those already in nature, particularly in and near cities. Chlorinated hydrocarbons including the DDT family of pesticides are among the most important of these. In this case the diffusion of toxic materials by wind and water works against, rather than for,

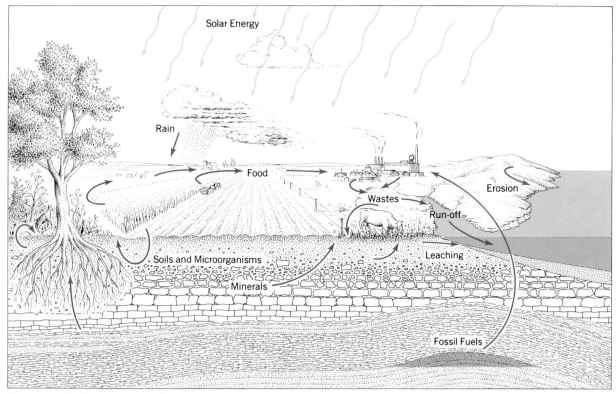

**Figure 9-1 Biologic cycles and energy cycles** The major inanimate inputs to biotic cycles are solar energy, water, and minerals from the lithosphere. The human food system is part of the biotic system and results in spatial transfers of nutrients, which must be taken into account in order to sustain viable agricultural production. The return to space of long-wave reradiation is not indicated.

man's benefit. The dispersion of local concentrations of industrial and urban waste does not mean getting rid of the dangerous substances. They often are reconcentrated in other locations and continue to be hazardous to all kinds of life. We will return to these problems in the chapters ahead. At this time we wish only to point out that the systems and processes which produce and sustain life can be contaminated as well as enhanced by human works.

In summary, the interdependence of organisms is a major characteristic of biological systems. Biotic processes are cyclic and tied inextricably to earth processes. Material is captured and processed by plants using energy from the sun. Animals live on plants, and both plants and animals die and return to the soil. Microorganisms in the soil complete the cycle by breaking down dead materials through decay into forms which serve as plant nutrients. The major inanimate inputs are solar energy, water, and minerals from the lithosphere. Losses from this system (Figure 9-1) occur when materials are transferred out of it either by wind and water or leaching and erosion or when man harvests and exports plants and animals to distant areas. Other losses or negative inputs can occur when materials are changed into unusable or dangerous compounds. In every case equilibrium must be maintained within the system. A lack of critical components or excessive losses at any geographic site will cause the decline of the

*Earth as the Habitat for Life: Vegetation and Soils*

**195**

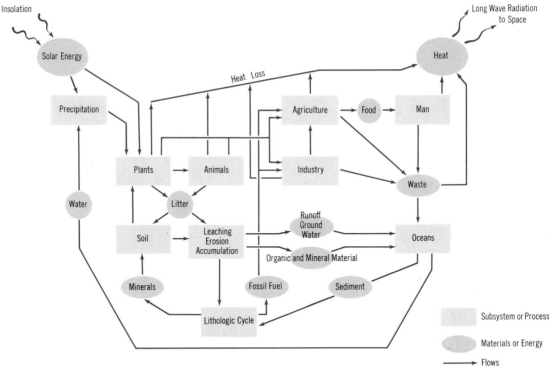

**Figure 9-2 Biotic cycles and energy flow** This systems diagram refers to the cycles and flows depicted in Figure 9-1. The human food system is part of the biotic cycle. The entire biotic cycle is driven primarily by solar energy, which flows through the subsystems and eventually returns to space as heat radiation.

ecosystem at that location. Conversely, adding missing elements can restore the system and allow increases in the biomass produced at a given site. These ideas give us a key to understanding why we find certain types of vegetation and soil in certain places, and also suggest strategies which can be employed to improve life environments.

### Nutrient Cycles

The illustration in Figure 9-1 tells only a small part of the story. When its parts are ordered and arranged as a system or set of subsystems (Figure 9-2), it is possible to distinguish the through flow of solar energy already described in earlier chapters. The hydrologic cycle also becomes apparent, sometimes as a loop or cycle

of short duration, sometimes as a much longer one involving a trip to the sea and back. Minerals also have their paths through the total system. Short paths lead from soil to plants to litter and back to the soil. Still longer routes return critical elements to the soil only after geologic spans of time. In the latter case, although the system is confined to the earth and therefore is a closed one, the materials are for all practical purposes lost to humankind.

Many elements are caught up in cyclical paths through the biosphere. Figure 9-3 indicates the journeys taken by nitrogen and phosphorus, two important plant and animal nutrients. Carbon, sulphur, and oxygen are equally important and could also be traced along similar routes. However, a brief description of the first two elements will make clear the importance of nutrient cycling.

Nitrogen production by artificial processes is an important part of modern agricultural systems. The 30 million tons produced annually by industry is equal to the entire amount fixed on an annual basis by natural systems. Mankind is further increasing the amount of fixed nitrogen by the increased planting of legumes and other nitrogen-fixing plants. *Denitrification*, the release of nitrogen back into the atmosphere, takes place most easily in waterlogged soils with the help of anaerobic bacteria, which do not depend upon oxygen for their life processes. Not enough is known about the nitrogen cycle to predict the total consequences of human interference with it. It seems quite certain, however, that the concentration of nitrogen in lakes and streams with accompanying *eutrophication*, the runaway growth of algae and other aquatic life forms, may seriously unbalance this cycle. The human organization of space in the form of regional sewage systems, as well as surface runoff from areas having intensive modern agriculture, plays a significant role in these matters (see Chapter 13). Thus it is important to know something about the geographical analysis of space as well as about the ecosystems involved when considering nutrient cycles.

Phosphorus follows much the same pattern. It is taken up from the soil by plant roots and by certain fungi which live on plant roots. Since bedrock is the major source of phosphorus, its removal from the soil by clearing vegetation, and/or the export of crops from the land without return through plant decay, could make this path a one-way street. The eventual accumulation of phosphorus in water bodies effectively removes it from existing biocycles, shifting it into the geologic time scale, where it becomes unavailable to contemporary generations.

The two diagrams to which we have just referred are in their own way geological rather than geographical in character. While their vertical cross sections or profiles tell us a great deal about such cycles, much is overlooked by this approach. Another, more geographical, way to consider the movement of nutrients is to view

their transportation across the surface of the earth from one region to another by international trade. The map in Figure 9-4 gives some inkling of this transfer of nutrients and the problems associated with it. To make this map, United Nations data showing the imports and exports of cattle, sheep and goats, swine; beef, mutton, pork; fresh, dried, and condensed milk, butter; wheat, barley, maize, associated flours; and fish products were converted to metric equivalents and then further reduced to the average amounts of phosphorus, nitrogen, and potassium those things contained. If exports exceeded imports, a negative balance was recorded; if imports exceeded exports, a positive balance resulted. Three types of countries are revealed: those that are net exporters, those that are net importers, and those that break even. This map does not comment on the overall nutritive levels within the countries, but that can be seen by comparing it with the one in Figure 9-5.

This nutrient shipment map may come as a surprise. The net exporting countries are those which are the most technologically advanced. Thus, modern high-energy agriculture tends to feed the world. Major importers are most of Europe and the third world nations which have some source of foreign exchange. Money to buy food is provided by oil revenues in the Middle East, special crops in Brazil and Indonesia, industrial products in Great Britain and Japan, and tourism or foreign aid in Greece and the east European countries. The least developed nations gain or lose little but eke out a general low level of nourishment. It is interesting to note that the drought-stricken nations of Mauretania, Niger, and Chad show up as net exporters of nutrients. This is explained by their production for export of peanuts and cotton despite the great need for food crops at home. It becomes evident from this map that artificial fertilizer production and the massive mining of phosphates is subsidizing many of the world's mouths. Given these flows of nutrients, who would be willing to judge the wisdom or propriety of diminishing agricultural production in the "have" nations at the present time? Is disruption of the environment too great a price to pay

*Earth as the Habitat for Life: Vegetation and Soils*

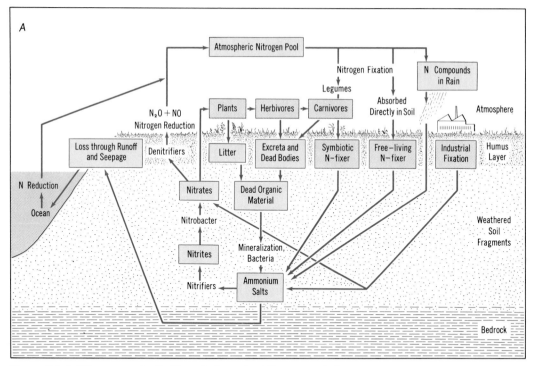

**Figure 9-3 A. Nitrogen cycle. B. Phosphate cycle** Nitrogen and phosphate are essential elements in biotic systems. Their cycles through the environment are complex and subject to modification by man. Most fertilizers consist of these two elements plus potassium. The practice of proper fertilizing is an example of beneficial direct intervention into plant systems by man. Unintentional and detrimental modification may also occur whenever the elements are sent into a long cycle which makes them unavailable to agricultural plants. An example is their loss into ocean sediments.

**Figure 9-4 Net change in nutrients due to international shipments of food** This map represents annual change in nutrients due to international shipment of nine foods (barley, wheat, corn, sheep and goats, cattle, swine, milk, butter, and fish) by measuring the net change in kilograms per square kilometer of arable land of nitrogen (N), phosphorus (P), and potassium (K). All countries showed a net loss or gain in all three nutrients except as follows: *Brazil*, gain N, P, no change K; *Columbia*, gain P, K, no change N; *Austria*, gain N, P, no change K; *Thailand*, loss P, K, no change N; *Denmark*, gain N, loss P, no change K. "No change" means no significant change. (Data compiled and analyzed by Nicholas W. Beeson. Data sources: FAO *Yearbook of Fishery Statistics*, vols. 34, 35, 1972, United Nations; FAO *Statistical Yearbook 1972*, United Nations; *Composition of Foods*, Agriculture Handbook No. 8, United States Department of Agriculture, Government Printing Office, Washington, D.C., 1964)

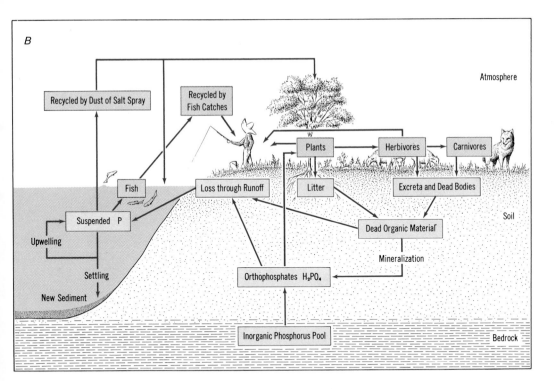

B

Recycled by Dust of Salt Spray

Recycled by Fish Catches

Atmosphere

Plants → Herbivores → Carnivores

Fish

Loss through Runoff

Litter

Excreta and Dead Bodies

Soil

Suspended P

Upwelling

Dead Organic Material

Settling

Mineralization

New Sediment

Orthophosphates H₂PO₄

Inorganic Phosphorus Pool

Bedrock

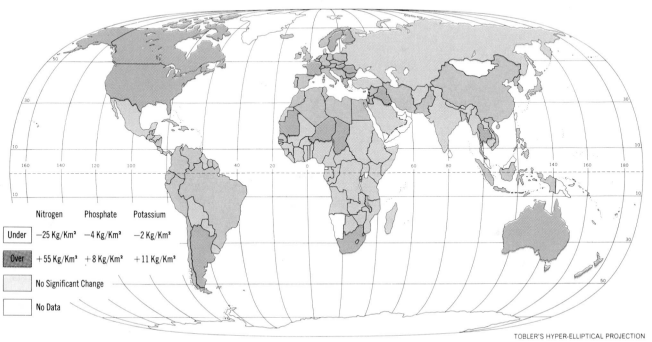

| | Nitrogen | Phosphate | Potassium |
|---|---|---|---|
| Under | −25 Kg/Km² | −4 Kg/Km² | −2 Kg/Km² |
| Over | +55 Kg/Km² | +8 Kg/Km² | +11 Kg/Km² |
| No Significant Change | | | |
| No Data | | | |

TOBLER'S HYPER-ELLIPTICAL PROJECTION

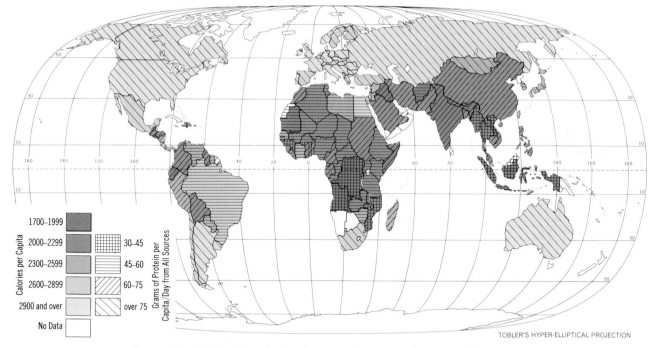

Figure 9-5 map legend:

Calories per Capita
- 1700–1999
- 2000–2299
- 2300–2599
- 2600–2899
- 2900 and over
- No Data

Grams of Protein per Capita/Day from All Sources
- 30–45
- 45–60
- 60–75
- over 75

TOBLER'S HYPER-ELLIPTICAL PROJECTION

**Figure 9-5 World food supply (net food supplies per capita at retail level)** Statistics on food consumption are for the latest year available, mainly from 1960 to 1970. They are not uniformly reliable, since they generally reflect over- rather than underestimates. They are also averages by country; great variations exist in almost every country, so that even in countries which have average surplus consumption there are undernourished people. (Prepared from data in *UN Statistical Abstract 1970,* Table 161)

when hunger threatens? Solomon himself might sit and scratch his head.

## Soil Systems and Classifications

The cloak of soil that rests so lightly on earth's shoulders is a strange garment. We take soil for granted, scarcely recognizing that it is a complex mixture of gases, liquids, weathered rock, humus or dead vegetation, and microorganisms. Soil is formed by the action *in time* of climate and living organisms—both plants and animals—on the rocky surface of the earth. Moreover, soil reflects variations in the relief of the surface upon which it is formed. Vegetation and soils are so interrelated that it is impossible to describe either one as the "egg" and the other as the "chicken" in answer to the ques-

tion of which develops first. We offer only as a matter of convenience this brief discussion of soils and their classification before talking of soils and vegetation in combination.

The processes which form soils are in large part represented by the vertical movement of liquids and gases through weathered materials. In humid regions, which we somewhat arbitrarily designate as those with more than 30 inches of rainfall annually, a layer of plant debris will cover undisturbed soil. This surface mat of organic material consists of recent leaves and grasses at its top and of *humus,* i.e. decomposed organic matter, lower down which gradually shades into weathered rock and the parent lithosphere as one digs deeper and deeper. Water filters down through these layers. The organic materials form weak solutions of carbonic and other acids which leach out the uppermost layer of weathered, rocky stuff. Still deeper in earth's unconsolidated mantle the dissolved minerals may precipitate out again, and microscopic, colloidal bits of altered clay from the leached stratum also may be deposited. Fully developed soils under these conditions often exhibit distinct strata or horizons from top to bottom, composed of an organic layer, a leached layer (called the *A horizon*), a layer of deposition (called the *B horizon*), and a layer of relatively unleached and uncontaminated but weathered materials (the *C horizon*).

These acidic and leached soils often have iron (fe) and aluminum (al) removed from the A horizon and reconcentrated in the B horizon and are called *ped*alfers to indicate this. They usually represent open systems in that energy, water, and some earth and organic materials move through them from top to bottom where either colloidal materials or materials in solution can be carried away by groundwater.

In regions with less than 30 inches (76.2 cm) of rainfall—or more accurately in those places where only the upper layers of the ground become wet and surface water does not seep down to the water table—a different process takes place. If the land is too dry for much plant growth, the surface organic layer may be sparse. Considerably less leaching takes place in the A horizon. Instead, the vertical movement of water in the soil is often upward and a layer of carbonates may be deposited when soil moisture evaporates near the surface. These precipitates can actually form a hard, sometimes impenetrable, crust anywhere from a few inches to a few feet beneath the surface. This *hardpan* or *caliche* is the extreme result found under these conditions, particularly where mineralized groundwater may continually move to the surface by capillary action in response to high temperatures. Soils of this type are called *pedo*cals in reference to the process of calcification which has taken place. Here the system is a closed one, at least for the mineral substances held temporarily in ionic form in the intermittently moist surface layers. Water and energy may move through such systems, but little else escapes.

If we consider the variety of climates on a world scale, as well as variations in vegetation and rock type, it is not surprising that a multitude of soils and soil regions can be distinguished. Defining and classifying soils, therefore, represent the same problems as do defining and classifying climates. We may distinguish very few soils and the areas they occupy, or we may divide and subdivide them into an almost infinite number of types. Typical of soil conditions which could allow impossibly small microregions is the *soil catena*. This term is used to describe the variation in soil types which occurs from the top to the bottom of a hill. The parent rock may be everywhere the same, but differences in soil moisture and the depth of the water table create a variety of moisture conditions which in turn affect soil chemistry and the intensity of microbiotic activity.

This raises the question of choosing a useful soil classification. Two systems are currently used to designate soil categories. The first, and oldest, was first suggested by V. V. Dokuchaiev, a Russian soil scientist, and was subsequently modified for use by the United States Department of Agriculture. This classification recognizes three basic orders of soils: the *zonal*, the *azonal*, and the *intrazonal*. Zonal soils have well-developed profiles and reflect the climatic and vegetative conditions under which they have developed. Soils with poorly developed profiles or with none at all are called *azonal*. Such soils are essentially recently deposited and were borne by wind and water to their resting places. Some specialists do not recognize these as true soils because of their undeveloped layering. Intrazonal soils reflect special local conditions such as excess water (*hydromorphic soils*), excess salt or alkali (*halomorphic soils*), and excess lime (*calcimorphic soils*). A three-level classification based on the three orders is shown in Table 9-1.

The new soil classification subsequently used by the Department of Agriculture is known as the *Seventh Approximation* in reference to the manner in which it was devised. It creates ten soil orders based on the morphology and formative processes of each. In view of our intent to

**Table 9–1 Zonal, Intrazonal, and Azonal Soils**

| Order | Suborder | Great Soil Group |
|---|---|---|
| Zonal soils | 1. Soils of the cold zone<br>2. Soils of arid regions | Tundra soils<br>Sierozem<br>Brown soils<br>Reddish-brown soils<br>Desert soils<br>Red desert soils |
| | 3. Soils of semiarid, sub-humid, and humid grasslands | Chestnut soils<br>Reddish chestnut soils<br>Chernozem soils<br>Prairie or brunizem soils<br>Reddish prairie soils |
| | 4. Soils of the forest-grassland transition | Degraded chernozem<br>Noncalcic brown |
| | 5. Podzolized soils of the timbered regions | Podzol soils<br>Gray wooded, or gray Podzolic soils<br>Brown podzol soils<br>Gray-brown podzolic soils<br>Sol brun acide<br>Red-yellow podzolic soils |
| | 6. Lateritic soils of forested warm-temperate and tropical regions | Reddish-brown lateritic Soils<br>Yellowish-brown lateritic Soils<br>Laterite soils |
| Intrazonal soils | 1. Halomorphic (saline and alkali) soils of imperfectly drained arid regions and littoral deposits | Solonchak or saline soils<br>Solonetz soils (partly leached solonchak)<br>Soloth soils |
| | 2. Hydromorphic soils of marshes, swamps, seep areas, and flats | Humic gley soils<br>Alpine meadow soils<br>Bog and half-bog soils<br>Low-humic gley soils<br>Planosols<br>Groundwater podzol soils<br>Groundwater laterite soils |
| | 3. Calcimorphic soils | Brown forest soils<br>Rendzina soils |
| Azonal soils | | Lithosols<br>Regosols<br>Alluvial soils |

Source: Charles B. Hunt, *Geology of Soils*, W. H. Freeman, San Francisco, copyright © 1972, table 8–1, p. 175.

deal with man's use of the environment we will refrain from the lengthy description necessary to make the Seventh Approximation's terminology intelligible to non-soil scientists. There is much to recommend this new classification, but Dokuchaiev's has the advantage of being relatively simple as well as being still widely used. Therefore, we will return to our discussion of zonal soils.

Zonal soils are most widespread and dominate any world map of soil distributions such as that in Figure 9-6. At this scale, azonal and intrazonal soils either occur in areas which are too small to be distinguished, or merge with nearby zonal soils, or are included among those soils designated as mountain and mountain valley types. Only nine categories are used, and a few of the semiarid to semihumid soils have been lumped together for convenience. Another way of organizing our perceptions of soil variations is to use a variation of the 3 × 3 matrix introduced earlier in Chapter 6. Figure 9-7 shows the distribution of nine soils in terms of their energy-moisture corollaries. In the next section we will describe the major humid soil sequence from the equator to the tropics and another one in the midlatitudes from humid to dry conditions. In Figure 9-7, notice how low energy conditions at high latitudes override moisture conditions and create world-circling bands of tundra and podzol soils. Notice too that moisture-energy conditions in the marine west coast and Mediterranean areas resemble east coast humid and prairie conditions sufficiently to produce similar soils.

*Major soil-vegetation sequences: The Arctic to the humid tropics*

To begin our examination of the combinations of soils and vegetation shown in Figure 9-7 and 9-8 we will look at the series of humid environments ranging from ice cap to tropical conditions. While the nature of the ice caps of the world is of considerable interest and significance—remember, three-fourths of all fresh water is locked up in glaciers and ice caps—they do not concern us here. Where rocky islands or *nunnataks* thrust out of ice fields, no soil exists,

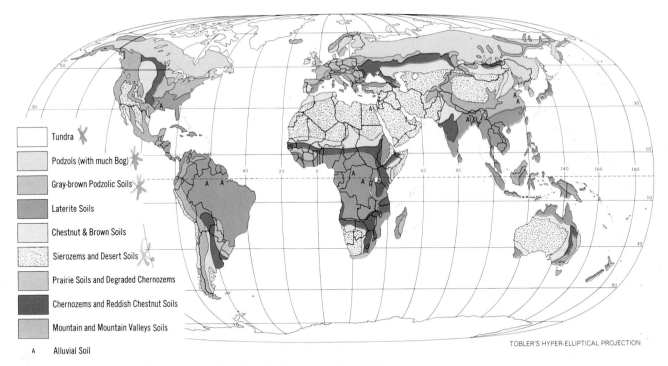

Tundra

Podzols (with much Bog)

Gray-brown Podzolic Soils

Laterite Soils

Chestnut & Brown Soils

Sierozems and Desert Soils

Prairie Soils and Degraded Chernozems

Chernozems and Reddish Chestnut Soils

Mountain and Mountain Valleys Soils

A    Alluvial Soil

TOBLER'S HYPER-ELLIPTICAL PROJECTION

**Figure 9-6 Major soil regions** Zonal soils reflect the climates under which they developed. Notice the similarity in patterns on this map and those in Figure 6-1, *"World Climates."* The classification shown here is very broad. Great differences in soils exist within the classes shown, depending upon parent material, time in place, vegetation cover, slope, and other conditions.

only sometimes a bit of pulverized rock. Let us pass on, for at the same time no vegetation beyond the rare occurrence of a few lichens and mosses graces such inhospitable surfaces.

**Tundra soils and vegetation** In arctic areas where ice does not prevail, a kind of soil forms which is almost like a pudding. Soils of this tundra region border the Arctic Ocean and

**Figure 9-7 The nine major soils in terms of energy/moisture content** Soil types reflect the moisture and energy conditions (shown as temperature) in which they develop. This diagram is not a map, but because moisture and energy conditions vary systematically over a continent, so too do the soil types. Each continent has a soil covering which resembles at least part of this sequence, depending upon its location on the globe.

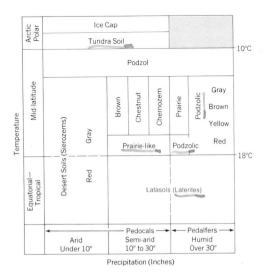

*Earth as the Habitat for Life: Vegetation and Soils*

| Temperature | Precipitation (Inches) | | |
|---|---|---|---|
| | Arid Under 10″ | Semi-arid 10″ to 30″ | Humid Over 30″ |
| Arctic Polar | Ice Cap. Tundra (Mosses Lichen and Low Bushes) | | |
| Mid-latitude | | Boreal Forest (Conifers)  —10°C | |
| Mid-latitude | Xerophytes | Short Grass Steppe / Long Grass Prairie | Conifers / Mixed Deciduous Forest |
| Mid-latitude | | Sclerophytes | Conifers  —18°C |
| Equatorial—Tropical | Xerophytes | Savanna (Scattered Trees and Grass—Thorn Scrub) | Selva (Perennial Tropical Rain Forest—Trees with High Crowns) |

**Figure 9-8 Major vegetation sequences** The vegetation sequences found on the continents are associated with the corresponding sequences of soil types shown in Figure 9-7.

extend inland for varying distances. There, despite scanty precipitation, what little water exists is locked up within the soil for most of the year by low summer temperatures and long, cold winters. When the weak summer sun finally brings 24 hours of daylight to the area north of the Arctic Circle, melting occurs and the surface is turned into a swampy maze. In many places it is difficult to tell whether it is land with a lacework of water or water with a lacework of land. Permanently frozen ground called *permafrost* underlies much of the area, sometimes to a depth of a hundred feet or more. Only the uppermost portion thaws each year during the brief summer. Then, as winter settles in, the surface freezes. This makes a sandwichlike structure with frozen materials both above and below a soggy filling. Sometimes hydrostatic pressure builds up and mud volcanoes and great blisters, or *pingos*, erupt on the surface.

The soil itself that forms under these condi-

Tundra vegetation alternating with open puddles. This is summer in the Arctic, and the surface layer of the permafrost has melted. (Photo by S. Outcalt)

tions is extremely shallow, perhaps only a foot or two deep. A third of its depth will be an acidic, humus-rich mixture of clay and sand particles with considerable undecomposed organic matter within it. This is particularly common because of the small amount of bacterial action possible under extreme cold. The remaining soil which rests upon the permafrost is gray and waterlogged, and if the rocky lithosphere is near the surface, it may contain chunks of rock shattered and freed by winter freezing. No normal trees can grow under these conditions, although sometimes dwarf willows less than a foot high exist. The usual plant life is composed of low arctic shrubs, reindeer moss, lichen on the bare rock, grasslike sedges, and a few flowering plants.

This combination of shallow soil and plants with poorly developed root systems presents particular problems for human use. No domesticated vegetation does well on tundra soils, and even better soils which are sometimes underlain by permafrost heave and buckle from differential heating if they are plowed and planted. More tragic is what happens when dangerous atmospheric pollution such as the dust from the Russian hydrogen bomb tests on the arctic island of Novaya Zemyla is blown onto these areas by the polar easterlies. Much of the resulting fallout rests upon bare rock; the remainder is caught in the shallow surface layer that thaws and freezes, for the underlying permafrost effectively prevents such materials from being carried to deeper levels or from being transported out of the region by underground waters. Mosses and lichens, which have no true roots, and all the other plants mentioned above, which have very shallow roots, thus quickly absorb the radioactive materials. Reindeer eat the moss and other vegetation and in turn become radioactive. Hunters and herdsmen of the Arctic depend upon these animals for survival and in Though official sources seldom discuss such matters, there is evidence that natives of the Arctic are perhaps the most radioactively contaminated people in the world, and dangerously so.

**Podzol soils and the boreal forests**   At the margin of the tundra region, dwarf vegetation gives way to true trees. Because plant growth ceases when temperatures go below freezing, these trees are dormant for great portions of the year. Despite the fact that they are frozen solid for part of their lives, they still survive. In some places these northern, or *boreal*, *forests* will colonize tundra territory underlain with permafrost. Small tundra vegetation has growth cycles adjusted to the long daylight hours of the short summers. The invading trees, however, are perennials which slowly grow larger until their roots extend down to the permanently frozen ground which they cannot penetrate. This places a limit on the size of the trees, and in some particularly cold years the soil does not thaw deeply. This means that the working root systems of the trees are too small to provide for the bulk of the aboveground portion. The trees cannot transpire sufficient water and die of thirst while water stands on the surface all around. As the result of this, sometimes large tracts of small, perfectly formed hemlocks or spruces, all the same size and all dead, can be observed on the boreal forest–tundra boundary. Tundra vegetation eventually reclaims the region, and the boundary is therefore a changing and irregular line (Figure 6-3). Forests of these

A boreal forest, or taiga. Aspen and balsam fir form a two-story cover of relatively small trees. (U.S. Dept. of Agriculture photo)

extreme northern regions are of no commercial value.

Well south of the arctic tree line but still within the subarctic region, the boreal forests consist of nearly solid stands of spruce and hemlock and cover vast territories in Canada and Siberia. Another name for these forests is the *taiga*, by which they are known in Russia. Several inches of needles and twigs usually carpet the ground under the trees. The soil beneath this cover is called *podzol* soil, a Russian word meaning "ash," for it is the color and consistency of ashes. Podzols are highly acidic and infertile. This is because both chemical and organic actions are slowed by the cold of these regions and the organic cover on the forest floor does not decompose very rapidly. Unlike the leaves of deciduous trees at lower latitudes, the pine needles of the taiga, or boreal forests, contain few nutrients and are mostly cellulose. Moreover, by their very nature they form acidic solutions when they do decay. Organic acids leach out not only calcium, potassium, and other nutrients, but also iron and aluminum oxides which normally color the soil with red to brownish hues. The silicon oxides that are left behind explain the gray ash appearance which gives the podzols their name.

Trees along the southern margins of the boreal forests can be used for the production of paper and wood pulp. Regrowth is slow, however, and the land is used for this purpose largely because it has practically no other agricultural value. The needle-leaf conifer forests of the subarctic need only small amounts of plant nutrients such as potassium and phosphate, for their bulk is primarily cellulose. This means, however, that their leaves are of no value as fodder and their organic debris is of no use as green mulch. This accounts in large part for the lack of northward expansion of the agricultural and dairy belts in the eastern United States and Canada.

In areas such as the Maritime Provinces of Canada where the weather is more moderate, these soils can be successfully farmed with proper treatment and choice of crops. Cranberries, which require acid soils, potatoes and other root crops, and some fruit trees are grown. The soil must be limed (calcium carbonate is added to neutralize the acid in the soil), and fertilizer with the primary macronutrients, nitrogen, potassium, and phosphate, must be added. This condition does not substantiate the popular image of an agricultural frontier waiting for settlers in the subarctic.

**Podzolic soils and mid-latitude deciduous forests** Underlying the mixed deciduous hardwood and coniferous softwood forests of the eastern United States are soils typical of the forests found south of the taiga wherever sufficient moisture is available. The hardwoods include the familiar maple, beech, oak, walnut, and hickory. Deciduous trees, which lose their leaves every winter, and especially nut producers bring nutrients up from the B horizons of the soil and the parent material. Such trees produce abundant organic material which each season is released as the leaves change color and fall to the ground along with the nuts and seeds. This organic debris is mineral-rich, especially the nuts and seeds, which have more nutrient value than leaves and stems. The soils under these forests are called the *gray-brown podzolics*. When the forests are cleared, the soils may be very productive, while in their natural state under undisturbed forest they support a sizable animal population. However, they are normally too acid for domestic crops and require liming. They are darker in color than a true podzol because of higher humus content. At the same time, warm summers and generally milder climates allow greater bacterial action.

Mixed livestock–feed grain farming practiced in the Midwest is suited to these podzolic soils. This farm system involves rotating small grains and corn with legumes, hay, and soybeans in order to help restore nitrogen to the soil. Livestock, in previous decades much more than at present, were kept on each farm and fed on feed grains as well as hay, and their manure was returned to the grain fields. These farmers also learned to put phosphate and potash (potassium oxide) on their fields. With such treatment podzolic soils continue to produce a wide range of products, but with poor farming techniques they rapidly deteriorate.

A

B

(A) Mid-latitude white pine forest in northern Minnesota. These trees are 200 to 300 years old and are remnants of the once extensive forests of the upper Midwest. (U.S. Dept of Agriculture photo)

(B) Mid-latitude hardwood deciduous forest in Wisconsin. Survivors of another age, the trees are, from left to right: maple, elm, black ash, and elm. (U.S. Dept. of Agriculture photo)

(C) Mid-latitude West Coast coniferous forest in northern California. The trees are redwoods. (U.S. Dept. of Agriculture photo)

The soils of the humid southeastern United States and of a similar area in China are noticeably red and yellow in color. They are also quite infertile, a condition typical of the *red and yellow podzolics*. The color results from high iron and aluminum content in the clay particles of which they are principally composed. Abundant rainfall, high temperatures, leaching, and bacterial action contribute to this condition. Also, in the southeastern United States, the

C

soils developed on deeply weathered ancient surface material undisturbed by the continental glaciers which never reached so far south. To the north all across the continent, fresh mineral-bearing materials were exposed by the action of the glaciers, either by scraping, by direct deposition, or through action of water and wind associated with the ice front.

The pioneers who came to the southern Piedmont on the seaward flanks of the Appalachian Mountains found a natural vegetation of mixed broadleaf and pine forests. Clearing the forests and raising crops invariably resulted in a very rapid decline in yields. The land was often abandoned after only two or three years. They then moved on to the west to repeat the process. This destructive land use practice is sometimes referred to as "grasshopper farming." Heavy rains caused severe erosion in those bare, abandoned fields. In some areas, pine forests eventually grew back in place of broadleaf forests. Old field shapes sometimes can still be distinguished in these second-growth pine barrens. This was because pine trees can manage better than broadleaf trees under poor soil conditions. But these podzolics are worthless for agriculture unless very heavily and carefully fertilized.

The chemical nature of the soils in this region can explain why this destructive and poverty-producing sequence occurred. We noticed before that nutrients must be in water solution to be available for plant use. In most soils less than 1 percent of the minerals present are in dissolved form, the bulk being in the solid clay or sand particles. The minerals in solution are contained in water films clinging to the soil particles. As plant roots or leaching action draw the dissolved ions from the water, more ions move into solution from the solid material. The direction and amount of such transfers depend upon the chemical balance available. Plant roots essentially offer hydrogen ions in exchange for other positive ions such as those of calcium ($Ca++$) and magnesium ($Mg++$). Such action is inhibited in acid soils, which by definition have an excess of hydrogen ions ($H+$) already in solution. A soil containing clay particles rich in calcium and other nutrients will slowly release ions into solution as the plants use them up. Such soils are chemically stable and are referred to as being "buffered." The red and yellow podzolics are not buffered at all. The soluble positive ions have moved out of the soil particles due to the long weathering to which they have been exposed, leaving the aluminum and iron oxides behind. The useful nutrients have been leached away.

Humus in the organic layer is normally a good source of recycled plant nutrients. Unfortunately there are also two problems associated with humus in red and yellow podzolics. The first is that there is not much organic material present, a fact that can be deduced from the color. Humus is dark, and soils rich in humus are dark brown or black. These are much lighter and yellower.

The second problem has to do with the quality of the nutrients available to animals, including humans, using the plants grown on these soils as food. Soil microbes, necessary for the decay portion of the plant-soil cycle, under these conditions compete with the plants for proteinaceous food. This is because they are made of proteins themselves. The plant litter under the trees, particularly pine trees in warm, moist regions, is quite carbonaceous. The trees are little more than cellulose and other carbohydrates which provide an excess of energy food for organisms but little or no growth material.

Therefore the infertility of podzolic soils has a differential quality about it. Fiber and leaf crops such as cotton, sugar cane, and tobacco will do relatively better than protein-rich fodder crops and small grains. Seeds and nuts, being the reproductive parts of plants, require protein nutrients to be viable. Animals fed on protein-deficient fodder grow slowly, do not reproduce well, and are susceptible to diseases. Again, the distribution of crop types in the United States and elsewhere is the result of a combination of both cultural and natural factors. Protein deficiencies, like those described above, also cause malnutrition in human populations. Conditions of this sort also become critical in association with the lateritic soils of both semitropical and tropical regions.

**Latasols of the selva** We frequently think of tropical rain forests as deep and luxurious. In places that have been relatively undisturbed by man, they are, but as with many other biotic communities, these lush forests rapidly decline under man's impact. Both subsistence and commercial farming contribute to the disappearance of the *selva*, another name for tropical rain forests. The huge biomass built up per unit area within undisturbed tropical rain forest despite its mass constitutes an extremely fragile plant community growing on meager soils. Perhaps only primitive agriculturalists have found a system by which the selva can achieve sustained food crop production. Certainly, attempts to convert the apparent plenty of the selva into commercial food crops are often disappointing. There is also evidence that past civilizations have fared poorly in rain forest regions.

Most of the nutrients in a tropical rain forest biosystem are held in the vegetation itself. The plants in this community are dominated by tall trees with high crowns. Little underbrush can grow on the dim forest floor. Our image of impenetrably thick equatorial forests is based in large part on photographs taken from rivers and trails, which are lined with dense vegetation, taking advantage of the additional sunshine penetrating the forest cover at those places.

Some vegetative cycles exist without benefit of soil in these forests—for example, epiphytes, which grow on other plants but which are not nourished by them, and parasitic plants which live high above the forest floor in the tree canopy. The trees are often evergreen in character and lose their leaves one at a time throughout the year. Thus the high green canopy remains unbroken from season to season. At the same time, very little litter covers the ground beneath, for high temperatures, abundant moisture, and intense bacterial action quickly dispose of any fallen plant materials. Excess water saturates the land, leaching out much of the parent material and weathering the rock to a great depth. The loss of clay silicates in this way creates a crumbly, porous soil rather than a sticky one when wet. In areas where wet and dry seasons prevail, saturation followed by drying out of the soil facilitates oxidation of iron

A tropical rain forest with squatter houses, gardens, and tea fields in the foreground. Notice that the solid tree canopy has been opened in the process of clearing the fields, thereby exposing the tree trunks to view. The scene is in Tanzania, East Africa. (Photograph by A. Larimore)

and aluminum compounds. This results in the characteristic red tones associated with the *latasols*.

Under such conditions, nutrients released in the soil are quickly taken up by the roots of existing plants or else are leached away. In places where plant roots cannot reach fresh mineral soil, the plants tend to be protein-poor and to reproduce vegetatively. That is, they send out shoots rather than reproducing by flower and seed. Clearing such forests and replacing them with crops breaks the smooth flow of plant food from roots to canopy to litter and quickly back to the roots. Soils formed under these conditions provide only protein-deficient foods for people and animals. Even the soil microbes, usually thought of as necessary and symbiotic partners in the soil-plant biosystem, begin to compete with plants for growth nutrients. This situation comes about in the following way.

The proportion of carbon to nitrogen in tropical forest litter is out of balance, and nitrogen is in short supply. Plants and animals, including the decay-producing organisms in the soil which break down litter into products usable by vegetation, all compete for the limited supply of proteinaceous minerals. Life processes are very active in humid, warm environments. Decaying carbohydrate plant parts provide ample energy food but little, if any, protein for growth. Plants requiring seed production on a large scale in their life processes do not do well under these conditions. For example, small grains do not flourish, and starchy crops are grown in their stead.

People who depend upon these tropical crops suffer from malnutrition and conditions such as kwashiorkor, a nutritional disease associated with a starchy diet. Young children in Africa and other tropical areas where carbohydrate-rich subsistence crops such as bananas, cassava (manioc), and maize predominate all suffer from these illnesses. It is significant that most of the original populations of the tropics were concentrated along the shores of lakes, streams, and oceans where fish provided necessary protein supplements. This is still true of places like the Amazon basin, where territories remote from the rivers are nearly empty of people. Since protein-rich crops do not do well under these conditions, it is likely that attempts to improve the health and nutrition of native populations by persuading them to grow imported proteinaceous crops will not succeed. Such crops will not do well without massive inputs to the soil. Such inputs are likely to rapidly wash away, and it may well be cheaper to import food rather than fertilizers directly to such places.

On the other hand, the plants of the tropics have considerable commercial value. Those best suited for this type of production reproduce vegetatively and yield large quantities of carbohydrates. Tropical lumber is one such crop, although the mixed stands of timber mean high costs on a per tree basis. This limits the present production of tropical woods to those that fetch high prices, such as mahogany and teak. The best commercial crops include sisal, jute, and hemp, which provide a variety of fibers; tea; coffee; sugar cane; and bananas. Spices, for which the tropics are famous, come from seeds

(cloves, nutmeg, pepper, and vanilla) as well as from bark and roots (cinnamon and ginger). Spices and medicinal products (quinine bark, for example) very often come from wild trees which must be quite large and old to be productive. Even then their yield is quite small per tree.

Tropical plantations try to overcome some of these problems by raising homogeneous stands of a particular species. However, the continual exporting of nutrients in the form of crops rapidly depletes the soil. Many operators, despite knowledge and use of fertilizers, find it necessary to use land for a few years and then to let it lie fallow for decades or to abandon it altogether. The "hollow frontier" in Brazil is a prime example of this. Coffee plantations are extended into previously unfarmed areas. After a few years the frontier moves forward again, and the land behind it is abandoned, thus making a thin line of active production between the forest and the depleted land.

Humidity and high temperatures encourage all manner of plant and human diseases. The onslaught of Panama disease, or wilt disease, among the banana plantations on the Caribbean shores of Central America in the period from 1890 to 1941 was responsible for the wholesale movement of production to the lowlands on the Pacific side of the isthmus. Similar diseases attacking coco trees in West Africa have also taken a high toll. But even if disease can be controlled, these commercial crops are not products which contribute to the balanced diets of local men and animals. Despite their vast stands of vegetation, the tropics of the world do not offer an easy answer to the search for new lands to farm.

So-called "primitive" systems of agriculture developed by local populations may actually be more sensible in both the short and long run. Such peoples employ variations on a common pattern of *shifting agriculture*, sometimes called *fire-field* agriculture. In this system, small plots are cleared in the forest. Trees are girdled and killed; smaller ones are chopped down and along with brush and grass are burned when they have dried out. The ashes from these fires are mixed in the soil, and crops are planted around the stumps and trunks of the larger trees. Small quantities of maize, cassava, and other plants are removed by harvest for up to three years, and then the fields are abandoned. Nearby plots are opened in the same manner, and the cycle of clearing and abandonment continues. Needless to say, even the nutrients consumed by humans find their way back to the soil as feces and the bodies of each dying generation. When fields are taken out of production, they are quickly reclaimed by grasses and eventually by brush until after 20 years or more the fragile nutrient cycle is restored. As long as populations do not become too large, it is then possible for the farmers to return to their fields or the ones of their fathers.

In the long run, though, even this low intensity of land use cannot endure. More and more of the virgin selva is depleted, and there is evidence that this technique of farming may have contributed to the creation of savanna grasslands in the wet-and-dry tropics.

## Laterization

Many tropical soils are characterized by concentrations of aluminum and iron. At the same time, they are low in silica and poor in plant nutrients. A typical soil profile shows practically no organic surface layer of litter. Instead the top inch or so may be a granular, dark-red clay. This gives tropical latasols their typical reddish hues. The rest of the A horizon is somewhat lighter. It is more coarsely granular and penetrated by roots and insect burrows. The B horizon is composed of blocky red clay that may be up to 6 feet deep.

These deeply weathered soils when exposed to the sun and air through removal of plant cover may turn into a bricklike substance called *laterite* (derived from the Latin word for brick). This material is so hard and durable that roads and buildings may actually be built from it. For example, the famous temple of Angkor Wat belonging to the former Khmer civilization in what is now Cambodia was constructed of laterite. This structure has endured unattended in its tropical environment since the sixteenth cen-

Savannas are grasslands with a scattering of trees. This vegetation type is found in regions subject to seasonal rain and drought. Herds of grazing animals such as these African antelope, called Topi, must migrate annually to obtain an adequate food supply throughout the year. The photograph was taken near the Seronera River in Tanzania, East Africa. (Photograph by A. Larimore)

tury and remains in reasonably good shape to the present day.

The Khmer civilization may have perished because of the laterization of the soils upon which its citizens depended for food. As their population increased, they may have pressed upon the fragile biosystem too hard by not allowing sufficient fallow periods between crop cycles. They may also have exposed large tracts to too much sun by removing the natural vegetative cover. In any event, their once populous area is now nearly empty. A similar example can also be found in Central America, where the Mayan civilization may have suffered the same fate. The Mayas occupied an area of lateritic soils, and some of their temples are also constructed of laterite. This may well be more than coincidence. These conjectures linking the decline of civilizations with the destruction of soil resources are not proven, but present-day people attempting to develop tropical agriculture

Typical depleted schlerophytic vegetation on the Mediterranean shore of Turkey. This was once a forested area. Now only thorny shrubs and herbaceous plants survive. The goat is one of the few animals which can utilize such vegetation. (Photo by Dick Scott)

should be concerned with the hazards of laterization.

### Relation of soil fertility to population distribution in the tropics

Not all tropical soils are poor. Notable exceptions occur on the flanks of recent or still active volcanoes. Ash and lava are mineral-rich and form the basis for a rich soil. The island of Java in Indonesia and the flanks of Mt. Kilimanjaro in East Africa are examples. Significantly, each of these places supports a very heavy population density. Commercial plantations also tend to cluster in places where rich parent material is available for soil formation.

In the middle latitudes, mountainous terrain is usually the most sparsely settled, with the low plains having more productive soils and carrying the heaviest rural population density. The opposite is often the case in the tropics. The population there is concentrated in the highlands and on steep slopes. One reason for this tendency in the wet-and-dry tropics is that the climate is cooler and the rainfall more predictable at higher elevations than closer to sea level. Also in the wet-and-dry tropics, the presence of

mountains may result in year-round orographic rainfall. Another reason is that the steep slopes and heavy rainfall cause erosion of surface layers at a rate which exceeds the rate of laterization. New mineral soils are therefore continually made available, and populations can depend on the renewal of vital soil resources.

### Savanna and Mediterranean vegetation

The savannas of the world are large tracts of open grassland that merge with the drier margins of the tropical rain forests. Scrub and thorn forest intermingle with patches of open ground, until grassland predominates. These grasslands, however, are unlike those of the middle latitudes in that they are subject to processes of soil formation and laterization similar to those in the selva. The tall grasses which predominate are full of cellulose and particularly when dry are unnutritious and unsuitable for fodder. Even the wild grazing animals of the area must migrate from pasture to pasture on an annual cycle in order to survive (Figure 2-7). Thus, a combination of climate, man, and special environmental conditions including those of soil-forming processes helps to create this particular type of environment with its unique problems.

Before discussing the problem of soil laterization in the tropics, let us consider the particular types of plant communities living where year-round high temperatures and wet-and-dry—rather than cold-and-hot—seasons prevail. The savanna areas of the tropics and the Mediterranean environment of the subtropical latitudes fulfill these conditions. Characteristically, the vegetation communities in either type of area would have widely spaced trees alternating with grasslands. Elsewhere, thornbush and scrub create patches of drought-resistant vegetation. While some typologists might object to our discussing the vegetation of these two environments in the same section, a strong similarity exists between Mediterranean and savanna plants. Both vegetative communities must adjust to annual periods of drought followed by rainy seasons of short to long duration. Both kinds of vegetation may become dormant during periods of high temperatures and aridity. This condition is called *estivation*, as opposed to *hibernation*, dormancy during cold, low-sun periods.

In Mediterranean environments, plants sprout new leaves in the autumn with the coming of the first rains. The main growth period when fruit and seeds form is during the milder, winter months. Then as the hot summer approaches, deciduous plants shed their leaves in order to reduce water loss. Such plants are called *schlerophytes*, of which evergreen oaks such as the cork oak are typical. Crops include the olive—the distribution of which defines the geographic extent of the Mediterranean climate in Europe. Where overcutting, overgrazing, and fires have degraded the stands of Mediterranean trees, a thick growth of thorny second-growth scrub often occurs. This is called *maquis*, and during World War II the famous French Underground Army which resisted the invading Nazis was given the name *maquis* after the thickets where they hid. When maquis vegetation is degraded still further until bare ground with patches of low herbs predominates, the degraded plant community is referred to as *garigue*.

Nearer the equator in the true wet-and-dry tropics, savanna woodland, thornbush and tropical scrub, and true savanna predominate. The woodland has an open, parklike appearance, although dense stands of flat-topped, medium-height trees frequently occur. This vegetation has small leaves and thick, fire-resistant bark, and is adjusted to a long dry season followed by a fairly long rainy one. Elsewhere, thornbush and scrub are even more tenacious and drought-resistant in character. Such stands are found where rains come in fairly large quantities but for very brief periods, with most of the year hot and dry. True savanna, with its large tracts of open grasslands and scattered, drought-resistant trees, borders on the steppe lands and desert margins of the subtropics and continental interiors.

## Major soil-vegetation sequences: Mid-latitude humid to arid environments

The preceding discussion of soils and vegetation from the Arctic to the humid tropics provides us with an introduction to soil-forming processes.

We can now take a somewhat briefer look at another basic series of soils and vegetation found in mid-latitude locations. These environments represent the transition from the humid to the arid mid-latitudes. In North America such a sequence occurs along the 40th parallel of latitude from central Ohio through Illinois, Missouri, Kansas-Nebraska, and on into Colorado. In the East, podzolic soils originally developed under hardwood deciduous forests. Even though the forests have been cleared away, with proper care such soils can remain fertile almost indefinitely and produce large yields from a wide variety of crops. Acidic and humus-rich, with proper liming and the application of phosphate, potash, and farm manure, they are among the best in the world.

**Prairie and chernozem soils of the semihumid-semiarid grasslands** West of the podzolics of Ohio and Indiana are soils formed under fairly moist conditions but without tree cover. Representative of these are the prairie soils found throughout the Corn Belt. On our transect they occur in parts of Illinois and Missouri and in eastern Kansas-Nebraska. These are formed on deposits left by continental glaciers which once covered the Midwest as far south as the Ohio and Missouri Rivers. The long grasses on these prairies send roots deep into the soil and have concentrated minerals and other nutrients in the A horizon. The flow of nutrients through the grasses, plus the contribution of the grass itself to the humus content of the prairie soils, is a critical factor in the fertility of these soils. Prairie soils are usually quite thick, with a characteristic dark-brown and loamy surface layer which shades gradually into a yellowish-brown B horizon overlying the original glacial clays and sands.

This region receives adequate precipitation throughout the year, and severe droughts are infrequent. While the soil loses some of its fertility once the grasses have been removed and replaced with farm crops, small inputs of phosphate and nitrogen, plus a system of crop rotation involving corn, grains, and nitrogen-fixing plants such as beans or alfalfa, can keep it fertile and farming profitable. The choice of possible crops is nearly as great as those which

*Earth as the Habitat for Life: Vegetation and Soils*

"Tall corn" country in the American Midwest. The prairie soils here are among the best in the world. (Photograph courtesy of Rockwell International)

can be grown on the podzolics, and the title "Breadbasket of the Nation" can apply here as well as to the chernozem soils immediately to the west.

The name *chernozem* means "black earth" in Russian. These are the soils of the Ukraine which have fed generations of East European people, people who in turn have died defending the land and the riches it represents. In America

A Maryland farm prospering on well-managed podzolic soils. (U.S. Department of Agriculture photo)

the same type of soil lies just to the east of the 100th meridian, which in turn marks the approximate location of the 20-inch annual rainfall isohyet on the Great Plains.[1] Only slightly leached, they are among the calcium-rich pedocals and need no liming and only small amounts of fertilizer. Adequate but relatively sparse rainfall and cold winters have kept bacterial action at a reasonable level, and the characteristic black color of the A horizon is the result of its rich humus content. A combination of moisture regime and soil type limits the number of crops that can be grown profitably in these parts, but wheat and other small grains do exceptionally well, and here is the source of much of the world's bread. It is also chernozem soils, interesting to note, that underlie the bountiful wheat fields of Argentina. While rainfall diminishes in a westerly direction across the Plains, the chernozems are still watered by nature with reasonable predictability. However, once beyond the 20-inch rainfall line, precipitation comes with less and less regularity. Thus, farming becomes more and more of a gamble as one leaves the

[1]This also corresponds to the "0" level moisture index in Thornthwaite's classification (see page 120).

black earth and enters into the area of chestnut and brown soils.

## Chestnut and brown soils of the short-grass steppes

The area of chestnut and brown soils denotes a true semiarid environment. Rainfall averages between 10 and 20 inches per year. As a consequence very little leaching has taken place. If there is groundwater present, it may move upward through capillary action, leaving behind an accumulation of carbonates in the lower layers. The grasses under which these soils form are shorter than those of the chernozems and long-grass prairies. This reflects the increased aridity with which vegetation must contend. Despite their lower organic content—a function of less moisture and less vegetative cover—these soils can still provide good crops in moist years. However, the general rule is that as the annual amount of rainfall decreases, the unpredictability of its occurrence will increase. This is an important point to which we will return very shortly. Few crops do well under the combination of these conditions, and grains, particularly winter wheat, are the best choice.

## Desert soils and xerophytic vegetation

In western Colorado, deep within the continental interior and barred by mountains from rain-bearing winds, we come to the end of this sequence. Here less than 10 inches of rain falls each year. The spatial occurrence of rain is haphazard at best, and its timing is unpredictable. Vegetation under these circumstances enjoys a "boom or bust" type of existence and is adjusted to it. Two types of plants exist: drought evaders and drought resisters. We will have more to say about these in the next chapter when we discuss the strategies men and other life use in their struggle with nature. For now, we will simply point out that desert plants space themselves farther apart than plants in other climates; that they often have thick, waxy skins and small leaves, characteristics which cut down water loss through transpiration; and that they may estivate, that is, go dormant during periods of high temperature rather than during the winter. These plants are highly specialized, so much so that they receive the special name *xerophytes*.

An absence of leaching makes desert soils

Rows of soy bean plants on a frame in southern Minnesota in what was the western end of the "crop belt." Soy beans, a rich source of vegetable protein, have been steadily replacing corn as the major crop on many Midwestern farms.

relatively rich in nutrients. On the other hand, scattered vegetation means much less humus in the surface layer. Altogether, no agriculture is possible under these conditions unless irrigation is practiced. Even then there is need for careful management of the land. Failures can come as frequently as successes, for the heavy concentration of carbonates and salts near the surface of desert soils can be easily dissolved and reconcentrated in quantities deadly to most plants. Many fields in the dry western part of the United States have been abandoned as the white signature of alkali writes "finis" to their productivity.

The above account of soils and vegetation carries a particularly geographic message. The distribution of vegetation across the earth's surface has been so changed by human activities that it is impossible to understand the geography of plants without reference to human geography. Soils are the same. Everywhere mankind has altered and even created soils. The nature of human ecosystems is therefore of critical importance to physical geography. In the next chapter we will examine man's organization of space and some of the environmental consequences of his efforts. We will also consider some of the strategies farmers and others use in their never-ending interaction with natural systems.

*Earth as the Habitat for Life: Vegetation and Soils*

# 10 | HUMAN ORGANIZATION OF THE ENVIRONMENT: ADJUSTMENTS TO ENVIRONMENTAL UNCERTAINTY

*There was a good gardener named Gunn,*
*Who was growing a cactus for fun,*
*But whenever his daughter*
*Tried to give it some water,*
*He said, "What it needs is more sun."*

More potted plants die from overwatering than from thirst. The care you give your windowbox qualifies you as one of the multitude that attempt to manipulate and relate to nature through various strategies. Where in the house is the best spot to grow a pot of chives?
Rain next Sunday for the wedding reception or should we wait another week? How much grain should the government of India stockpile against possible famine? Thousands of questions like these illustrate why humans constantly play games of chance and strategy with their environments.

For example, a New York apartment dweller owned a cactus. She knew that it shouldn't be watered too frequently, but when? Her solution was to subscribe to an Arizona newspaper and read about the local weather. Whenever it rained there, she watered her plant. It seemed like a great idea, but the cactus still died from overwatering. What our apartment gardener failed to take into consideration was the spatial as well as the temporal character of rainfall in

the desert. The newspaper reported rain whenever it fell anywhere within a large area in or near the city. But our friend didn't consider that convectional showers in the desert are of brief duration and highly localized. By noting the occurrence of rainfall within too large an area she overlooked the true nature of an arid environment; the apartment cactus received far too much water; and our gardener lost her game with nature.

During the thousands of years of his existence man has devised numerous strategies to help him survive natural conditions which are often difficult or dangerous. The purpose of this chapter is to outline some of the ways in which we organize our lives and environments to accomplish this. Let us first consider an example drawn from the vast panorama of prehistory and history. The story concerns the buildup of increasingly complex ways of social interaction and resource use and is an example of a deviation-amplifying process.

## Resource Use and the Organization of Space: The Tehuacan Valley

Twelve thousand years ago or more a band of Paleo-Indians stood on the hills overlooking the Tehuacan Valley in what today is the state of

Puebla, south of Mexico City. The small group probably consisted of two or three couples and their children joined into some type of extended family. They had few material possessions, and their technology at best was able to produce chipped flint tools and weapons: scrapers, gravers, leaf-shaped knives, and some projectile points. Though sometimes described as "big game hunters," they depended mainly on wild plants as well as on small game. Birds, rabbits, and turtles were among the animals cooked at their fires. They seldom were successful in killing anything larger, and as one archaeologist has put it: "They probably found one mammoth in a lifetime and never got over talking about it."

Those people were part of the several migrations of ancient hunters who generations after crossing the Siberian-Alaskan bridge to North America eventually found their way to Central and South America. The Tehuacan Valley before their coming was unpopulated. The climate there in the Mexican highlands at the end of the Pleistocene was a dry one much like that of western Texas today. The valley, while not lush, offered some possibilities for hunting and collecting wild foods; and that group of hunters, or another much like it, descended into the valley and stayed on to become its first inhabitants. The tangible remains of their occupance are like faint shadows on the sand and rock. Eleven of their early hearths or campsites, some in caves, some in the open, have been found by modern scientists.

Nearly 12,000 years later, on another morning, with the clank of light body armor and the neighing of horses, a Spanish patrol also stopped on the hills overlooking the Tehuacan Valley. Their presence marked the end of a distinct and independent Indian culture in that area, as in the rest of Mexico and all of what was to become Spanish America. But that patrol of invading conquistadores did not find a few small and simple groups of hunters and collectors. Instead, the valley presented to them a collection of little kingdoms, each with its towns surrounded by villages and hamlets. Farmers produced a wide variety of domesticated crops,

among them maize, squash, tomatoes, peanuts, and avocados. There were few domesticated animals—only dogs and turkeys—but valley industries produced cotton and salt for export to other regions. An elaborate hierarchy of priests and administrators controlled a population of perhaps a hundred thousand or more people within the valley. The Indian civilization encountered there and elsewhere by the Spanish was nearly as elaborate as their own. Unfortunately for the Indians, the crafts in which the Spanish excelled were iron metallurgy and weaponry, including gunpowder. The destruction of the Indian culture, and its replacement by one based in large part on European norms, is a matter of history.

### Spatial and social organization for resource use

An amazing sequence of events in that 12,000-year period transformed a small handful of wandering hunters and gatherers into a complex sedentary, agricultural society with thousands of people living in villages and towns. The Indians found by the Spanish possessed a high degree of material civilization, although many of their customs were perhaps still as savage as those of the conquistadores who so thoroughly disrupted them. The whole story of their odyssey through time cannot be told here. What concerns us as geographers is the changing pattern of the way in which those people spatially organized the Tehuacan Valley and its resources.

A major problem faced by all groups of people is how to organize their lives spatially to achieve satisfactory resource use. People must live far enough apart to ensure the areas necessary for their well-being. Such areas include room for farms, for recreation, and for adequate amounts of fresh air and sunlight. On the other hand, people crowd together in settlements in order to maximize the interaction and the spatial convenience upon which they also depend. A kind of tension thus exists between these two ways of organizing space.

*Human Organization of the Environment: Adjustments to Environmental Uncertainty*

## Spatial organization without domesticates

The earliest people to reach the Tehuacan Valley were well along the road of human technological development. They knew the use of fire and could make chipped stone tools and weapons. They also had considerable knowledge about edible wild plants as well as hunting and trapping skills. This assemblage of skills and knowledge helped them organize their activities in time and space in order to ensure group as well as individual survival. With each seasonal change a new set of resources became available, although not necessarily in the same locality. Overhunting or overcollecting also might exhaust the plants and animals within walking distance of a campsite. To meet these and other circumstances, ancient man from time to time had to relocate himself in the space in which he lived. Carl Sauer in his discussion in *Agricultural Origins and Dispersals* neatly summarizes one interpretation of the basic geographic rationale of early man:

*We need not think of ancestral man as living in vagrant bands, endlessly and unhappily drifting about. Rather, they were as sedentary as they could be and set up housekeeping at one spot for as long as they might. In terms of the economist, our kind has always aimed at minimizing assembly costs. The first principle of settlement geography is that the group chose its living site where water and shelter were at hand, and about which food, fuel, and other primary needs could be collected within a convenient radius. Relocation came when it was apparent that some other spot required less effort, as with seasonal changes in supplies. Consumer goods were brought to the hearth and processed there. Women were the keepers of the fire, and there prepared the food and cared for the children. They were the ones most loath to move, the home makers and accumulators of goods. The early hearths recovered by archeology are not casual camps, but fire places used so long and sites so significantly altered as to have withstood the obliterating effects of time. The normal primitive geographic pattern is that of a community, a biologic and social group, clustering about hearths at the points of least transport, holding a collecting*

*territory for its exclusive use, and relocating itself as infrequently as necessary.*[1]

The natural resources of the Tehuacan Valley then as now imposed certain constraints on the activities of its inhabitants. The valley receives from 500 to 600 milimeters (19.7 to 23.6 inches) of rainfall during its wet season, which begins sometime in April or May and extends through the summer months. The rest of the year is dry. As a result most of the valley is either arid or semiarid in character. In some places streams bring water from the mountains, but many of these flow only intermittently or during the rainy season.

Nature's larder used by early people did not include any domesticated plants or animals. It is hard for us to imagine living under such conditions. But early man all over the world survived for at least $2^1/_2$ million years before gaining the skills and knowledge which led to domestications. The history of the development of the Tehuacan people is closely tied to their gradual mastery of the plants and animals in their environment.

Before 5000 B.C. the people depended entirely upon wild foods. The plants and animals constituting their diet were adjusted to a subtropical wet-and-dry seasonal cycle: a rainy season from May to September alternating with a cooler and very dry period from October to March. Human activities were organized spatially and chronologically to meet such fluctuations. The first signs of plant domestication appeared there about 5000 B.C. In the interval from then until about 200 B.C. a permanent system of irrigated agriculture using numerous species of domesticated and hybrid plants was established. The thing that interests us as geographers during this gradual shift from wild to domesticated sources of food is the development of connectivity and spatial hierarchies within the resource-utilizing system described.

## From wandering microbands to semipermanent villages

In the period before 6800 B.C. the total population of the valley was small. Perhaps three tiny

[1]Carl Sauer, *Agricultural Origins and Dispersals*, American Geographical Society, New York, 1952, p. 12.

groups, or microbands, wandered through the area collecting and hunting such available food as their technology permitted. The basic moves necessitated by this relatively simple pattern of existence are shown in Figure 10-1*A* as a spatial pattern superposed upon a circular calendar of the year. Points represent campsites or hearths, and the lines connecting them represent the connectivity of the system. At this stage there was no hierarchical ordering of the space used by the people of Tehuacan. Perhaps the several bands met occasionally by prearrangement or by chance in their wanderings. Perhaps those contacts meant simple trade and friendly exchange; perhaps they were times of hostility. No record of such meetings remains for archaeologists and others to read. And yet the pattern of life within the valley was slowly changing into new and more complex forms.

In the 1,800 years that followed, slight but significant differences appeared in the resource-using pattern of the Tehuacan people. The space-time chart of this period is shown in Figure 10-1*B*. Macrobands gathered together in larger campsites during the early rainy season with its abundant food supplies. As the dry season got underway in the fall, the group split into microbands in order to hunt deer and to forage for "starvation foods" like cactus fruit. The resource system was still dependent upon three or more camps in different localities at different times of the year, but in this case certain hearths had precedence over others as sites for macroband camps. Single-family units probably utilized dry season hearths while extended families or related lineages congregated at spring camps. At that time social organization and interchange increased along with opportunities for group ritual. We can now think of two levels of campsites, with dry-season hearths being subordinate to those of the wet season. It is also likely that some idea of territoriality, that is, a social group's right to occupy a particular area in order to exploit its resources, came into play at this time.

The years between 5000 and 3000 B.C. saw even more changes. By the end of this period the population of the valley numbered ten times the original. Chilies, squash, corn, beans, and gourds were being cultivated during the rainy

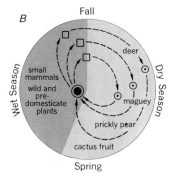

*A* Wandering Microbands
1–2 Persons Per 100 Sq. Mi.
Before 6800 B.C.

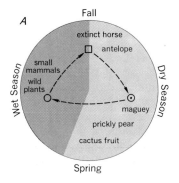

*B* Micro-Macro Bands
3–7 Persons Per 100 Sq. Mi.
6800/5000 B.C.

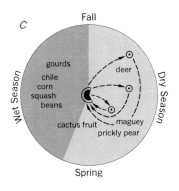

*C* Semisedentary Macrobands
9–17 Persons Per 100 Sq. Mi.
5000/3000 B.C.

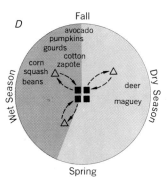

*D* Semipermanent Village
340–680 Persons Per 100 Sq. Mi.
3000/1500 B.C.

**Figure 10-1 Seasonal movement in the Tehuacan Valley, Mexico** Archaeological evidence found at ancient campsites has been used to piece together probable movement patterns of these early people. See Figure 10-2 for key to symbols.

season, and domesticated foods constituted about 30 percent of the people's diet. With increased food supplies the smaller microbands could coalesce in the rainy season into semisedentary macrobands, which in the final centuries of this stage lived in villagelike clusters of pit houses. However, it was still necessary for the large groups to separate into microbands during the dry season. Figure 10-1*C* indicates the annual time span of these semipermanent camps with the complementary scattering of microbands during the dry season. From a

*Human Organization of the Environment: Adjustments to Environmental Uncertainty*

**219**

spatial viewpoint the wet-season hearths were becoming more and more dominant compared with the dry-season camps, which had scarcely changed from previous times.

The next 1,500 years until 1500 B.C. saw the establishment of full-time agriculture, growing among other domesticated plants hybrid corn, beans, squash, chilies, avocados, pumpkins, and cotton. The population of the valley had increased by then to some forty times the original. As the Indians' ability to support themselves increased, it allowed them to reorganize their lives for greater convenience and intensified social interaction. The spatial expression of such reorganization was a contracting of the scattered population about better locations. This phase, which ended about 1500 B.C., is the last one represented by the time-space diagram in Figure 10-1D. Thereafter, populations were more or less permanently fixed at one location throughout the year.

### Permanent settlement hierarchies

In the Tehuacan Valley following 1500 B.C., the number of settlement sites again increased (Figure 10-2). Such an increase represents the prospering of a larger and more sedentary population. Full-time agriculture which utilized more productive hybrid plants and increasing amounts of irrigation stabilized the population at fixed locations through the annual cycle of climatic events. Improved agricultural technology, which was responsible for this stabilization, can be viewed as ways of rearranging resources in time and space. For example, the availability of water, which normally was found only in a few streams after the rains stopped, was extended in time by storing runoff behind dams and was relocated in space by leading it in canals to distant fields. Soil nutrients and solar energy made available to humans by plants became more accessible through the efficient production and storage of surplus crops. Even the development of markets and marketing affected the use of resources. As systems of exchange and barter developed in the valley—just as they have done elsewhere in the world at earlier and later times—resources were not only redistributed in

space but also redistributed among different groups within the total society. Thus, the period of permanent settlement that the Tehuacan people entered into was very different from the thousands of years of wandering and near-starvation from which they emerged. Additional resources in greater quantities became available to them. In the same way, new spatial and temporal arrangements of those resources and the communities which consumed them appeared with an increasing tempo.

By 3,000 years ago a rich religious life had begun to develop. Priests and chiefs attained more and more power. Artisans learned to make new styles of pottery, and trade or some kind of cultural exchange with areas beyond the valley occurred with greater frequency. The growth of these interregional contacts is inferred from similarities of design and decoration found in many parts of Meso-America including the Tehuacan Valley.

In the years between 900 and 200 B.C. the people became full-time farmers living in small villages which were in turn the satellites of larger settlements containing central religious structures. At first these ceremonial centers were little more than large villages, but change continued after 200 B.C. By A.D. 700 these same settlements had become elaborate hilltop sites with a variety of streets, plazas, courts, and pyramids. New crops were added to the growing list of domesticates: among them peanuts, lima beans, and tomatoes. Irrigation systems were becoming larger and more complex and undoubtedly required managers and engineers. Just who those specialists and administrators were is not certain, but priest-kings may have filled the highest positions. More and more similarities appeared between the valley and other sites in Mexico, indicating increased contacts with the outside.

Finally, in the seven centuries before the arrival of the Spaniards, an elaborate system of secular cities, ceremonial centers, villages, and camps or hamlets grew up. Urban centers and the lesser settlements which focused upon them apparently constituted the realms of minor kings. Only about 15 percent of the people's diet came from wild sources, and industries produc-

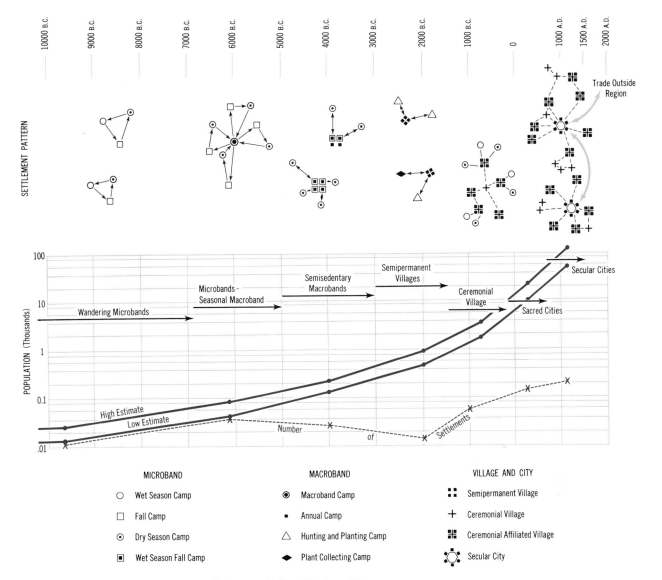

**Figure 10-2 Rise of civilization in the Tehuacan Valley, Mexico**  The migratory habits, population density settlement pattern, and social interaction were all functions of the improved reliability of food supplies that accompanied domestication of plants and animals.

ing cotton textiles and salt for trade outside the region added to the general prosperity. Much remains to be learned from the records of the Spanish regarding what they found within the valley, and much more remains to be learned through archaeological methods. Nevertheless,

the society lacked a detailed written record of its own, and after A.D. 1500 more of the Indian culture was destroyed than preserved by the conquering Europeans.

If we cast back over the 12,000 years encompassing this story of development and seek the

*Human Organization of the Environment: Adjustments to Environmental Uncertainty*

geographic features contained therein, we will plot patterns of increasing complexity. Of primary importance to geographers is how the organization of space offered a variety of solutions to environmental problems of resource use. A major element of this organization was the development of a hierarchically ordered group of settlements with a few large cities, more middle-sized towns, and numerous villages and hamlets.

## A Theory of Agricultural Land Use

Just as the Indian civilizations of Mexico learned to organize their environments for their greater benefit, so too does modern society manipulate and arrange its own resources. We may think at first that the crops we grow are located solely in response to conditions of soil and climate. But this is only partially true. The human organization of space is a vital factor in the distribution of domesticated crop production. It also influences those parts of the world which are still relatively untouched. Thus, our knowledge of environmental systems will remain incomplete without some understanding of human ecosystems and their spatial organization.

In order to understand this point, one of the least appreciated and most critical in physical geography, we must turn aside from strict environmental processes for the moment and consider a theory explaining differences in land use proposed more than a century ago by Johann von Thünen.

A pilot flying over the cities of the North China Plain would see each settlement surrounded by a ring of green fields. The circles shade from bright green nearest the cities to more yellow hues and finally merge with browns and russets of the open farmlands far away from any urban place. This is the *green ring effect* and results from the use of human fertilizer on lands adjacent to the settlements. Soil nutrients concentrated in agricultural products are shipped to the cities, passed through human consumers, and returned in part to the fields as human waste, called *night soil*, which is collected and spread back on the land. Transportation costs are high for night soil, and it cannot be shipped long distances. Thus, the

rapidly diminishing ring of green vegetation reflects the friction of distance on the return flow of nutrients to China's land.

You're on your way west. Interstate 80 stretches 2,000 miles ahead of you to San Francisco. You and your friend are driving shifts, and the road peels away at a steady 55 miles per hour. The suburbs of Chicago with their plumes and fumes drift back behind you. Tall corn rises to the right and left as Illinois swings past. You started late, and as night comes down you are still moving through the green fields of Iowa. Morning comes up behind you and you stop to walk around a rest stop, stretching and looking. Western Nebraska now. Wheat fields. Dry and getting drier. Grasslands ahead with *Danger: Livestock on Road* signs whipping by. The farms turn to ranches and windmills set back against the hills, and then you reach the Continental Divide with little towns hot under the high sun and sheep like pillows far off where pines begin. . . .

Chinese cities or American landscapes, the great earth stretches on; band upon band of different crops, different uses for the land; each region shading into the next. Is it chance the way farm activities are distributed? Does nature alone dictate that corn grows in one place and wheat in another? Why corn, then wheat, then cattle, then sheep? What structures the country as it is? In this chapter we discuss how geographic theory helps to explain the distribution and location of different types of agriculture. In the following sections we suggest ways in which world land use as a whole is organized. More specifically in this chapter we will be talking about a well-known body of thought referred to as *agricultural location theory*. Such theory was formalized by Johann von Thünen in the nineteenth century. Using his ideas as a starting point, we hope to illustrate how the interaction of natural and cultural systems operating in the context of nature accounts for the worldwide pattern of crop types and farm regions.

## The regionalization of farm types

When we speak of agricultural regions, we necessarily deal with abstractions born of our own imaginations. Nevertheless, we divide the world around us into homogeneous areas for convenience in classifying and understanding it. We speak casually of the *Corn Belt* or the *Cotton Belt*. We write learned papers about the types of agriculture, their numbers and spatial distribution. Even a simplified map of world agriculture identifies nine types of farming scattered across the globe (Figure 10-3).

The most interesting thing, however, is not the number or complexity of the regions we perceive, but rather the homogeneity within them and the singleness of choice which their farmers exercise. Farm numbers are limited by the availability of land, and within broad limits

by the character of the physical world. A conservative estimate gives us more than 280 million individual farms throughout the world, with at least 2 million in the United States. If we were to examine these farms in more detail than shown in Figure 10-3, in order to acknowledge the variety we observe around us we might increase the number of farm types to 100 or more. The possibility of 280 million farmers choosing from over 100 types of farming leads to astronomical combinations, but similar farms are found clustered together. There is a distinct regional effect which creates the Corn Belt in the United States, the Rice Bowl of China, the

**Figure 10-3 World agriculture**

Intensive Subsistence, Rice Paddy

Intensive Subsistence, Small Grain

Commercial Wheat

Specialized Horticulture (Commercial) and Mediterranean Agriculture

Commercial Feedgrain/Livestock and Dairy

Tropical Export Crops, Plantation/Small Holders

Tropical Subsistence

Stock Raising

Nomadic Herding, Hunting, Fishing, Collecting, Forestry

Little or No Economic Activity

TOBLER'S HYPER-ELLIPTICAL PROJECTION

dairy districts of Scandinavia, and the cattle region of Argentina, to name but a few.

Why do people making independent choices end up making the same one? Why does one farmer in Iowa decide to raise maize and soybeans and to fatten hogs, and his neighbor decide in favor of the same combination, and his neighbor, and his? These are important questions, for if the nature of farm decision making is understood, then some part of the complex biosphere surrounding us will have been explained.

## Variables determining farm production

The reasons for the choices farmers make can be subdivided into four categories. These are (1) site characteristics, (2) cultural preferences and perception, (3) available technology and organization, and (4) geographic situation or relative position.

**Site characteristics** Site characteristics are the *in place* attributes of a particular area viewed at large, local, or neighborhood scales. Thus, the amount of rainfall and average annual temperature of an area are considered important site characteristics. Soil type and fertility, slope, drainage, and exposure to sun and wind are also used to characterize the physical geography of each and every site. These things all relate to the amount of energy available in the physical system within which the location is incorporated. Other site characteristics could include the number of insect species, their populations, and their potential for destroying crops. The same is true for plant, animal, and human diseases. At still another level of abstraction, the human population density of an area can be considered one of the characteristics helping to determine the qualities of site. The type and intensity of pollution, the amount of built-up area, and the nature of land ownership and property fragmentation could also be included in this category.

**Cultural preference and perception** Perhaps the least known and possibly the most important of all the conditions that help to determine the type of agricultural activity which takes place at a given site are the cultural, psychological, and emotional characteristics of the people involved. For example, we do not eat everything which is available; sometimes people starve rather than consume perfectly edible but taboo food. Muslims abhor pork; certain Hindus abstain from eating all meat, but particularly beef; many Africans will not eat chickens or their eggs. The Chinese and some other peoples of East and Southeast Asia refuse to drink milk or eat milk products.

The refusal to eat certain foods places real constraints upon the agricultural systems possible within an area. Maize is scarcely considered human food in much of Europe, and therefore its production is restricted to animal fodder in all but a few places. Americans consume large quantities of meat despite its expense. A diet with greater emphasis upon vegetables and cereal grains would be just as healthful and cost much less. Most nationalities can be characterized by their food preferences and prejudices.

The way we perceive the resources around us is also important. For example, the European settlement of North America was largely from the northeast to the west. The firstcomers were Anglo-Saxons and other Europeans accustomed to a moist, mild climate and a tree-covered landscape. Those yeoman farmers equated trees with fertility. To them, land to be suitable for farming should, in its wild state, have a cover of trees. New England and the East Coast met their expectations when they settled there. But as subsequent waves of migrants pushed west to the edge of the central prairies and Great Plains, they encountered treeless, grass-covered areas. This lack of trees failed to meet their perception of truly fertile land. They referred to this area as the Great American Desert—which according to some people began at the Mississippi River —and pushed across it to the valleys along the Pacific Coast. There again they had to clear the land of timber before farming it, but they were satisfied. Those early farmers failed to see in the grasslands the latent fertility of what was to become the Great American Breadbasket. It took a later generation of migrants, this time people from the steppelands of Eastern Europe, to take advantage of the rich, grass-covered soils of Nebraska and Kansas. Thus, the way in which those immigrant groups perceived the environments which they encountered colored their subsequent use of the resources available to them. Many other factors influenced the pattern of settlement on the Great Plains. Certainly technological developments such as the moldboard plow and barbed wire fencing also were important. However, it is not our purpose in this section to explore these topics in great detail. We want, rather, only to identify some of the important elements which complicate reality and make simple explanations so difficult. As part of this we should not overlook intangible but significant human interpretations of the environment.

**Available technology and organization**
Since the end of World War II and the subse-

quent creation of international development programs, whole libraries have been written about technology and organization in agriculture. It is convenient to summarize this general category by describing two types of farms and farmers located at opposite ends of the developmental spectrum. Modern commercial agriculture as it is practiced in the United States might be one case; subsistence-level farming in an emerging economy would be the other.

The modern commercial farm is characterized by the large amount of capital necessary for its operation. We describe this as *capital-intensive*. Farms substituting human labor in place of all the conveniences and mechanical aids that money can buy are called *labor-intensive*. The investment necessary to operate a viable, capital-intensive farm unit in the United States is impressive. The average value of the property and buildings on "first class" American farms in 1959 was 135,000 dollars. Table 10-1 shows the

Table 10–1   Size, Investment, and Returns by Type of Farm, United States, 1963

| Type of Farm* and Location | Size of Farm in No. of Units | Total Farm Capital, 1/1/63 | Gross Farm Income† |
|---|---|---|---|
| Dairy, Central Northeast | 32.2 cows | $ 43,400 | $ 14,475 |
| Dairy, western Wisconsin | 23.8 cows | 37,410 | 10,267 |
| Hog and beef fattening, Corn Belt | 153 acres | 98,920 | 31,024 |
| Cash grain, Corn Belt | 246 acres | 137,020 | 24,581 |
| Cotton, southern Piedmont | 101 acres | 30,750 | 7,153 |
| Cotton (nonirrigated), Texas, High Plains | 445 acres | 84,950 | 19,584 |
| Cotton-specialty crops (irrigated), San Joaquin Valley, Calif. | 329 acres | 305,450 | 112,987 |
| Tobacco, North Carolina Coastal Plain | 47 acres | 27,640 | 12,581 |
| Spring wheat, small grain, livestock, northern Plains | 588 acres | 57,540 | 12,384 |
| Winter wheat, sorghum, southern Plains | 684 acres | 125,910 | 16,632 |
| Cattle ranches, intermountain region | 149.5 cows | 95,550 | 17,460 |

*All except cotton farms in California are family-operated.
†Includes both income from farming and government payments.
Source: *Farm Cost and Returns Commercial Farms by Type, Size and Location*, Agricultural Information Bulletin 230, Economic Research Service, U.S. Department of Agriculture, June 1964, p. 4.

*Human Organization of the Environment: Adjustments to Environmental Uncertainty*

**Table 10–2  Energy Inputs (kilocalories) in Corn Production**

| Year / Input | 1945 | 1959 | 1970 |
|---|---|---|---|
| Labor | 12,500 | 7,600 | 4,900 |
| Machinery | 180,000 | 350,000 | 420,000 |
| Gasoline | 543,400 | 724,500 | 797,000 |
| Nitrogen | 58,800 | 344,400 | 940,800 |
| Phosphorus | 10,600 | 24,300 | 47,100 |
| Potassium | 5,200 | 60,400 | 68,000 |
| Seeds for Planting | 34,000 | 36,500 | 63,000 |
| Irrigation | 19,000 | 31,000 | 34,000 |
| Insecticides | 0 | 3,300 | 11,000 |
| Herbicides | 0 | 1,100 | 11,000 |
| Drying | 10,000 | 100,000 | 120,000 |
| Electricity | 32,000 | 140,000 | 310,000 |
| Transportation | 20,000 | 60,000 | 70,000 |
| Total Inputs | 925,500 | 1,889,000 | 2,896,000 |
| Corn Yield (output) | 3,427,200 | 5,443,200 | 8,164,800 |
| Kcal return/Input Kcal | 3.70 | 2.88 | 2.82 |

Source: David Pimentel, et al., "Food Production and the Energy Crisis," *Science*, vol. 182, No. 4111, pp. 443–449. The above data are drawn from Table 2, page 445; a complete explanation of the values and their derivation can be found in footnotes to that table and to Table 1, page 444, as well as in the text of the article.

average size of and the amount of capital invested in a variety of American farms in 1963. When we compare the capital invested with gross farm income for the same properties, it is easy to see why farmers prefer to leave the countryside and take jobs in urban areas. Remember, the gross farm income must compensate the farmer for his annual investment of labor as well as capital.

*Modern farming is energy intensive.* Mechanized agriculture utilizes energy not only to run its machines and to transport its crops to market but also to produce fertilizer and pesticides for its fields, to acquire irrigation water, and to carry out essential management functions. Fossil fuels are used for all these activities, and the growing energy crisis promises to influence modern farming everywhere in the world. We must take into account, like it or not, the fact that one characteristic of "developed" nations is their use of very large amounts of energy. For example, the United States in 1970 accounted for more than one-third of the energy consumed in the world, while only about 6 percent of the earth's people lived there.

All this is reflected in conditions on modern American farms. From 1950 to 1971 the number of farm tractors increased 86 percent from 2.4 million to 4.5 million. Fuel consumption rose accordingly by 4.3 billion gallons between 1940 and 1969 to a total of 7.6 billion gallons in the later year. In the case of corn, actual fuel use by farm machinery increased from 15 gallons per acre in 1945 to an estimated 22 gallons per acre in 1970. Table 10-2 shows the energy equivalents required for all the other tasks and items associated with corn production. Altogether, in 1945 kilocalories equal to the amount contained in 26 gallons of gasoline were used to produce one acre of corn, while in 1970 the equivalent amount of gasoline was 80 gallons. There was a compensating increase per acre in corn yields from 34 bushels per acre in 1945 to 81 bushels in 1970. However, energy inputs increased 3.1 times, while the yield per acre in corn food calories increased only 2.4 times. This was a decrease of 24 percent in the production of corn calories per kilocalorie input of fuel.

As long as fossil fuel is plentiful and inexpensive, such figures can be justified. But there are many indications that the lavish use of fossil fuel for farming cannot continue indefinitely. In 1970 it took about 112 gallons of gasoline per person to feed the population of the United States. If similar technology were used to feed a world population of 4 billion at the American nutritional level, 488 billion gallons of fuel would be required each year. One estimate of known world petroleum reserves is 546 billion barrels—which, if used only for farm-oriented activities, would last an estimated twenty-nine years given present conditions.

Another way of viewing modern agricultural production is to contrast the cost of producing 1,000 kilocalories of plant product in America and in India. In the former, the cost is about 38 dollars; in the latter, about 10 dollars. However, it is not our purpose here to suggest alternatives to the dilemma implied above. The point we wish to make is that modern farming requires very large amounts of energy and that energy is becoming increasingly expensive. The implications of all this give considerable food for thought, if little else.

*Modern farming is quick to change* under the necessity to return profits on such sizable investments of money, material, and energy. Fluctuations in the market are watched closely by farmers, and for example, the number of animals they raise varies dramatically from season to season. Figure 10-4 illustrates the rapid fluctuation in the number of hogs butchered over a 19-year period. The enormous variation from year to year reflects the uncertainty of farmers' efforts to anticipate market demands and shifting wholesale prices. Animal types also change rapidly. American hogs were once fat porkers heavy with lard. Consumer tastes changed rapidly in favor of lean bacon and ham at the same time that vegetable oils provided a cheap substitute for cooking fats. Innovation plays its role here, as well. Now pigs being fattened for market are sometimes fed from raised troughs which they must reach by standing on their hind legs, thus producing lean, well-developed hams.

Not only does the market fluctuate widely, but new markets for new crops bring about abrupt changes in farming. Hybrid corn and soybeans have both made dramatic entries into American farming in recent years. Figure 10-5 illustrates the nearly geometric increase of the area devoted to soybean production. In the period between 1960 and 1965 alone, more than 10 million additional acres were sown in this crop. Changes result not only from market fluctuations but also from competition with new sources and substitute products. Rubber was originally produced from trees growing wild in South American jungles. By 1920, 90 percent of all rubber came from trees grown in orderly plantations halfway around the world in Southeast Asia. World War II reduced all production of rubber trees, both wild and plantation-grown, to only 16 percent of the total. In the decades that followed, synthetic rubber was less important, but in every case, wild rubber production remained an insignificant proportion of the total (Figure 10-6).

The modern farmer's qualities match the demands of the system within which he operates. He must be an agronomist able to assess the physical requirements of crops both new and old. He must be receptive and able to understand and accept the advice given by the Department of Agriculture's county agents as well as that from state university experimental farms and from seed, chemical, and machinery salesmen. He also must be able to think like a market economist. If his evaluation of the market is wrong, he will be left with an unsalable surplus of crops or animals. To avoid this he must utilize every source of information available. City people would be surprised if they listened to the farm programs broadcast in the early morning over many stations. Market futures are given regularly, as are the number of animals delivered, bought, and butchered at major stockyards. The farmer listens to all this and more and must decide for himself how to operate his enterprise. At the same time, the commodities produced on modern farms travel long dis-

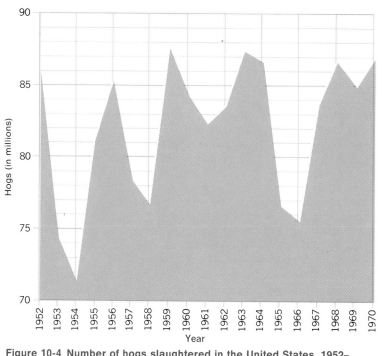

**Figure 10-4 Number of hogs slaughtered in the United States, 1952–
1970** The number of hogs slaughtered in the United States varies greatly
from year to year. From this record, it appears to take a few years to recover
from a decline, which can be as much as 20 percent lower than the previous
peak year. The fluctuations are due, in large part, to uncertainty regarding
future price and the fact that a farmer must start his production cycles months
ahead of when he plans to market the products. (Data from *Historical Statistics of the U.S. Colonial Times to 1957: Continued to 1962 and Revisions;
Statistical Abstract of the U.S. 1963–1971,* U.S. Dept. of Commerce, Bureau of
the Census, Wash., D.C.)

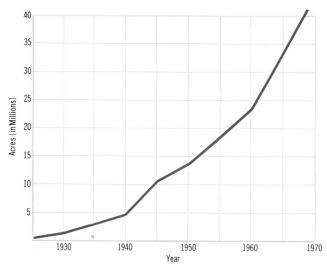

**Figure 10-5 Introduction of soy-
beans in the United States**
Soybeans, a source of vegetable oils
and proteins, have become a major
crop in the United States in recent decades. The consumer has shifted to vegetable oils for cooking and in marga-
rine and has preferred leaner meats.
These changes in market demand have
encouraged farmers to switch from corn
to soybeans, which are processed directly into foodstuff as well as providing protein supplements for animal
feed. (Op. cit., Figure 11.3, and *Bulletin #951,* Dept. of Commerce, Bureau
of the Census, Wash., D.C.)

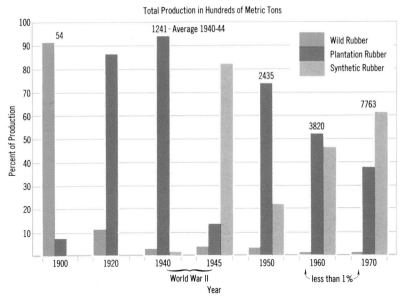

Total Production in Hundreds of Metric Tons

Figure 10-6 **Percent of world rubber production in wild, plantation, and synthetic rubber, 1900–1970** (Data from: Rubber Study Group, *Commodity Year Book 1956 and 1971*; wild rubber percentages, Jones and Darkenwald, *Economic Geography*, Macmillan, New York, 1954, and Jean le Bras, trans., *Introduction to Rubber*, Hart New York, 1969)

tances, sometimes to the other side of the world, to reach the consumer. This means that the farmer must follow not only the fortunes and activities of his neighbors but also those of farmers producing similar crops in distant states or other countries.

All in all the modern commercial farmer must be a superb manager if he is to survive. He must assess and juggle all the things described above and many more while balancing his decisions against a capricious and often uncooperative Mother Nature. To do this he has a variety of information sources available: radio and television, farm journals and magazines, special newspapers and stock market reports, government farm agents, and advisors sent by private industry. At the same time, his aspirations are almost the same as those of urban dwellers. He wants the same conveniences, the same family transportation, the same schools for his children, and the same high level of living. In the final analysis, the modern farmer in all ways resembles his city cousins much more closely than he

Sugar beet harvest in California. American agriculture is capital intensive; that is, great use is made of equipment, chemicals, and high-quality but expensive breeds and seeds. This results in high labor productivity but at the cost of great energy inputs. Such a system is not easily exported to less developed countries, where the economy may not be able to supply the financial and industrial services required. We are also becoming increasingly aware of undesirable ecological results from capital intensive agriculture.

*Human Organization of the Environment: Adjustments to Environmental Uncertainty*

Subsistence agriculture. This man is cultivating maize with a hoe on a steep mountain slope in Colombia, South America. His farming methods are labor intensive; that is, most of the value he applies to the land is his own labor. He buys little or no mechanical equipment or chemicals. As a consequence, his inanimate energy and financial inputs are modest to nil. However, production is also low, perhaps not even enough to feed his whole family throughout the year. From an ecological point of view the evidence is mixed. Some subsistence agriculture is clearly destructive of the long-run food producing capacity of the environment; others seem nicely balanced with nature.

own the property he tills nevertheless works the land with his own labor and without the use of modern machines. *Labor-intensive* agriculture requires that he and his family toil long hours in order to farm fewer acres of ground than his modern counterpart. It is difficult to make direct comparisons between these two types of farming, for virtually no data exist which show the capitalization of subsistence-level farms. We can, however, make some qualitative comparisons. The Near Eastern farmer, for example, may own one or two oxen which he uses to pull a wooden plow or solid-wheeled cart. In some cases fodder is so scarce, because of the extreme scarcity of irrigation water, that only the very richest farmers can afford to feed draft animals. Lacking animal power, the farmers must spade their fields by hand. Similar conditions exist in South and Southeast Asia, where more fortunate farmers rely on oxen or water buffalo while the poor depend on the sweat of their own brows. The picture is repeated again in Latin America, while in much of Africa sleeping sickness and rinderpest disease have kept the horse and cattle population at a minimum, unavailable even to those farmers who might afford them. Particularly in Africa, *hoe culture* is common, with humans using those tools in place of plows pulled by animals.

In some parts of the emerging world absentee landlords living in the cities may own hundreds of villages. In these cases, the landlord may provide seed, equipment, land, and water, while the tenant farmer invests only his own labor. The harvest, however poor, is divided into five portions; each part is allocated to one of the five subdivisions just mentioned, with the landlord receiving four-fifths and the tenant one-fifth of the produce.

All these conditions are reflected in the low per capita incomes found in the predominantly agricultural nations of the world. Figure 3-9 shows the world distribution of income. It is no coincidence that the distribution of predominantly rural populations shown in Figure 3-8 matches that of the low per capita income countries.

Subsistence-level farming with its lack of cash or surplus crops presents few opportunities for experiment and change. We should not con-

does subsistence-level farmers in emerging economies.

By subsistence-level farming we mean agriculture which provides food, shelter, and the basic necessities for its participants but which allows little or no surplus with which to enter the commercial market system. It would be difficult to find a completely self-sufficient farm anywhere in the world today. Here, however, we present a description of such a system in order to clarify the traditional characteristics which in various combinations with those of modern farming go to make up the middle and lower range of the agricultural spectrum. For our present purposes we will use the terms *subsistence-level* and *traditional* agriculture interchangeably.

The traditional farmer fortunate enough to

sider this as completely bad, though, for subsistence-level agriculture throughout the world is remarkably resilient and able to survive all manner of disasters. We should not ask the question, "Why are traditional farmers so inefficient?" but rather, "How have such impoverished methods of farming survived thousands of years of drought and flood, heat, cold and storms, unfair taxation, war, pillage, and looting? Indeed, why does traditional agriculture continue to resist the well-intentioned, well-financed, and highly trained technicians who have tried to change it in recent years?"

The answer can be summed up in a short sentence: *Tradition is wise.* Subsistence-level or traditional agricultural systems lack the flow of information so necessary for rapid change. Poor education and ignorance of modern farm methods are everywhere apparent in the emerging nations. But we should not think that the participants in these systems are either stupid or lazy. Lacking capital, outside information, and scientific methodology, the farmers have learned farming strategies by trial-and-error methods over hundreds and hundreds of years. Their inherited culture, which provides them with techniques and attitudes necessary for survival, is their most valuable asset. For example, wooden needlenosed plows without moldboards are used everywhere in arid regions by subsistence-level farmers. While they are seemingly far less efficient than our own moldboard metal plows, which turn a deep furrow, thereby exposing the soil to sun and air, needle plows stir the earth without severely disturbing the surface. By not exposing the underlayers, valuable soil moisture is preserved for subsequent plant use. The simple needle-plow is also less expensive, and can be made and repaired from local materials. When modern farm technicians first attempted to introduce the iron moldboard plow into the Near East, it was not readily accepted by local farmers, who knew more of their own environment and pocketbooks than did their would-be helpers. Similarly, when tractors were used to replace oxen in some areas, a major conflict arose. Plowing was easier and the timing of crop planting was improved, but the departure of the oxen deprived the villagers of their major source of fuel. In those treeless areas, dried dung mixed with straw was in many settlements the only material available for the cooking fires. The tractors were an improvement but necessitated an additional investment in kerosene cookstoves. In the words of one enlightened developmental technician, "There's no fuel like an old fuel." Thus farmers in the emerging world are slow to change their ways for fear of overtaxing themselves, their pocketbooks, and their resources. Given enough slack and the opportunity to change, they are as willing to accept new developments as are our own farmers. It is simply because they already have a system that works reasonably well that they are cautious about experimenting with irreplaceable materials and money. Tradition tells them what will succeed and how to survive, albeit at a low economic level. In other words, don't rock the boat.

In summary, the traditional and the modern farmer *viewed as stereotypes* have contrasting characteristics and skills. The modern farmer is a specialist in technology, money matters, and management. The traditional farmer is able to provide himself and his dependents with food, shelter, clothing, and equipment made with his own hands. He is at a disadvantage in the modern market system but could probably survive a major catastrophe like war as well as or better than his modern counterpart. This is particularly true when we consider the elaborate supply system which provides the modern farmer with necessities. If his communication lines were cut, he would soon run out of fuel, spare parts, store-bought foods, and clothing purchased off the rack. Traditional communities, on the other hand, depending on the outside world for fewer things, would miss it far less if cut off from central places ranked above them in the settlement hierarchy.

**Geographic situation or relative position**
The emphasis placed upon communication and organization in the above section brings us to the central point of this chapter. Wherever the movement of energy, goods, and information is important, so too will be the friction of distance and the relative location of the farms in question. We have already described three sets of variables which help to determine the form

that agricultural land use will take. Let us now consider our fourth set, relative location expressed particularly in terms of distance.

## The Land Use Theory of Johann Heinrich von Thünen

Relative position is important in agriculture at all scales from world patterns to patterns of production surrounding a single settlement. The analytical principles underlying this statement were first demonstrated in 1826 by Johann Heinrich von Thünen, a north German landowner and farmer who wrote on the economics of production. Thünen had observed that various types of farming occurred with surprising regularity in circular bands or rings around his own settlement. The pattern was not always clear, but in his book, *The Isolated State,* he presented a logical scheme which explained what he had observed. The importance of Thünen's work, however, lies not in his explanation of the world in which he lived but rather in the fact that *his methods may be applied to other situations with other sets of data, with results differing from what he observed but consistent with the geographic theory which he outlined.*

Thünen's ideas are of particular interest to geographers because they deal with geographic rather than nominal locations. In the words of Michael Chisholm:

*His argument started from the premise that the areal distribution of crops and livestock and of types of farming depends upon competition between products and farming systems for the use of any particular plot of land. On any specified piece of land, the enterprise which yields the highest net return will be conducted and competing enterprises will be relegated to other plots where it is they which yield the highest return. Thünen was, then, concerned with two points in particular: 1. The monetary return over and above the monetary expenses incurred by different types of agriculture; 2. Such net returns pertaining to a unit area of land and not to a unit of product. For example, if a comparison is being made between potatoes and wheat, we will not be concerned with the financial return obtained per ton of produce but with the return which may be expected from a hectare of land in either crop. Thus, at certain locations wheat may be less profitable than potatoes because, although the return per ton on wheat is higher than on potatoes, the latter yield perhaps three times the weight of crop to a hectare of land. In this case, potatoes will occupy the land.*[2]

### Characteristics of the isolated state

In order to explain the world in which he lived, Thünen had to simplify and restrict the conditions describing his model of farm production. To do this he assumed six characteristics for his agricultural region. (1) At the center of the area was a single, isolated market town. No links connected it to other settlements or to the outside world in general. Movement was only to and from this one place, with its population being the only urban one and all other people being rural farmers. (2) The area in question was a homogeneous plain having equal fertility in all its parts and neither hill nor valley to vary its surface. (3) All labor costs were everywhere equal on this plain. Nowhere were there fewer laborers or more skilled workers. No cost differential could occur as a result of competition for employees. (4) Transportation costs were the same everywhere and in every direction. This required an initial roadless condition, since roads of any kind would focus transportation into a radial pattern centered on the town. Thus, he assumed that all carts could go to the central market by the most direct route. (5) Within this region there existed a static economy. The entire system was in equilibrium, with no long-range trends leading inevitably to lower or higher prices, nor were there sudden shocks within the system such as depressions or inflations. (6) Finally, he assumed that the market price of any commodity was fixed for any single farmer and that farmers could not form combines or cooperatives in order to manipulate the market by holding back crops to raise prices or by dumping them to ruin their competitors.

[2]Michael Chisholm, *Rural Settlement and Land Use,* Hutchinson, London, 1962, pp. 21–22.

## The isolated state as an energy system

Thünen showed a city and its hinterland in an isolated and very stable condition. His model, however, cannot be considered an isolated system. Although he did not consider his isolated state in modern systems terminology, it may be convenient for us to view it in that way. Energy in the form of sunlight constantly entered the area with which he dealt. Foodstuffs were shipped to the central settlement and were reduced there to waste materials and heat. The waste in turn might further decompose, releasing more heat; some of the waste would be returned to the fields. (In this latter case Thünen concerned himself only with the return of horse manure as fertilizer to the land. In the early nineteenth century the major form of power for urban transportation was horses, which required large amounts of fodder and produced equally important quantities of manure as a by-product.) Eventually, the energy which had entered the system as sunlight would escape from it as some form of reradiation back to the heavens. This kind of system, which exchanges energy but no mass with its surroundings, is a *closed system.*

It was this flow of energy through the system which helped organize its many parts into a recognizable structure. Much as logs floating on a stream become aligned with each other as a result of the flowing water, so too do all man–environment systems reflect the particular characteristics of the energy flows which they utilize and in turn help to create. We have introduced this idea of energy movement within various systems in order to familiarize and prepare the reader for more detailed discussion in later portions of this book. In the meanwhile, it will help us to remember that just as a steady stream of water maintains the logs in a given orientation, so too does a steady flow of energy in the form of farm products through Thünen's model maintain it in a *steady state* or single form without change. Should new conditions be introduced into the model, adjustments to new steady states will result.

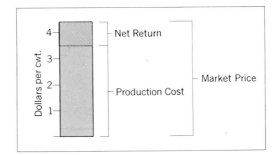

**Figure 10-7 Price and cost condition at market** The market price $p$ of a commodity at the market less the production cost $c$ equals the net return $r$; that is: $p - c = r$.

## Unit commodity concepts

Before we examine Thünen's model for its areal characteristics, we should define some of its basic terms. These deal with unit measures of commodities such as bushels of corn or hundredweights (cwt) of milk, liters of wine, and kilograms of butter. We also need to introduce the notation which will be used to indicate other elements such as distance. As soon as our definitions are clear, we will transform our thinking into its areal form.

Let us begin, for example, with milk. It has a market price we can call $p$. That would be the per unit price for any commodity, in this case, a hundred pounds (cwt) of milk. Land and labor and the cost of cows, barns, and fodder are all investments of capital that must be repaid. The total expenses necessary to produce our cwt of milk must be subtracted from the market price. Market price $p$ minus production costs $c$ leaves a net return $r$, that is, the profit for each unit of produce, in this case milk (Figure 10-7).

The above relationship assumes that the market is located at the production site. This might be true if we lived next door to a dairy and could buy our milk by leaning out the window, but in most cases the produce has to be shipped some distance $d$ to market. This can be measured in miles or kilometers. Milk is perishable, and the glass-lined, chilled tank trucks, sanitary milksheds, and everything else that it takes to get the milk from the cow to you contribute to transportation costs.

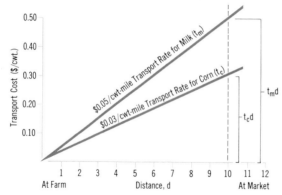

**Figure 10-8 Transport rate per mile and transport cost related to distance from market** The transport cost for delivering a commodity to the market is the product of the distance to the market and the transport rate per mile.

Now let us consider another commodity with different shipping characteristics. Bulk corn is much easier and cheaper to move from farm to mill or market. It can be shoveled or sucked up with vacuum hoses. It will not spoil if it is kept dry. High temperatures within the normal

**Figure 10-9 Net return relative to distance from market** Net return is market price $p$ minus production costs $c$ minus transport costs $td$. Beyond distance $x$ from the market, transport costs exceed the market margin (price minus production costs), and net return is negative. Farmers located this far away have no incentive to enter the market. In this diagram, transport rates and cost are exaggerated relative to production costs in order to show the relationship clearly. Today milk is often shipped over 100 miles to market.

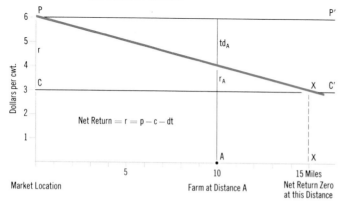

range will not damage it. Thus, the cost of transporting corn will be much less than the cost of moving fresh milk to market. Large or small, the cost of transporting a unit commodity a given distance (one bushel or cwt per mile) is called the *transport rate* and is given the symbol $t$.

Let us assume that we ship a hundredweight of milk and a hundredweight of corn 10 miles. For each mile that the commodities are shipped we must add one increment of transport cost $t$, but each commodity will have a different value for $t$ depending upon its perishability and general transportability. This rate times the distance shipped gives the total transportation cost $td$. The lines in Figure 10-8 representing the cost of shipping milk and corn have different slopes, a steep one for milk and a more gentle one for corn.

We can put the terms $p$, $c$, and $td$ together in order to see their interrelationship when the point of production is not located at the marketplace. This relationship is stated: The net return on a unit of a commodity is equal to the price at the market less the cost of production less the transport cost. It is written thus:

$$r = p - c - dt$$

We can graph this relationship for the crop being considered. Figure 10-9 shows the unit commodity price $p$ at the marketplace. This price is extended across the graph (line $P-P'$) to suggest the *market price* which any farmer would receive once he got his goods to the market. Line $C-C'$ represents the production costs for a commodity unit; the difference between $p$ and $c$ illustrates the net return $r$. However, this presentation shows the value of $r$ as being everywhere the same and does not take the cost of transportation $td$ into account. We know that transportation costs increase in direct proportion to the distance from market and must be subtracted from the net return. To show this in Figure 10-9 we have taken the transportation costs off the top of the net return. If production occurred directly at the market, the distance between market and farm would be 0 and therefore the value $td$ would equal 0. At point $A$ the distance would be 10 and *td* would be 10*t*. When this is subtracted off the top of the net return *r*, a new value $r_a$ results.

If the sloping transport cost line is projected outward from the market, it will eventually intersect the production cost line *C—C'* at point *X*. At that intersection, the cost of transportation will equal the original net return *r*. In other words, all the profit earned if the market and the farm were in the same place will have been eaten up by transportation costs. Beyond that distance from market there would be no more profit, and production would stop.

## *Areal concepts*

To change our comments on unit commodities into ideas incorporating space we must multiply all the elements in the basic equation *r = p − c − dt* by the *yield Y* or output per unit area. For example, instead of talking about bushels of wheat we must now discuss bushels of wheat per acre. We must convert measures like gallons of milk into gallons per acre. Since milk will be produced every day on a dairy farm and, on the other hand, wheat is harvested but once a year, we also need to consider production over some reasonable period of time, usually 12 months. Multiplying our original expression by yield *Y* we obtain:

$$Yr - Y(p - c) - Ydt$$

We can simplify this by substituting single capital letters in place of subgroups in the above equation. Let *P* = *Y(p − c)*; that is, *P* equals the *market margin* or profit on the amount of a crop produced per acre. For example, if farmland can produce 20 bushels of wheat per acre, then the profit per acre f.o.b. the farm is 20 times the market price of 1 bushel less 20 times the cost of producing 1 bushel.

Let *T* = *Yt,* in other words, the transportation rate on the amount of crop produced on 1 acre of land. If it costs 1 dollar to ship 1 bushel of wheat 1 mile, then in our example *T* will equal 20 dollars.

After we have made these substitutions, it remains for us to use *R* in place of *Yr.* In this case, *R* represents the net return per unit area, or the rent. This is all expressed by a new

A. Rent and Transport Cost per Acre of Crop

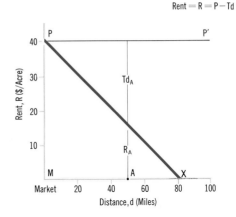

B. Rent Surface and Limit of Marketable Crop

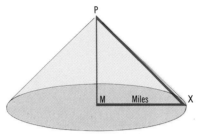

**Figure 10-10 Bid rent as a function of distance from the market** *(A)* Rent and transport cost per acre of crop. Rent is the net return of a commodity per unit area, e.g., $/acre. In order that the rent equation be in the proper units, one must think of the transport rate as the cost of moving an acre's worth of crop a unit distance. *(B)* Rent surface and limit of marketable crop.

equation very similar to our first one:

$$R = P - Td$$

Note carefully that distance *d* does not change.

We may now redraw the graph in Figure 10-9 in a simpler form showing the relationship between the market margin *P,* transportation costs *T,* distance *d,* and rent *R.* This is shown in Figure 10-10*A.* Again, if the farm is located at the market, distance and transportation costs

*Human Organization of the Environment: Adjustments to Environmental Uncertainty*

**235**

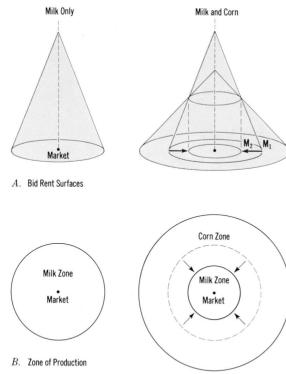

Milk Only

Milk and Corn

$M_2$  $M_1$

A. Bid Rent Surfaces

Corn Zone

Milk Zone

Market

Milk Zone

Market

Market

B. Zone of Production

**Figure 10-11 Two-crop system** (A) Bid rent surfaces. (B) Zones of production. The effect of adding an additional crop to the commercial agricultural system can be seen in the diagram. Corn has a lower bid rent near the market, but because this bid rent falls off more slowly with distance than the bid rent of milk, beyond a certain distance farmers receive more for corn than for milk. They will switch to corn production, and additional farmers beyond the limit of milk production will enter the market. There will then be a reduction in the size of the milk zone, with the result that less milk will reach the market. Milk prices will rise, and the market boundary will adjust outward somewhat.

the distance between $M$ and $X$ becomes the radius of a circle with the market at its center (Figure 10-10B). This is the areal extent of crop production.

We can now draw our first two conclusions from Thünen's work. First, we see that *rent R and transport costs Td are inversely related.* As transportation costs increase, land rent decreases. Second, *given a single market taken as a point on a homogeneous plain, there will be a limit to commercial farming.* No one beyond the radius where transport costs completely eat up profit will want to try to enter the market. Thus, an agricultural region with definite limits will be formed around the city.

## Agricultural interdependencies

What *adjustments* occur when a second commodity, like corn, is added to a one-product system? In the diagram 10-11 milk production extends from the central market to point $M_1$. When corn is added, the superior profits for milk end at point $M_2$. Under the competition from corn the milk-producing area will be forced to shrink inward from its original boundaries (Figure 10-11). If we add a third crop, let us say wheat, the boundaries of the agricultural regions again adjust under the impetus of competition among unequal rent paying abilities of the different crops (Figure 10-12A). The amount of wheat grown on an acre has the lowest price in the market but is least expensive to ship. It will be found growing farthest from the marketplace because of the slow rate at which its transportation costs use up available profits. Now it is corn's turn to draw in its boundaries from $c_1$ to $c_2$, which mark the intersection of the corn and wheat transportation slopes. When things have settled down, wheat will be found growing from line $c_2$ as far out as line $w_1$ beyond which no profits can be made by milk, corn, or wheat. We come, with this observation, to the third and fourth conclusions provided us by Thünen's analysis. *At any given distance from market the crop with the highest rent paying ability is chosen and agricultural land use forms rings of homogeneous activity around the market.* This is illustrated in Figure 10-12B for a three-crop system. Fourth, *agricultural industries (crop types) are interdependent.* If you change

are reduced to 0. This means that the market margin $P$ and the rent $R$ on a unit of land are the same. (Be sure to note that we have now included production costs in a single expression with the market price, and have eliminated the cost line $c$—$c'$ from our second diagram.) The point $X$ where the sloping transportation rate line intersects the line of 0 profit marks the distance from market beyond which production of the crop will not be found. If we take this sloping line and rotate it around its vertical axis,

one part of the system, you will affect all its parts.

Figure 10-13 shows the land use surrounding Thünen's original central place. Once we have seen the pattern, it is a simple exercise to reconstruct the relative value of crop types in the market and their varying degrees of transportability. Notice that the distribution of land use types in Thünen's day was somewhat different from our own. We again turn to Michael Chisholm for his commentary on this circumstance:

*A point which many writers have seized upon is the fact that Thünen put forestry as the land use occupying the zone second from the central city, whereas certain types of agriculture were put at greater distances. This arrangement accords so ill with the reality of location patterns in the developed parts of the world in the mid-twentieth century that people are often tempted to reject the whole analysis. A few explanatory words are therefore in place. At the time Thünen wrote, forest products were in great demand for building and, more particularly, for fuel. Large quantities of timber were required for these purposes, and consumers were not willing to pay high prices. A hectare of land produced a very large quantity of lumber, even though few inputs were applied; the bulky material incurred high transport costs. Thus, the advantages of proximity to the market were such that all other types of agricultural use, except the innermost zone of intensive production, were displaced by forestry; it produced a higher Economic Rent than any other product in the second zone. For the time at which he wrote, this arrangement was entirely logical.*[3]

### Intensity of land use

In much the same vein we may add a fifth and final conclusion: *Intensity of agriculture increases toward the market.* We have seen that a crop to be competitive must pay a high rent or profit per acre. This means that centrally located land with a subsequent transportation advantage will be high-priced. People will compete for property near the center to avoid high transportation costs. The actual price of land

[3]Michael Chisholm, op. cit., p. 30.

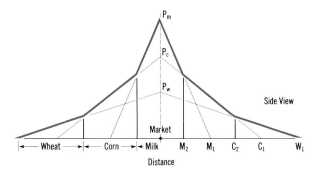

A. Bid Rent Surfaces for a Three Crop System

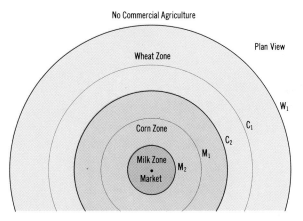

B. Agricultural Land Use Rings

**Figure 10-12 Three-crop system** (*A*) Bid rent surface for a three-crop system. (*B*) Agricultural land use rings. Adding additional crops to the system will cause market boundaries to adjust according to which crop has the highest rent-paying ability. Individual and independent farmers with profit motives for farming will come to the same crop decisions, depending upon their relative distance from the market, and land use rings centered on the market will form.

may be bid up and up until any advantage given to it by its centrality may all but disappear. If land becomes high-priced, it will then pay the farmer to shift more and more of this total investment from actual land to other factors of production. (By *factors of production* economists mean the three basic elements which in various combinations make farming possible: land, labor, and capital.) He must increase his yields, and to do so he must invest in more and more fertilizer, greater care by men and ma-

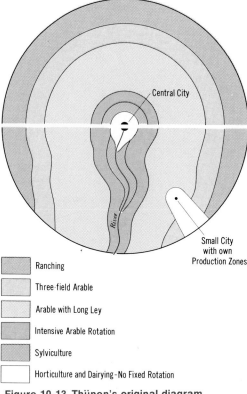

Central City

River

Small City
with own
Production Zones

Ranching

Three-field Arable

Arable with Long Ley

Intensive Arable Rotation

Sylviculture

Horticulture and Dairying - No Fixed Rotation

**Figure 10-13 Thünen's original diagram showing land use rings** Half of the diagram shows the effects of lower transport costs along a navigable river and Thünen's ideas concerning the effects of a satellite city in the region. (Michael Chisholm, *Rural Settlement and Land Use*, Hutchinson, London, 1962, p. 29)

chines, better seeds of livestock, and dozens of other improvements. This will result in higher intensities of land use in areas nearest the market. Conversely, farms on the periphery of things will utilize more and more land with less and less investment per acre. At the same time, perishable goods will be limited in space to locations near the market unless some way can be found to make them more durable and cheaper to ship.

### The nation and its agriculture

Now let us look at the United States as a whole in order to see how von Thünen's principles apply to modern agriculture and land use. The

sequence of American land use with which this section began took us from urban landscapes, through fields of corn, to wheat ranches, and finally to open range where sheep and cattle grazed. A moment's reflection on this sequence of land uses will suggest Thünen's rings to us again. The central city with its market; high-priced but perishable corn; cheaper, more transportable wheat; and finally cattle grazing on acres and acres of unimproved range are similar to other sequences of land use already described. (We should be careful to note that the cattle referred to here are not the sleek, corn-fattened beasts kept in feedlots and destined for immediate butchering. These are, instead, yearling steers getting their first growth on cheap grass before being shipped cityward for final conditioning on special diets. Range cattle are called *stockers* and provide new stock for city-oriented feedlots, where they became *feeders* before becoming steaks and hamburgers.)

Given the concentration of industry and urban markets in the northeastern United States, we would expect that area to be the focal point for a Thünen model conceived on a national scale. Adjacent to this core area we should find perishable, high-value commodities raised for city use such as dairy products and fresh fruits and vegetables. Next we might expect high-yielding field crops, followed by less and less intensive uses of the land. Corn, wheat, and rangeland for grazing match our expectations there. Thus a completely theoretical United States would look like the map pattern in Figure 10-14. Another way of putting all this would be in terms of the rent paid by various land uses. Urban and industrial rents are unquestionably highest; beyond them we would expect land uses, whatever they might be, to provide less profit per area the farther we traveled from urban centers. Now let us look at the actual distribution of crops and rent earned by agricultural activities in the United States.

We should not expect reality to match the smooth rings, neat homogeneous areas, and abrupt transitions allowed us by the models we build. This becomes evident when we look at the maps in Figures 10-15 and 10-16. The first of these maps shows the average value per acre in dollars derived from all types of agricultural

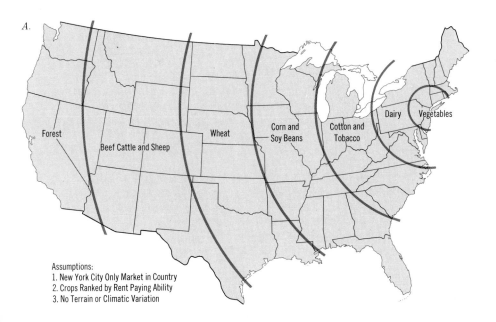

A.

Forest

Beef Cattle and Sheep

Wheat

Corn and Soy Beans

Cotton and Tobacco

Dairy

Vegetables

Assumptions:
1. New York City Only Market in Country
2. Crops Ranked by Rent Paying Ability
3. No Terrain or Climatic Variation

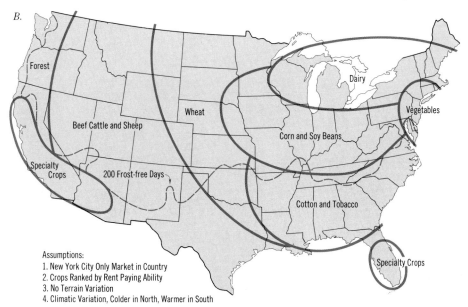

B.

Forest

Beef Cattle and Sheep

Wheat

Dairy

Corn and Soy Beans

Vegetables

Specialty Crops

200 Frost-free Days

Cotton and Tobacco

Specialty Crops

Assumptions:
1. New York City Only Market in Country
2. Crops Ranked by Rent Paying Ability
3. No Terrain Variation
4. Climatic Variation, Colder in North, Warmer in South

**Figure 10-14 Theoretical land use rings in the United States** Theoretical land use rings in part (*A*) become more realistic by recognizing North/South temperature variation as shown in part (*B*). Rings shift around because different crops respond differently to temperature variation. Pasture for dairying can grow in short, cool seasons; corn and soybeans require 150 frost-free days and a hot summer. Cotton cannot grow north of the 200-frost-free-day line. Specialty crops exist in regions with mild winters. Adopting other assumptions, such as accounting for soil and terrain differences, would result in a more complex and realistic pattern.

*Human
Organization
of the
Environment:
Adjustments to
Environmental
Uncertainty*

**239**

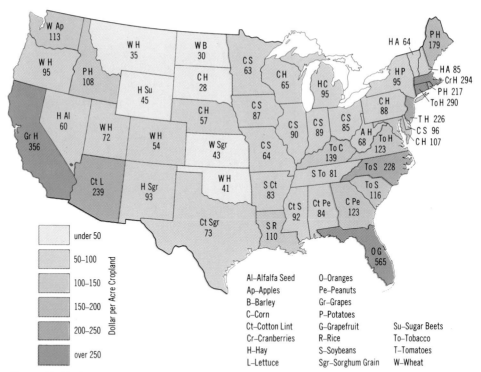

Figure 10-15 **Average value ($) per acre of cropland derived from all types of agriculture, excluding livestock and poultry 1970** (Compiled from *Statistical Abstract of the United States, 1972*, U.S. Dept. of Commerce Table 1011, "Principal Crops")

activity excluding livestock and poultry production. The two letters inside each state indicate the two leading money-earning crops in order of their importance. The pattern shown here is interesting both for the way in which it meets our theoretical expectations and also for the several exceptions for which Thünen's theory of land rent does not prepare us. As we expected, the highest returns per acre of agricultural land come from the part of Megalopolis between Boston and New York City. New York State itself has lower values, since the shape of the state means that most of its area is farther removed from its major city than areas of New Jersey, Connecticut, Rhode Island, and Massachusetts. Farm values fall off regularly to the west away from the urbanized Atlantic seaboard. In West Virginia and Pennsylvania the

effect of the Appalachian Mountains and Appalachia can be seen in lower values that pick up again as Illinois with the major urban focus of Chicago is approached. Going south along the Atlantic coast from New York City we encounter a ridge of high rents which runs inland from the Carolinas to Kentucky. This matches the southern Piedmont and is easily explained by the presence of tobacco. Another high point over Louisiana comes as more of a surprise, although rice and sugar cane account for this rise in the topography of rents. Viewed as a whole, the United States shows a steady, if slightly irregular, decline in rents westward across the Great Plains to the Rocky Mountains.

Idaho, Washington, Arizona, and California all seem much too high compared with expected

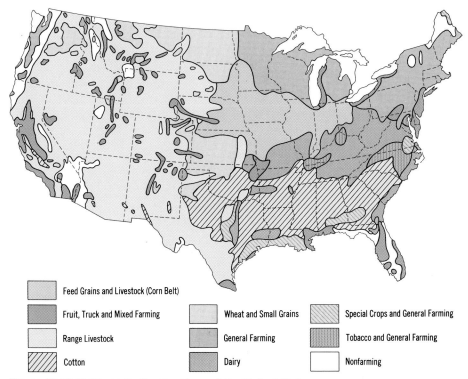

Feed Grains and Livestock (Corn Belt)

Fruit, Truck and Mixed Farming

Range Livestock

Cotton

Wheat and Small Grains

General Farming

Dairy

Special Crops and General Farming

Tobacco and General Farming

Nonfarming

Figure 10-16 Major agricultural regions of the United States

values. Irrigation plays a significant part in raising land rents. As part of national policy, major inputs of capital have reclaimed large tracts of dry land in our Western states. In the Southwest mild winters combined with irrigation water help to produce bumper crops of cotton, fruit, nuts, and vegetables. These conditions are enhanced by modern transportation developments with refrigerated freight trains and diesel trucks which carry fresh produce to the Northeastern states. The high value in Florida also represents areas of special crops such as citrus which are possible in the subtropical climate of the area. Wheat from the rich hills of the Palouse country in eastern Washington and Idaho potatoes raise returns in the Northwest. At the same time, the mild winters and green, rain-drenched pastures of western Oregon and Washington create conditions favorable for dairy farming although the immediate

market for fresh milk is very small by national standards. This problem is solved by turning fresh milk into condensed milk, cheese, and butter for shipment to the East and Southern California.

If we regionalize production by crop type for the country as a whole (Figure 10-16) and compare this map with the idealized one in Figure 10-14, we see that the general pattern is the same. However, dairy products dominate across parts of New England, northern New York State, and the upper Midwest. This is because these areas have poor soils and are north of the 90-day frost-free growing season. As a result, they are better suited for the production of hay for fodder. But even if we consider only the price structure for milk, we find our general expectations for the United States are fulfilled. Wholesale buying prices for fresh milk diminish steadily from New York State

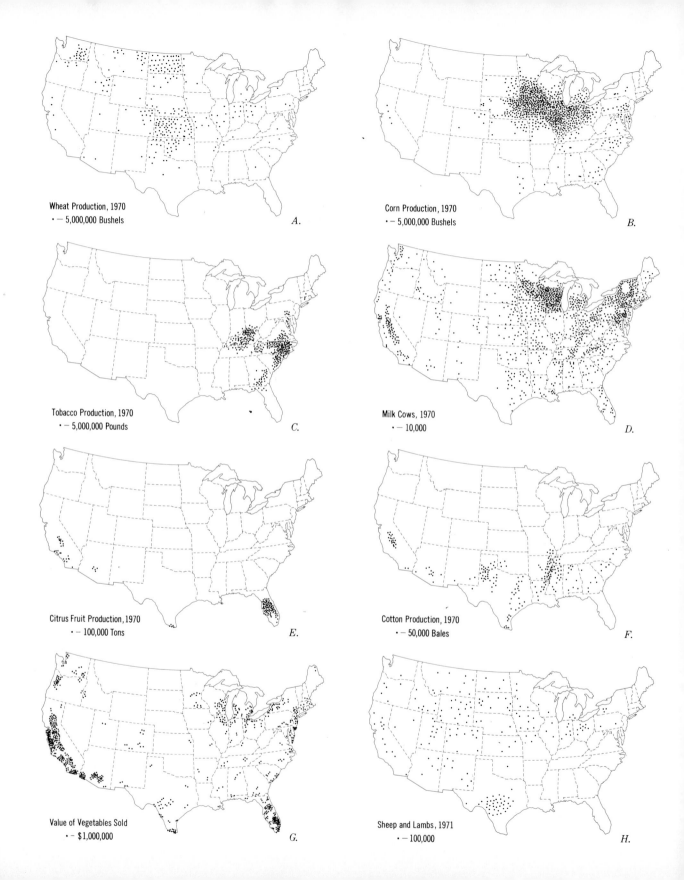

Wheat Production, 1970
• — 5,000,000 Bushels

*A.*

Corn Production, 1970
• — 5,000,000 Bushels

*B.*

Tobacco Production, 1970
• — 5,000,000 Pounds

*C.*

Milk Cows, 1970
• — 10,000

*D.*

Citrus Fruit Production, 1970
• — 100,000 Tons

*E.*

Cotton Production, 1970
• — 50,000 Bales

*F.*

Value of Vegetables Sold
• — $1,000,000

*G.*

Sheep and Lambs, 1971
• — 100,000

*H.*

west to Minnesota in the Northeastern dairy region. Since these prices do not include transportation costs, dealers can offer less for milk the farther they are from the central market area. In the South, where high temperatures make caring for dairy herds and producing fresh milk difficult, demand exceeds supply and prices are again higher.

Returning to our idealized map, we find exceptions where high-yielding cotton, tobacco, and peanuts tend to force out corn. The same is true for the specialty crops (Figure 10-14) which provide alternate choices for farmers distant from the major markets of the Northeast.

Perhaps the best way to summarize these several descriptions is to turn to the maps in Figure 10-17 which show the actual distribution of production for different categories of dairy products, corn, wheat, livestock, and specialty crops. The myriad dots scattered like confetti across these maps could be confusing, but we hope that by now the basic processes underlying their distribution have become clear. Naturally, one theory and one approach cannot fully explain anything as complex as agricultural production in a country as large as the United States, but a geographic point of view can help. We also hope that the next time you drive across the country, your trip will be enriched when you see and understand the sequence of crop production and other activities.

*The United States, Europe, and the world*

If the ideas expressed by von Thünen work for the United States, will they also help explain world agricultural production? Space does not allow us to investigate European agriculture in detail, but it can be said that the high level of industrial activity associated with Western European countries indicates a parallel intense level of agricultural activity. Figure 10-18 puts both Europe and the United States in global perspective. Here we see the average caloric yield from

both small grains (wheat, corn, barley, rye, etc.) and potatoes. Pounds and bushels per acre are converted to a more universal caloric value in order to overcome the problem of comparing unlike crops. The results of our analysis confirm what we have already anticipated. The bulls-eye pattern of intensive agricultural production centers now on the North Atlantic. Some details have been lost—for example, in the Soviet Union and America—but the general pattern of distribution focuses upon the *have* nations of the world. Outliers do exist in Argentina, South Africa, Australia, and Japan, but those places are all industrial and commercial centers which have grown up at the opposite end of the major sea routes which supply Europe.

In addition to the rings of production focused upon northwestern Europe, it is useful to note the natural limits of agriculture imposed by a short growing season in high latitudes, particularly in the Northern Hemisphere, and elsewhere by extreme aridity.

## Strategies for Adjustment to Environmental Uncertainty

The preceding discussion of the Tehuacan Indians and of modern agricultural land use on both a national and an international scale emphasizes the increasingly complex use of space by society. Embedded in humankind's major solution of environmental and production problems through spatial organization are a number of adjustments to the natural environment needing further explanation. We know by now that natural systems cannot be ignored. In fact, human beings have worked out a number of strategies to counter the capriciousness and unpredictability of nature. Let us present a systematic review of some of these devices.

To do this we will refer to the idea of *operators,* that is routines, techniques, or devices which act to maintain or change conditions within a given system. Let us examine some

**Figure 10-17 Distribution of selected farm products in the United States** (Compiled from *Statistical Abstract of the United States, 1971*, Dept. of Commerce, Wash., D.C., Tables 968, 970, 977, and 989; *Annual Crop Summary*, Crop Reporting Board, SRS, Dept. of Agriculture, Wash., D.C., January, 1972)

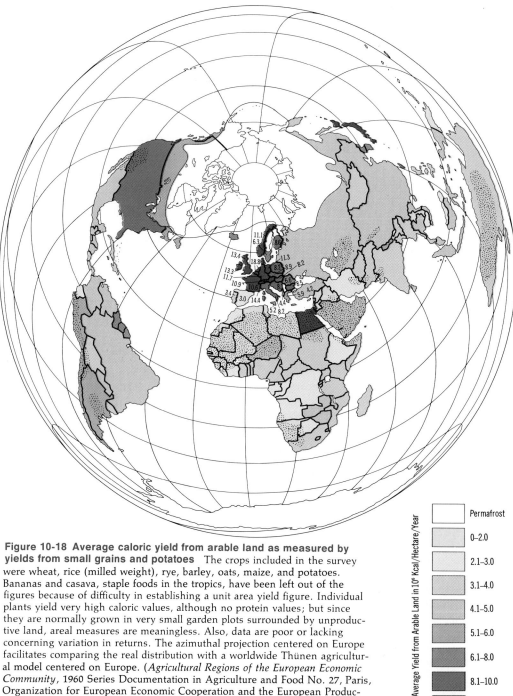

**Figure 10-18 Average caloric yield from arable land as measured by
yields from small grains and potatoes** The crops included in the survey
were wheat, rice (milled weight), rye, barley, oats, maize, and potatoes.
Bananas and casava, staple foods in the tropics, have been left out of the
figures because of difficulty in establishing a unit area yield figure. Individual
plants yield very high caloric values, although no protein values; but since
they are normally grown in very small garden plots surrounded by unproduc-
tive land, areal measures are meaningless. Also, data are poor or lacking
concerning variation in returns. The azimuthal projection centered on Europe
facilitates comparing the real distribution with a worldwide Thünen agricultural model centered on Europe. (*Agricultural Regions of the European Economic
Community*, 1960 Series Documentation in Agriculture and Food No. 27, Paris,
Organization for European Economic Cooperation and the European Productivity Agency; *Annual Abstract of Statistics*, United Kingdom, 1970, London,
Her Majesty's Stationary Office, Central Statistical Office; Tarimsal Yapi ve
Uretim (*Agricultural Structure and Production*), Ankara Turkish State Institute
of Statistics Publication No. 564, 1968; and materials from the Foreign Market
Information Division, Foreign Agricultural Service, USDA; *Production Yearbook 1968*, FAO of the UN, Statistics Division, 1969; *Statistisk Arbog 1970*, vol.
74, Copenhagen; *Statistisk Arsbok 1970*, vol. 57, Stockholm)

Average Yield from Arable Land in 10⁶ Kcal/Hectare/Year

| | |
|---|---|
| | Permafrost |
| | 0–2.0 |
| | 2.1–3.0 |
| | 3.1–4.0 |
| | 4.1–5.0 |
| | 5.1–6.0 |
| | 6.1–8.0 |
| | 8.1–10.0 |
| | 10.1 & over |

Desert Climate

scale operators, time operators, and site operators.

## Scale operators

As with any scale-oriented problem, we must choose the size of the area within which we plan to work. In this case, it is not convenient to divide up the continuum of reality into a great many levels or scales. Instead we can talk of local and global environmental hazards and of the *local* and *global operators* by means of which farmers and others can survive them.

**Local operators** Local hazards are those which occur either at points or along lines throughout the environment. The path of a tornado is linear at a regional scale. Actually its swath of destruction can be hundreds of feet wide, but this strip of area reduces to a line in terms of an entire region. This nonareal aspect also would be true of a well which goes dry or of house fires, explosions, and acts of violence, to name a few point-type hazards. In modern cities individual robberies and car crashes can be considered local events.

There are two strategies by which local hazards can be overcome. The first is the *sharing of risk* by a group, the larger the better. Thus, most tribal groups maintain elaborate rituals for dividing up and sharing throughout the community the food found or killed by every single person. In this way, a hunter with a run of bad luck can still count on the food provided by others until his luck changes and he, in turn, must share his kill. It is likely that the early bands of Paleo-Indians in the New World had such arrangements to ensure group survival. In modern society, insurance, family ties, and community obligations help to offset local disasters. Insurance is a good case in point. Many people buy fire and theft insurance with the expectation that only a relative few of the purchasers will need to use it. By paying a small amount we are thus insured in the event of a much larger loss. It is also possible to buy tornado insurance, since when a tornado strikes, only relatively few people are killed or ruined along its path, while the population over

a much wider area who feel threatened by the possibility of experiencing a tornado help through their insurance payments to offset the damage. On the other hand, hurricane insurance is not available. This is because hurricanes cover broad areas and affect almost everyone in a community. In that case everyone needs help, not just a few, as with tornadoes.

The second strategy used against local hazards is *spreading the risk.* In this case, a subsistence-level farmer may own several fields scattered across a wide area. Local occurrences of wind, hail, or insect damage may wipe out part of his harvest, but with luck most of his crops will go unscathed. Again, we do not know enough about Tehuacan society to speak with confidence about their pattern of field distribution, but if other modern village systems are any indication, there, as elsewhere, fields were scattered for greater security. In another example, the Amish people of North America do not buy tornado or fire insurance on sale to the general public. This is part of their belief in removing themselves from the world at large. However, Amish communities all over the United States will come to the aid of each other in the event of disasters. This private tithing and the donation of food, supplies, money, and labor by widely scattered people ensures the well-being of the group as a whole.

**Global operators** Hazards that cover entire regions present a different kind of problem for the people involved. It is generally impossible to avoid a drought by having two fields a short distance apart, particularly if rainfall fails across an area as large as southern India or the American Midwest. Another way to imagine global hazards is that they extend beyond the range or effective limits of the social institutions of the system involved. The strategy here is to *discriminate* among minor variations in the macroenvironment. For example, there are areas of higher winter temperatures and milder summers on the lee side of lakes. This results from the greater insulating quality of water and the downwind drift of the air from above the water body. Freezing temperatures are less likely to occur under these conditions, and it is common

*Human Organization of the Environment: Adjustments to Environmental Uncertainty*

**245**

A hurricane viewed from space. Hurricanes may contain destructive high
winds 50 to 100 miles distant from their centers and, when they move onshore, may damage an area extending over several states. Strategies useful
for counteracting local hazards are not effective in such circumstances. In the
United States one response is to declare a national disaster area. In this way
restorative energies of the entire nation may be brought to bear on the
problem. The social/political institution is enlarged to match the scale of the
natural phenomena. International relief agencies may offer similar aid to
smaller nations in which the storm may be as big as the whole country.
(Photograph by NASA from Apollo 7)

to find localized areas specializing in apples, berries, flowers, and other frost-sensitive crops in these situations. The eastern shore of Lake Michigan is a good case in point. Another famous example is the vineyards of France. French grape growers recognize tiny areas with exactly the right conditions suitable for the best grapes which will produce wines near perfection. Such areas are known as *climat* and occur as a precious string of fields along the hill flanks of Burgundy, the Rhone Valley, and in parts of the Bordeaux region (Figure 10-19). The hazard there takes many forms: unsuitable soils, improper soil moisture, too much or too little sun, and dozens of other things. The strategy is to find the best spot and to lavish care and protection upon it. Thus, to return to the Tehuacan example, by the time sedentary agriculture was well established, trial and error had shown the farmers which fields within the generally dry valley retained soil moisture longest. They had also learned to concentrate their efforts upon a relatively small group of reliable and durable food plants selected from a much larger number of wild species. In this latter, nondimensional sense domestication became a special kind of discrimination and selection which ensured the food supply of those people. In other words, the system involved deviation-amplifying processes.

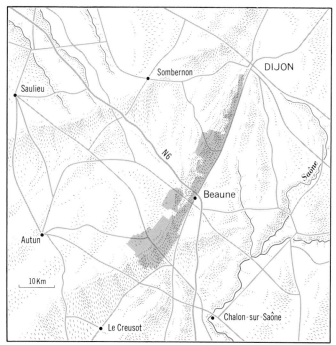

**Figure 10-19 Vineyards of Burgundy** The best wines of France are produced from grapes grown on small fields possessing a combination of soil, climate, slope, and other conditions which have proved optimum for the purpose. Such areas are known as *climat*. In Burgundy, for example, they occur in narrow bands along hill flanks called *cotes*.

## Time operators

Just as space can be organized to ensure group welfare, so can time-dependent processes be manipulated for the commonweal. *Storage* of all kinds clearly illustrates this concept. We have already spoken of *food processing* and storage and their role in preserving nutrients and energy for periods of low availability in the annual cycle. In a similar vein, we can consider preserving, storing, and transporting foods as one of the linchpins holding urban society together in city locations. But other kinds of storage are equally important. The damming of streams and ponding of water is really nothing more than holding back the flow for use later in time. Certainly, this technique ensured the farmers of Tehuacan larger harvests and more predictable results.

Another technique useful in temporal events is the seasonal or cyclic adjustment of activities. For example, the farmers of the Flint Hills in Kansas sell their herds before the late summer drought comes on. Wheat farmers in the drier parts of the West will "roll with the punch" in dry years and simply write off their crop as a loss early in the season rather than pour good money after bad by attempting to harvest a submarginal crop. The early hunters and gatherers of Mexico also matched the intensity of their activities to the resources available in their environment when their macrobands broke up into smaller units during the dry season. Still another time adjustment that can be made is to compress the time during which certain activities take place. This requires the concentrated effort of everyone involved, but it often pays off in better results. It is a common sight to see harvest work continuing day and night in order to avoid the onslaught of autumn rains. This can be the case whether the farm is a modern one and the work is done under floodlights or a traditional one where only the flickering light of lanterns helps the laborers. During other seasons other kinds of insurance also exist. If the coming of the rainy season is uncertain, the farmers cannot tell if the first or second rain may also be the last. To counter this, a few seeds may be planted with each rainfall, although some may not survive because of premature germination before adequate soil moisture builds up to see the crop through. On the other hand, the first rains may be the most important, and if they are, some crop is assured.

An example of a way in which the same field and the same time period can serve several purposes takes place during the transition from a subsistence- to market-oriented farm system. Typical of this are the modernizing farmers of southern Turkey who at the present time are changing from an age-old pattern of goat herds and grain fields meant for local consumption to citrus orchards supplying major Turkish and European cities. It takes an orange tree four to six years to grow large enough to produce a profitable crop. If the ordinary farmer had to wait that long for an income, he would starve. But citrus seedlings are interplanted between rows of maize or wheat, and for the first few years both species grow side by side until just as the young trees are large enough to begin shading out the grain crop, the first oranges can be picked and shipped to market.

### Site operators

Some survival strategies involve not so much scale differences or the timing of events as they do taking advantage of the special *in-place* characteristics of a given site or creating special conditions at certain spots. For example, in some farm systems *intercropping* is practiced, where different species of crops are grown together. A famous example of this originated in Mexico. There, maize and beans were grown in the same plots. The beans helped fix nitrogen

in the soil for the corn, while the corn provided poles, as it were, for the beans to climb upon. But the mutual effectiveness of the two plants did not end with that. *Zein*, the principal protein in corn, in order to form a complete protein suitable for human needs, requires the amino acid *lysine*, which is supplied by beans. Thus, the combination serves not only plants but man as well. It is unlikely that this combination is the result of coincidence. Thousands of years passed during the domestication of these plants, and that time span provided ample opportunity for many plants to be tried and rejected before an adequate dietary combination was found. In any event, this is typical of strategies for human survival where the key is manipulating the elements at the site of operations.

Another reason for mixing the number of species in a garden is the disease control offered by isolating individual plants from others of the same kind by scattering them among unrelated species. The spread of disease through homogeneous stands of vegetation is well known to those American communities which have lost all their shade trees because of Dutch elm disease. The American chestnut was another victim of disease which wiped out the homogeneous groves formed by that species.

Perhaps the most effective site strategy employed by traditional farmers is to seek out multiple environments on steep gradients. On a local scale this can simply mean taking advantage of differences in soil moisture and soil types from the top to the bottom of a hill. But more important are those farmers and herders who utilize entire mountain slopes in their activities. In southern Turkey nomads traditionally wintered in the Mediterranean lowlands near the sea. Then, as summer came on with high temperatures and endemic malaria, they packed their tents and moved in easy stages to high summer pastures with plentiful grass, pure water, and cool temperatures which discouraged mosquitoes. In the same area, semi-nomadic dirt farmers still occupy villages on the coastal lowland as well as others high in the mountains (Figure 10-20). Winter is spent down below in relative warmth where early spring crops can be planted or where winter wheat grows throughout the rainy season. Again, as summer conditions make the lower elevations inhospitably hot, entire villages pack up and retreat to fields at heights where lower temperatures prevail. Sometimes even more than one supplemental village is occupied for short periods en route. The overall system has many advantages: summer diseases are avoided; two or three planting and harvest periods can be spaced out instead of one, thus making more labor available; and if crops fail at one elevation, others lower or higher on the mountain can stave off hunger. The similar use of multiple sites in the Tehuacan Valley is an obvious feature of that place.

Counter to the above examples are the homogeneous sites sought by modern farmers using large machines which cannot easily compensate for local variations in vegetation, topography, soil type, and other environmental features. In this case fields with only one kind of crop must be planted. To protect these homogeneous plant communities from epidemics of disease and insects, poisons and medicines must be applied in profusion. High yields are the result, but once a disease does break loose, often the only solution is to abandon one site and to move the entire operation to another location. The spread of Panama disease and the abandonment of the banana plantations of the Caribbean have already been mentioned.

## Other Strategies: Magic and The Farmer's Almanac

Hunting and gathering, farming and herding, the day-to-day business of all our lives present situations in which it is impossible to decide the exact, optimum time to do something or the best possible location for an activity among many very similar choices. Sometimes seeking the absolute optimum time or place is not practical. In other words, the best strategies people can devise may still leave them with unresolved conflicts and decisions. It is at this point that magic, astrology, *The Farmer's Almanac,* and other such devices become useful extensions of rational thought. The seasons do

not wait for any man, and indecision may be disastrous. Action, any action, is usually better than none at all. Let us illustrate this with two widely separated examples.

The Naskapi Indians of northern Labrador occupy one of the most difficult environments of North America. They are hunters and gatherers who live in camps which are moved from time to time as the availability of game dictates. Like all other hunters, they possess a large store of knowledge about the animals they seek and about the local habitat and act upon such information in a rational manner. But when game becomes scarce and the vast and bitter landscape offers no clues to the whereabouts of animals, they must nevertheless continue their endless search for food. Now if the hunters always searched in areas where they had most recently been successful, they would soon "overhunt" the game and sensitize it to the point where the animals would hide or run away. It is, therefore, useful for the Naskapi to introduce an element of randomness into their search patterns.

Psychological experiments indicate that it is nearly impossible for humans to act in a purposefully random manner without the aid of some randomizing device. The Naskapi are able to introduce randomness into their hunting patterns by the use of magic. They take the shoulder bone of a caribou and hold it in a forked stick over a low fire. Cracks, checks, and charred spots appear on the bone in a random fashion. The bone is then oriented to the landscape according to another ritual, and the pattern on it is read for the whereabouts of game much as we would read a map. Since the Indians change camps frequently, the randomness of the orientation is further assured. The end result is that the search trips of the Naskapi hunters are varied enough to cover the entire territory and at the same time lack repetition which might drive away needed game. Now let us consider a modern example of the successful use of nonrational or magical strategies in agriculture.

Tomato farmers in the modern southern United States face an interesting problem each year. They can take advantage of warmer annu-

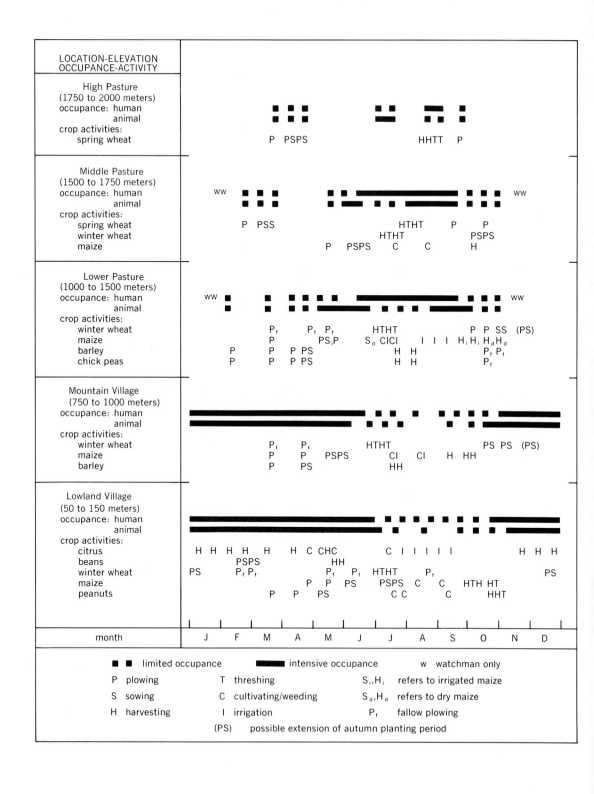

al temperatures and easily beat Northern farmers to the market with their produce. The earlier their tomatoes are harvested and shipped, the higher the price they will command in Northern cities. Meanwhile, farmers outside the South are also trying to get their plants into the ground and their produce to market as early as possible. Transportation costs on these perishable vegetables are such that when summer blankets the entire country, locally grown Northern tomatoes force Southern tomatoes out of the Northern markets. Thus, tomato farmers must decide upon the earliest possible date that they can plant their tomato seedlings and still avoid possible killing frosts. This is true for Southern as well as Northern farmers, although the chances of late frosts are considerably less in the South. Tomato growers use three means of predicting the best time to plant. They rely upon their own knowledge of their local environments accumulated from years of experience. This serves them well but leaves them uncertain about global events such as major cold waves originating outside their immediate ken. To anticipate these potential disasters they turn to National Weather Service reports and long-range weather forecasting. But the science of forecasting the weather is still in its infancy, besides which the scale upon which those predictions are made is so gross that the Weather Service's prognostications often do not serve the needs of a particular farmer using a field with a specific microclimate.

At this point the farmer may hit a plateaulike area of knowledge. He can tell that spring is well on its way; he "feels" that the time to plant has come; but no additional amount of rational effort can tell him if the odds are right for him to gamble. Just exactly when is the best day, late enough in the season to be reasonably safe, early enough to beat everyone else to market? What a quandary! It is at this point that many a farmer, frustrated by indecision, turns to that age-old friend *The Farmer's Almanac*. Many different editions of these books exist, and while they contain useful information based on fact, they also give advice on the exact days to plant and other things, all of which, quite frankly, are the creation of the editors' lively imaginations. The advice is not wildly foolish but no better and no worse than any knowledgeable farmer's. However, the air of authority surrounding the *Almanac* is such that the advice is followed when other, more rational, sources of information are inadequate. Action is taken; the crop is planted; most often with a little luck the farmer gains some degree of success and the *Almanac* is vindicated. While the *Almanac* may seem a trick to trap the gullible, actually its function is a good one in a magical sense. It has helped the farmer move off dead center and given him that little additional bit of confidence that everyone needs when the going gets rough. In this fashion, some of the old folkways of doing things can be effective even in modern times.

Figure 10-20 **The seasonal cycle of agricultural activities and occupance patterns as practiced in two seminomadic villages** Traditional societies use complex strategies in dealing with the environment. On the Mediterranean coast of Turkey, seminomadic peasant farmers use seasonal variations in conditions on the mountain slopes to optimize their return on labor. (J. F. Kolars, *Tradition, Season, and Change in a Turkish Village*, Research Paper No. 82, Dept. of Geography, University of Chicago, Chicago, Ill., 1963, fig. 14)

The farmers with whom we ended the last chapter are like football coaches. They are very often pessimistic and feel that nature will "do its worst." But even the phrase "do its worst" implies a kind of consciousness, a humanizing of the environment into something with which we can deal as an individual. How do such attitudes affect our view of resources? We have already commented on the role played by perception in human affairs. For example, the transformation of the "Great American Desert" of the mid-nineteenth century into the "Great American Breadbasket" of the twentieth was a matter of how emigrant farmers interpreted the treeless steppelands of the Great Plains.

Even more complex and subtle than those farmers' views of the environment are the philosophies which provide a basis for all our interpretations of nature. This chapter deals with some of the attempts to explain individual and group behavior with respect to the natural enviornment. We also look at the role played by perception and attitude in the management of resources. In other words, we view natural resources as cultural appraisals.

## Approaches to the Study of Man and Nature

In considering the various interactions between man and his natural environment, we are really referring to two types of studies. The first treats *homo sapiens* as a creature living in a physical environment which satisfies his biological needs and imposes physical limitations on his health and well-being. This type of study tells us such things as how many quarts of water are needed per person per day at certain temperatures, or the manner in which high elevations affect the human body and mind. All such investigations can be grouped under the heading *human ecology*, that is, the biological considerations of man-evironment systems. These studies consider human reactions in the same manner as they treat the reactions of all other animals found in ecosystems. However, in considering human use of the earth it becomes difficult to distinguish the purely physical environment from the social one. Therefore, more complex, but to most geographers more interesting, is the study of man-environment systems from the viewpoint of *cultural ecology*. This considers human ecosystems in which man is no longer a simple animal, but rather a carrier of culture with its accumulation of tools, techniques, beliefs, and prejudices.

Our view of the man-environment relationship is similar to that of Harold and Margaret Sprout:

*So far as we can determine, environmental factors (both non-human and social) can effect human activities in only two ways. Such factors can be perceived, reacted to, and taken into account by the human individual or individuals under consideration. In this way, and in this way only, environmental factors can be said to "influence," or to "condition," or otherwise to "affect" human values and preferences, moods and attitudes, choices and decisions. On the other hand, the relation of environmental factors to performance and accomplishment (that is, to the operational outcomes or results of decisions and undertakings) may present an additional dimension. In the latter context, environmental factors may be conceived as a sort of matrix, or encompassing channel, metaphorically speaking, which limits the execution of undertakings. Such limitations on performance, accomplishment, outcome, or operational result may not—often do not—derive from or depend upon the environed individual's perception or other psychological behavior. In many instances, environmental limitations on outcome or performance may be effective even though the limiting factors were not perceived and reacted to in the process of reaching a decision and initiating a course of action.[1]*

In other words, the way we feel, whether we are optimistic or pessimistic, energetic and industrious or lazy and slothful, deceitful or honest, *does not derive* from the type of natural environment in which we live. On the other hand, our crops may fail, our bodies may deteriorate, we may starve or grow fat according to our ability to cope with conditions in the natural environment which surrounds us. Furthermore, we need not always know the forces and events coming our way in order subsequently to be affected by them.

### Nonenvironmental deterministic theories of human behavior

In all attempts to explain human behavior one or more modes of thought are always apparent. The first can be described as *deterministic*. This means "simply that all empirical phenomena of

[1]Harold Sprout and Margaret Sprout, *The Ecological Perspective on Human Affairs with Special Reference to International Politics*, Princeton University Press, Princeton, N.J., 1965, p. 11.

the system under consideration (be it mechanical, biological, or social system) can be predicted by reference to some set of causal laws." Let us first consider some common patterns of nonenvironmental deterministic thought which attempt to explain human behavior. Among these, racial determinism is most prevalent. Thus the idea states that heredity determines all major modes of behavior and that some groups of people sharing a common genetic inheritance will be either superior or inferior to other groups with different genetic legacies. Contrary to uninformed belief there is no evidence that the genetic inheritance of a group of people determines the average intelligence of the group.

We further agree with the recent statement of the American Association of Anthropologists adopted at their 1971 annual meeting that "there is no scientific warrant for ascribing to genetic factors the oppressed conditions of classes and ethnic groups."

Another nonenvironmental type of determinism centers upon the concept of *economic man* and *economic determinism*. This concept views all human activities in terms of profit maximization. No longer taken seriously in its purest form, economic determinism assumes that each person has perfect knowledge and perfect reasoning ability and that everyone acts to earn the highest possible return on his investment of the money, energy, time, and materials available to him. The overly simple notion of economic man has generally been rejected, for while profit maximization seems logical, no one can claim to know everything essential in order to make a completely rational economic decision. Moreover, many people, if not the great majority, will pass up profit in order to attain other kinds of satisfaction. One brief example of this is mountain rescue teams, which are composed almost always of unpaid volunteers. Very few people would tolerate the danger and discomfort of mountain rescue missions if an hourly wage, no matter how much, were attached to such work. Behavior of this kind allows us to substitute the concept of *satisficing man* for economic man. In doing so, however, we introduce so many psychological variables into the picture that simple cause-and-effect rela-

*Resources as Cultural Appraisals: Man's Changing World View*

tionships become obscured. It is reasonable to say that humans always perceive a limited range of economic possibilities and will often make their selections or choices of activity so as to satisfy their total needs rather than simply to maximize their incomes.

### Environmental determinism

Environmental determinism has always been, and remains, a popular explanation of human behavior. We have already given a general definition of deterministic thought. In the present case, the basic idea is that human behavior and human attitudes can be predicted by referring to the characteristics of the natural environment in which the participants are living. Examples of environmental determinism are available from all periods of history.

About 420 B.C. Hippocrates wrote, "The inhabitants of the colder countries are brave but deficient in thought and technical skill and as a consequence of this they remain free longer than others but are wanting in political organization and unable to rule their neighbors. The peoples of Asia on the contrary are thoughtful and skillful but without spirit, whence their permanent condition is one of subjection and slavery." We scarcely need to point out that our Greek philosopher has bracketed his own land and people, the implication being that Greeks are skillful and thoughtful but also freedom loving and quite capable of ruling their neighbors.

Nearly 24 centuries later the French philosopher Soulavie commented, "The inhabitants of basaltic regions are difficult to govern, prone to insurrection, and irreligious." At about the same time, the pithy statement "basalt is conducive to piety" was published in Germany. These two statements suggest that a person's religious attitudes are in part the result of the rock underlying the place where he lives. The trouble is that the two authors, typical of those who held this idea, cannot agree on which rock generates which mood and end up contradicting each other.

Geography has had two eloquent spokesmen for environmental determinism. Ellen Churchill

Semple, writing at the turn of the century, echoed Hippocrates: "The northern peoples of Europe are energetic, provident, serious, thoughtful rather than emotional, cautious rather than impulsive. The southerners of the sub-tropical Mediterranean basin are easygoing, improvident except under pressing necessity, gay, emotional and imaginative." A moment's thought provides easy contradictions to the above rule. Swedish movie starlets and Old Testament prophets seem to have reversed their assigned roles. The most cautious and persuasive of the American determinist geographers was Ellsworth Huntington. His major interest was climatic change and the effect of climate on man. He proposed that "a certain type of climate prevails wherever civilization is high. In the past the same type seems to have prevailed wherever a great civilization arose." We feel it only fair to point out that when England with its marine west coast climate was inhabited by savage Picts, Rome with its Mediterranean climate was the center of world knowledge. Now Northwest Europe surpasses Italy in research and technology. The jungles of Southeast Asia harbor great ruins which represent civilized societies that grew up under rainy tropical conditions. On the other hand, high-order indigenous civilizations never developed in the jungles of the Amazon or in the marine west coast climate of the Pacific Northwest. With such poor correlations between climate and achievement we must look elsewhere for explanations of human behavior.

What all this stems from is a kind of nonscientific selection of examples to prove the author's point. You look at a man or a nation and decide what you think. Then you engage in what might be called *retrospective inevitabilism*, in which you set out to prove that the end result was predetermined by—what could be simpler?—the climate. We do not for a moment deny that high temperatures and humidities can encourage endemic disease in tropical areas, or that poor soils can lead to malnutrition and slowed human responses. Those conditions are typical of the real effects of the environment on humankind. But what is uncomfortably cold or dry for one human may be too warm or wet for another. In other words the environment offers an incomplete explanation of the causes of human achievement.

## Possibilism

The antideterministic arguments given above were worked out in great detail by the generation of geographers following Semple and Huntington. The French geographer Vidal de la Blache, along with Americans like Isaiah Bowman, argued that within a given environmental setting there are a number of choices that humans can make about their activities. The history of any spot in the United States will demonstrate this point. Nowhere has permanent climate change altered the environment in the last four centuries. And yet within that span, or much less in the western part of the country, Indians and settlers and urbanites; hunting, then farming, then industry; camps, villages, towns, and sprawling metropolitan areas have all occurred in rapid sequence. If only one location in space or time were considered, a strong deterministic argument could be made for the particular activity found there. But a step backward from the scene will show many other possible uses or interpretations of any given site. Thus, the message of possibilism is that the environment offers not one, but many, paths for human activities and development. The major caution that must accompany possibilistic thought, however, is the one given by the Sprouts: "Possibilism is not a frame of reference for explaining or predicting decisions. . . ."

## Probabilism

Two events are seldom exactly alike no matter how similar the conditions preceding them. The world is too various for exact copies of things or processes to be other than rarities in themselves. Instead, most of us expect the world to usually live up to our expectations but are not too surprised if the unusual occurs. In other words, we have notions of the *normal* behavior of people and the environments they occupy. These are based on our past experiences and our accumulated knowledge. If the unexpected happens—or for that matter, the less expected—it does so with much less frequency. From this idea we can construct theories of normative behavior. The events which fill our lives and the way in which people will react to different environments can be described in terms of the probability of their occurrence. If we have enough examples, and if the environment has not changed significantly since our preceding observations, we should be able to predict the frequency or expectations of something's happening again.

Much of our activity is based on normative models of human behavior. For example, most people will step off the curb into the path of an oncoming car as long as the traffic light is with them and against the automobile. In most cases the car will stop; in a few instances it will run the light; in a few others it might suddenly turn into a driveway or make a U-turn. The pedestrian assumes that the driver will behave in a normal manner, and he assigns a high probability to that possibility. The application of this idea to geography was apparently first introduced two decades ago by the British geographer O. H. K. Spate when he suggested that a probabilistic view of the world might resolve the argument between determinists and those who advocated free will in human affairs. In any event, predictions based on probabilities can be no more accurate than our knowledge of past events.

## Cognitive behavioralism

Our frequent references to the *views* people have of things imply a more dominant role for human thought than deterministic theory allows. We are in great part the product of our education and experience. Twin children, separated at bith and raised in radically different cultures, may share many physical attributes, but their views of the world will depend primarily upon their experiences while growing up and what their foster parents and peers believed. Humans almost always act in terms of what their own cultures value most, and see only those things to which they have been sensitized. For example, coal was only a black rock in ancient times before man learned to burn it as fuel. Similarly, some of the richest deposits of radioactive minerals have been found in recent years by consulting old geological survey reports which simply listed their occurrence as curiosities in the days before atomic technology.

The concept of cognitive behavioralism goes far to explain human actions as long as we are familiar with the minds of those whom we are considering. However, there seems to be a randomness that intrudes into real world events and skews our expectations. Disruption by population growth, migration to new environments, the depletion of finite natural resources, the discovery of new ones like atomic energy, and events like war all twist our lives in unexpected ways which exceed our ability to behave according to our expectations and conditioning. Thus human behavior and attitudes cannot be explained in simplistic terms.

### Who Owns the Earth?

If we accept the idea that much of our behavior depends upon what we have been conditioned or taught to recognize and value, we come face to face with the current debate on religion and ecology. This issue was brought into the spotlight by Lynn White, Jr., in his discussion of "the historical roots of our ecologic crisis." White's argument is that much of the attitude of Western man (we recognize our use of the nominal term) toward resource use and exploitation stems from Christian theology. He reasons that the ecologic crisis that looms before us results from combining the traditionally speculative and aristrocratic sciences with more pragmatic and action-oriented technology. The result is a peculiarly "Western" approach to the use of the earth. Moreover, this revolution has swept the world and has so speeded up the pace at which fuels and other resources are being used that it actually threatens the well-being of the next, if not this, generation. White argues further that the roots of such attitudes extend back into history well beyond the Industrial and Scientific Revolutions of the seventeenth and eighteenth centuries. He says:

*What people do about their ecology depends on what they think about themselves in relation to things around them. Human ecology [we would say cultural ecology] is deeply conditioned by beliefs about our nature and destiny—that is, by religion. . . .*

*The victory of Christianity over paganism was the greatest psychic revolution in the history of our culture. It has become fashionable today to say that, for better or worse, we live in "the post-Christian age." Certainly the forms of our thinking and language have largely ceased to be Christian, but to my eye the substance often remains amazingly akin to that of the past. Our daily habits of action, for example, are dominated by an implicit faith in perpetual progress which was unknown either to Greco-Roman antiquity or to the Orient. It is rooted in, and is indefensible apart from, Judeo-Christian teleology. . . .[2]*

According to White, in pre-Christian times every item in nature, be it tree, lake, or the landscape itself, was inhabited by spirits which had to be placated and appeased if those things were to be used by man. When Christianity overcame the old gods, man gained the right to use all of nature without thought of the consequences of his actions. In this way he assumed mastery over nature. Thus, our views of natural resources, based as they are on Christian beliefs, are indifferent to any appeal but that of immediate and maximum use.

This argument is countered by that of the geographer Yi-fu Tuan. He reminds us that although early and medieval Christians may have felt that the earth was theirs to use in whatever manner they chose, lack of technology prevented them from seriously disturbing the ecologic balance of man and nature. At the same time Tuan points out that serious destruction of environmental resources had already occurred in China. He quotes songs, instructions to officials, and general written warnings by Chinese living well before the Christian era that the landscape must be saved by careful husbandry or the people would suffer. Deforestation, the manipulation of bodies of water, the

[2]Lynn White, Jr., "The Historical Roots of Our Ecologic Crisis," *Science*, vol. 155, no. 3767, Mar. 10, 1967, p. 1205.

creation of entire landscapes for aesthetic purposes were all indications of the Orient's ability to change and destroy the natural order of
the world. It is almost a truism among
geographers—particularly in grade school
courses—to talk about the destruction of the
Chinese landscape through deforestation and
erosion.

The point of these arguments, it seems to us,
is that it is relatively unimportant what beliefs a
people or society may have if, whatever those
beliefs, the end result is the same. In this case,
the issue of Christian versus Oriental philosophies and their subsequent effect on the landscape is a moot one. Instead, the practical
problems are those of population pressure,
technological developments, and a modern
life-style that continues to use up all kinds of
resources at an increasing rate.

## Resources as cultural appraisals

Resources of all kinds are defined by a complex
set of conditions and not by a few lines written
in a dictionary. All but one of the elements in
this complex definition depend upon the
learned behavior of humankind. As we have
already said, coal is simply a black rock, a chunk
of neutral stuff, until man has learned to burn it
and put its energy to use. Herein lies the key to
the definition of resources given by Erich W.
Zimmermann. "The word 'resource' *does not
refer to a thing or a substance but to a function
which a thing or a substance may perform or to
an operation in which it may take part*, namely,
the function or operation of attaining a given
end such as satisfying a want. In other words,
the word 'resource' is an abstraction reflecting
human appraisal and relating to a function or
operation."[3]

When we refer to resource *use*, at least four
subsets of conditions have to be met. These
relate to (1) physical presence and human
awareness, (2) technological availability, (3)
economic feasibility and managerial skills, and
(4) individual and social acceptability. First and

obviously, the material must be physically available and brought to the attention of those
people who would use it. This may be more
complicated than it appears, for nations like
Japan with few natural endowments of their
own are able to reach out, nevertheless, and
bring raw materials from around the world to
their factories. In the same manner, the utility
of a substance must be known. That radioactive
deposits were only curiosities in the nineteenth
century geological literature is a good example
of this. Second, the means for acquiring and
using the materials must be developed. Fossil
water beneath the Great Plains and the Sahara
Desert was useful to no one until the technique
of drilling deep tube wells was perfected. Similarly, aluminum—one of the most common
elements in the earth's crust—was too expensive for common use before the electrolytic
refining method was discovered.

Third, and following closely on the heels of
our second subset, people must have the financial means to set up the extractive, refining, and
manufacturing processes necessary to make
something available. More subtle, but just as
important, is the need for managerial skills by
means of which the factors of production can be
brought efficiently together. Many a firm or
factory with all the tangible attributes leading to
success has failed for want of good management. Fourth, and finally, both individuals and
the society to which they belong must find the
use of a given resource acceptable. Muslims
cannot eat pork although pigs could provide
meat to protein-poor areas of the Near East and
South Asia. Most Europeans (including ourselves) refuse to eat insects although locusts and
grubs are a tasty and welcome supplement to
the diets of many people. Alan Moorehead, in
his book *Cooper's Creek*, describes how expedition after expedition to the interior of Australia
starved because their members would not eat
the local and nutritious foods relished by the
aborigines. Land tenure and attitudes toward
land ownership also fall within this class of
restrictions placed on resource use. Land may be
held by individuals, as in the United States and
Western Europe; it may belong to the state as in
Communist bloc nations; or it may be the

[3]Erich W. Zimmerman, *Erich W. Zimmerman's Introduction
to World Resources*, Henry L. Hunker, ed., Harper and Row,
Publishers, Incorporated, New York, 1964, p. 8.

*Resources as
Cultural
Appraisals:
Man's Changing
World View*

legacy of the tribe or kinship group, as in much of Africa. In each case, the use of the soil and its products is limited or biased by the system involved. At another level, a community may object to the presence of an atomic pile or nuclear reactor in their neighborhood. Objections of this sort rarely depend upon the real issues of safety and cleanliness. Instead, vague fears may be enough to keep power plants of this kind away from urbanized areas. Another example of the sometimes violent reactions to resource use in a modern community is when it is suggested that the drinking water be fluoridated to cut down dental cavities. People will drink substandard water filled with industrial waste or treated sewage, but let someone attempt specifically to add chemicals to the water supply and the trouble starts.

We have emphasized material resources in the above discussion, but less tangible things can also be thought of in the same manner. The climate of a region may allow a variety of uses, but human beings must interpret energy availability, temperature regime, humidity, precipitation, and numberless other things before they can speak of a "vacationland" or an "attractive location for industry." Finally, we emphasize again that from the geographic point of view expressed in this book, *any material or process must enjoy a certain minimum locational advantage in order to become a resource in the complete sense of the word.*

### Perception of Environmental Hazards

We have repeatedly used the term *perception* in our discussion of man-environment systems. Much confusion surrounds the use of this word by geographers, psychologists, and others, but for our purposes we rely on the definition provided by Robert Ward: "Perception is the cognizance of the real world environment as it has been assessed by an individual. This includes the physical, social, and economic complexities of past, present, and future events, and their meanings as they relate to the decision making processes."

The apparent difference between psychologists and geographers in the use of the word *perception* is that psychologists feel that the perceiver "must come into *direct* contact with the stimulus in order to perceive the stimulus." Sometimes the subject reacts to his memory of the stimulus. This memory of his experience and his reactions to that memory are what geographers usually investigate. A geographer would want to find out what memories the inhabitants of a shoreline—across which hurricanes move—have about past disasters and how their evaluations of past events affect their use of the land and their preparations for the next possible storm. "Perception" in this geographic sense is something more than stimulus response and corresponds to what psychologists call *cognition.* This act of cognition, according to psychologists, relates the physiological awareness of the stimulus to a chain of *memories* which are in turn flavored by the *goals, decisions, actions*, and *consequences* of past events. In accordance with current geographic literature we will use the term *perception* as synonymous with the above definition of *cognition.*

Perception studies in geography cover a wide range of topics. Kevin Lynch, Julian Wolpert, and others have attempted to discover how people view the cities in which they live. Their studies show the different values placed upon elements within the urban environment by its inhabitants. Most important, urban perception studies indicate that modern cities seem fragmentary and uncoordinated to those who inhabit them. The conclusion is that more research is needed in urban design before we can build cities that are perceived and used as "remarkable and well-knit places" in which to live.

Far removed from the city are studies by T. F. Saarinen of the perception of drought hazard in the Great Plains and by Gilbert White, Ian Burton, Robert Kates, and others of storms and flooding. The results of these studies indicate that the adage "familiarity breeds contempt" does not always hold true. In general, people who have had the longest exposure to environmental hazards of this sort score best when asked to estimate the frequency of past floods, storms, and droughts. In the same way, people living in high-hazard areas tend to be more sensitized to the problems facing them than

**Table 11–1  Common Responses to Uncertainty Concerning Natural Hazards**

| Eliminate the Hazard | | Eliminate the Uncertainty | |
|---|---|---|---|
| Deny or Denigrate Its Existence | Deny or Denigrate Its Recurrence | Make it Determinate and Knowable | Transfer Uncertainty to a Higher Power |
| "We have no floods, here, only high water" | "Lightning never strikes twice in the same place" | "Seven years of great plenty. . . . After them seven years of famine" | "It's in the hands of God" |
| "It can't happen here" | "It's a freak of nature" | "Floods come every five years" | "The government is taking care of it" |

Source: Ian Burton and Robert W. Kates, "The Perception of Natural Hazards in Resource Management," reprinted with permission from *Natural Resources Journal* 412 (1964), University of New Mexico School of Law, Albuquerque, New Mexico, table 5.

those who encounter such things less frequently. On the other hand, it is human nature to downgrade or discredit the danger of events over which they have little control. People have lived for thousands of years on the slopes of active volcanoes or in low-lying coastal areas subject to typhoons and tidal waves. Their staying in such places may be enforced in part by circumstance and lack of alternatives, but not always. Table 11-1 lists some of the common responses to natural hazards.

The pragmatic importance of hazard perception studies rests in the realm of public policy making. Should millions be spent to protect summer cottages along open shores? Or would it be better to leave such places unbuilt upon? Once flooding is controlled in a river basin, how best can farmers be persuaded not to make long-term investments in bottomlands despite their perceptions of reduced hazard? Average hazards may be reduced, but exceptionally high flood waters can still occur, topping the levees and causing immense damage and loss. Flood protection is almost always designed to counteract average peak conditions. The cost of protecting against "once in 50 year floods" is prohibitively high.

In the next chapter we will discuss many of the ways in which environments deteriorate, sometimes because of man, sometimes because

of agents beyond his control. But before discussing the geographic implications of ecologic deterioration we need to review another significant aspect of human perception.

The world view which each of us nurtures can influence our different theories of human behavior. If we see the world as a series of isolated places and events, we are forced to seek explanations in terms of site conditions such as environmental characteristics. But if we see the world as a complex network of spatial interrelationships, situation explanations become important. For this reason, we have waited until late in our text to discuss one of geography's central activities, cartography and map making. We feel that one of the most important aspects of geographers as cartographers is the way in which maps reflect mankind's changing world view. Map making can be considered one of man's oldest attempts to record his perceptions of the world in which he lives and of its resources.

## Maps and the Changing World View

One basic activity of all animals is that of orienting themselves spatially within their environments. That is, they make mental maps and place themselves in them. Even with humans this kind of map making is often intuitive.

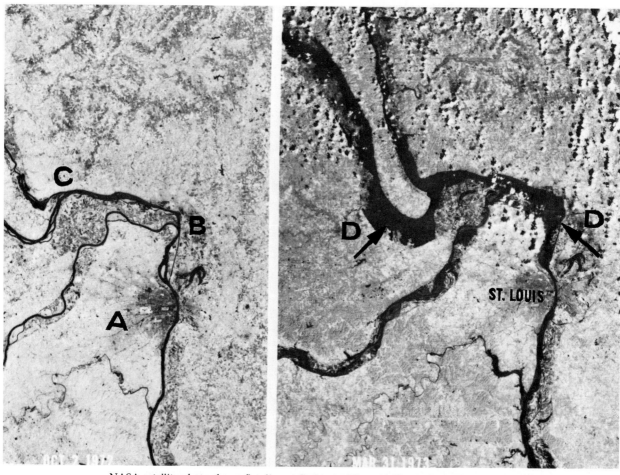

NASA satellite photo shows flooding at St. Louis, Missouri. The St. Louis area as seen by NASA's Earth Resource Technology Satellite shows the extreme flooding along the Illinois, Missouri, and Mississippi Rivers. In the photograph on the left, taken Oct. 2, 1972, St. Louis can be located by the letter *A*. North of St. Louis, the Missouri River joins the Mississippi River at point *B*, and farther upstream, the confluence of the Illinois and Mississippi Rivers is noted by the letter *C*. The photograph on the right, taken of the same region on Mar. 31, 1973, shows areas under water (letter *D*) as a result of the flooding. In this near-infrared wavelength view, the darkest tones indicate areas of deepest water. At the time this picture was obtained from space, the Mississippi River at St. Louis was at a stage of thirty-eight feet, the highest since 1903. In this frame, about 300,000 acres were already covered by water, and the river stage was still rising. This flood wave is slowly making its way downstream, threatening cities and agricultural lands along the entire length of the Mississippi River. (NASA photo)

The child psychologist Piaget points out that babies and small children spend much of their time orienting themselves within their environments. Anyone who has watched an infant crawling or toddling is immediately impressed with the small one's role as an explorer. As adults each one of us carries within himself many maps of the world in which we live. Most of these maps are mental ones, and relatively few are actually drawn on paper. But mapping as such is not limited to modern society. Fragments of ancient maps exist, as do maps drawn by tribal and preliterate people.

People map what interests them. Micronesian charts show the direction of wave fronts (represented by curved palm fronds) and islands (intersection of the fronds) (Figure 11-1). Eskimo maps are among the best made by preliterate peoples. They emphasize the true shapes of land masses and bodies of water, although distances seem to be of less importance (Figure 11-2). For example, the Netsilik Eskimos drew maps for the arctic explorer Rasmussen. Altogether 532 place names were listed, and of those 498 designated islands, bays, streams, lakes, and fjords. Only 18 showed scattered rocks and ravines, while 16 indicated mountains and hills. For people who earn their living essentially from the sea this is not too surprising. American Indian maps emphasized time rather than spatial relationships. A day's journey was given the same unit length whether the spatial distance was great or small. Travel effort was considered most important, just as modern geographers make maps of travel times. American Indian maps also frequently showed places where important events occurred.

In the final analysis, maps reflect the world view of the people who make them. They can either show limited horizons or actually attempt to chart the universe. Maps also reflect the perceptions and value systems held by their makers. In the United States many maps show where struggles between Europeans and Indians took place; if the white men won, such locations are labeled "battles;" if the red men won, they are called "massacres." We may

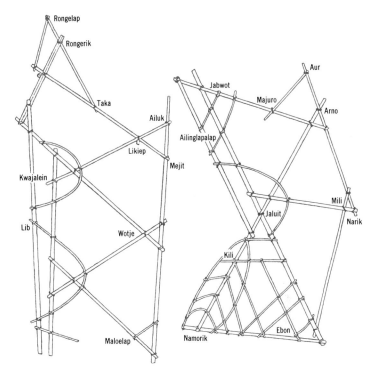

**Figure 11-1 Micronesian navigational stick charts** The charts were used for navigating in open boats between islands in the South Pacific. (E. H. Bryan, Jr., *Life in the Marshall Islands,* Pacific Scientific Information Center, Bernice P. Bishop Museum, Honolulu, 1972)

similarly consider how Canadians map much smaller settlements as towns than do people from the United States. This clearly reflects the Canadian's assessment of their smaller and scattered population.

Map making depends as well upon the technology of each group of map makers. At the same time, the need for more accurate maps has stimulated cartographic invention. Much of the continuing earth satellite program is concerned with mapping. It is, in fact, safe to say that the history of cartography reflects mankind's increasing knowledge and use of the world it occupies as well as gives clues to the value systems it chooses.

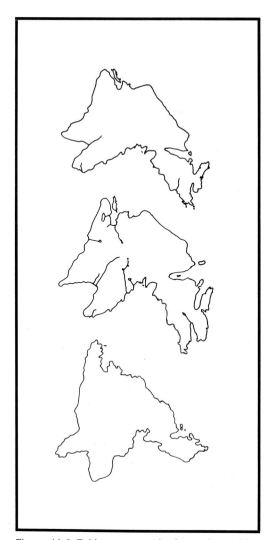

**Figure 11-2 Eskimo maps** The figure shows different maps of Southampton Island in the Canadian Arctic. The two topmost maps are sketches independently drawn from memory by two Eskimos. The lowest figure is a modern map prepared with the aid of areal photographs. The Eskimo maps are amazingly accurate and, just as interestingly, display similar distortions. The peninsula in the southeast was a favorite hunting ground, and it has been enlarged on both maps. Other locational shifts may relate to differences in travel times between positions. These deviations may reflect Eskimo realities better than does the modern map. (Edmund Carpenter, *Eskimo Realities,* Holt, Rinehart and Winston, Inc., New York, 1973, pages 10–11)

*Physical Geography: Environment and Man*

## Early beginnings

The role of maps in human affairs becomes evident if we review cartography in a historical perspective. One of the oldest known maps dates from 4,500 years ago and shows the location of a rich man's estate recorded on a baked clay tablet. Found about 200 miles north of the site of ancient Babylon this fragment is not important for its cartographic merit. But it does show the Babylonians' concern with the land and real estate. Other more important contributions which we have inherited from the Babylonians are the duo-decimal system of counting based on 12. They divided circles into 360 degrees; each degree into 60 minutes; and each minute into 60 seconds—measurements still used in surveying as well as in keeping time. Another somewhat dubious legacy was their concept of the universe with a disk-shaped earth floating on the ocean with the sky above and the firmament over all. This idea was accepted by many but not all Greeks, and thereafter by the Romans and Israelites. Through the Scriptures it came into medieval Europe where it was dislodged only with great difficulty from scholarly thought.

Other examples of early map making come from Egypt and China. In the former case, maps were less important than land surveys carried out in order that the pharaohs might levy grain taxes. A systematic land survey of the Egyptian Empire was carried out under the rule of Ramses II (1333–1300 B.C.). We will refer to this survey again in a moment as well as to some of its consequences. Map making in China was a case of parallel invention. The earliest Chinese literary reference to a map dates from 227 B.C. A famous cartographer, Pei Hsiu (A.D. 224–273), gave five sound rules for drawing useful maps. They should have rectilinear divisions by means of which relative locations might be shown. They should be properly oriented in order to show the correct direction from one place to another. They should have some accurate indication of distances; that is, they should have a scale. They should show higher and lower elevations, and they should pay attention to changes in the shapes of things such as roads. In other words, they should be topologically accurate. It might be said that this independent development of cartographic thought indicates the universality underlying mankind's concern with space.

## Greek and Roman contributions

The Greeks were the first great cartographers and developed a system of map making that was not equaled again until the sixteenth century A.D. The earliest existing Greek map appears on a silver coin from the fourth century B.C. and shows about 90 square miles on the shore of western Asia Minor (Figure 11-3). Some high points of the Greek achievement include the work of Anaximander of Miletus (611?–547? B.C.), who is credited with the first map. Hecataeus (ca. 500 B.C.) wrote a systematic description of the world which showed the flat earth surrounded by oceans and divided into Europe and Asia. In this description the habitable world was called the *oekumene*. The idea of a spherical earth was introduced by an unknown Greek as early as the fourth century B.C., although this was related more to the idea of the sphere as a perfect form than to any direct astronomical observations. By 350 B.C. three additional points were recognized by many of the Greeks. These were the concepts of the equator and the poles and the inclination of the earth's axis. Eratosthenes of Cyrene (276–196 B.C.) measured the circumference of the earth in a classic exercise which used the distance from Aswan to Alexandria in Egypt as it had been recorded centuries before in the land surveys of Ramses II. Because of compensating errors his figure for the circumference was 252,000 stadia (28,000 miles), accurate within 14 percent of the true figure. Posidonius repeated the experiment about a century later using the Island of Rhodes and Alexandria. His figure was reported by Strabo to be about 18,000 miles, which was 25 percent too small. This estimate was later accepted by Ptolemy in his *Geographia* and was used by geographers as late as the fifteenth century. Therefore, Columbus underestimated the size

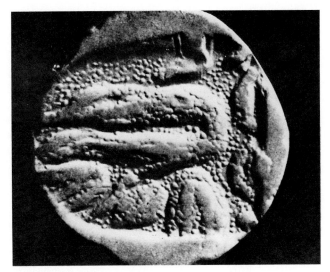

**Figure 11-3 Map on an ancient coin**   An Ionian coin, 350–730 B.C., bears a relief map of the area around Ephesus, south of the present city of Izmir, on the western shore of Asia Minor. (British Museum; George Kish, *History of Cartography*, Harper & Row, Publishers, Inc., New York, 1973, page 3, map 6)

of the earth and mistook America for Asia when he first arrived. Hipparchus as early as 150 B.C. suggested the utility of a regular grid system on maps.

Ptolemy summed up ancient geographic and cartographic knowledge in his eight-volume *Geographia*, which described and mapped the world from the Canary Islands to China. This work was subsequently lost to the West until the fifteenth century, although it was retained in the Arab world. Western Europe relied mainly on Roman cartographers, who emphasized practical maps for military and administrative purposes. They also maintained the notion of a circular and flat world, the *Orbis Terrarum*, on their maps. Their road maps have come down to us in the form of the Peutinger Table, which is the work of a twelfth century A.D. monk but is really a copy of a third century map. This scroll is about 1 foot in width and 21 feet in length and shows distances as well as military posts. We have already discussed part of the Peutinger

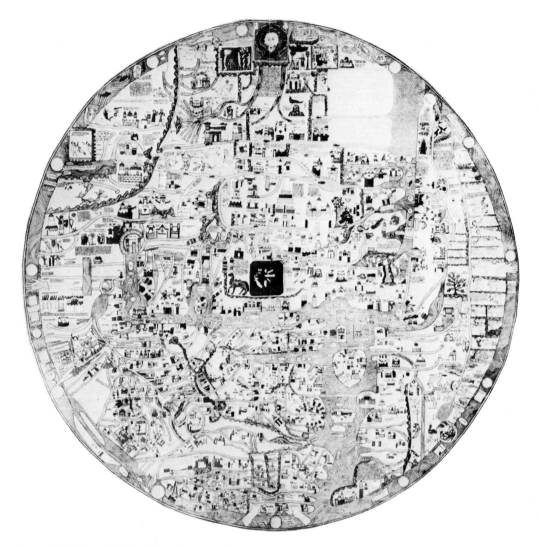

**Figure 11-4 Ebstorf "T in O" world map, thirteenth century** The original map was in the Benedictine abbey of Ebstorf, in northwest Germany. It is oriented with east at the top and with Jerusalem in the center. The head, hands, and feet of Christ are seen around the circular frame. The map combines the geography of its time with religious and mythical elements. (Op. cit. Figure 11-3, map 24, page 8)

Table in Chapter 1, Figure 1-6. The Romans have left us little of importance cartographically speaking, although their custom of placing east at the top of a map has given us our term "orientation" and the phrase "to orient a map."

*The Middle Ages*

Medieval map makers illustrate our viewpoint that cartographers show what interests them most. The world depicted by them was symbolic rather than even as they knew it to be. The basic shape for their world was taken from the Orbis Terrarum maps of the Romans. A typical map was circular and had Asia at the top, while Europe and Africa (Lybia) equally divided the lower half with the Mediterranean Sea up the middle. The effect of the circle and the inner seas, including the Aegean and the Red, was that of a T inside an O. In fact, maps of this type are often called "T in O" maps (Figure 11-4). Most important of all was Jerusalem, which occupied the center, thus fulfilling the scripture (Ezek. 5:5) "This is Jerusalem: I have set her in the midst of the nations, and countries are round about her." T in O maps were the work of Christian monks, and those devout fellows often placed the hands and feet and head of Christ at appropriate places around the periphery of the world they drew. Also prominent were places like the Garden of Eden and all manner of mythical creatures. Accuracy, or the

lack of it, probably reflected not only the monks' limited knowledge of the world but also their much greater concern with the journey of all men to salvation.

Little else of cartographic importance occurred during the Dark Ages and early part of the Middle Ages in Europe. Matthew Paris (1200–1259) began to make more accurate maps of England, and navigation charts created by officers of the Genoese fleet aided sailing on the Mediterranean Sea. These latter *Portolan Charts*, as they are called, were based on surveying by compass and showed harbors, capes, and coastlines. The interiors of the land masses were left blank. In both these cases, the maps were used as travelers' aids, with strong mercantile and military motivation underlying their production.

Elsewhere, the Arabs kept systematic and mathematic geography alive (Figure 11-5). The Arab cartographer Idrisi (1154) produced a

**Figure 11-5 Idrisi map, 1154 A.D.**   Idrisi was a Moslem geographer and cartographer born in Morocco. His maps are rich in detail and notably accurate. On this map, which is oriented to the south, notice the use of a rectangular grid. (Op. cit. Figure 11-3, map 18, page 5)

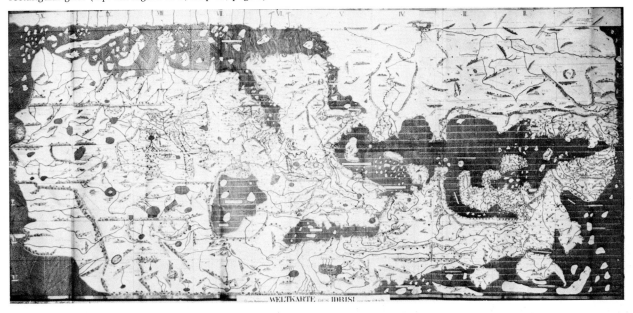

world map which was rich in information and based upon a rough rectangular projection. Remember, too, that it was the Arabs who provided European geographers with the works of Ptolemy and aids to navigation like the sextant and the compass. Anyone at that time who wished to learn about the world would have done well to have visited the Arab schools at Seville and Cairo rather than the dim cloisters of Europe.

### The renaissance of map making

The sixteenth century ushered in a revival of systematic cartography in Western Europe. This was the result of at least three major developments. Ptolemy's works were translated into Latin about the year 1405. Although they contained many errors and misconceptions, they enlarged the world view of Europeans and led them into thinking about far Cathay (China) and other places. Just about the same time, the art of printing and engraving reached new levels, which allowed many copies of a map or text to be prepared. Thirdly, the Age of Great Discoveries dawned upon Europe. The Portuguese rounded Africa and reached India in the second half of the fifteenth century. The names of Prince Henry the Navigator and Vasco de Gama are among the earliest. Columbus and Magellan are the most famous, but there were literally hundreds more. The end result of all this was that the errors in the old maps were pointed out and the need for new and better ones grew rapidly. Cartography, with the help of printing, became a lucrative business, and the demand for more and better maps and map-making techniques grew apace.

At about this time (ca. 1522) Juan de la Cosa drew a map which showed parts of the New World, including Cuba, which Cosa showed as an island despite Columbus's insisting that it wasn't. The important thing about Juan de la Cosa's map was that it attempted to depict the round earth on flat paper. That is, it utilized the first real *map projection*.

Once it was clearly established that the earth is essentially spherical in shape, the central problem of cartography, both as an art and as a science, became how to show the surface of a sphere on a flat or planar surface. This was originally accomplished by geometrically *projecting* a grid system drawn on a sphere onto a plane. In carrying out this exercise, one or more properties of the spherical grid system became distorted on at least some portion if not everywhere on the flat map. This violates the notion that accurate renditions of distance, direction, shape, and area are all desirable when making a map. Thus, a map may show the true shapes of areas, but their sizes become disproportionate. Conversely, the proportions from one part of a map to another may be maintained, but directions from point to point become skewed from their true compass bearings. For the purposes of navigation, true bearings are essential, and it was this property which Gerhardus Mercator (Gerhard Kremer, 1512–1594) incorporated in his maps. The Mercator projection invented by him shows all compass directions as straight lines. One feature of this projection is that parallels and meridians intersect each other everywhere at 90°. This allows navigation routes to be drawn directly on such a map using compass bearings.

This was extremely useful for the early navigators and remains so to the present day. For example, the maps in NASA headquarters which show the curving flight paths of orbiting earth satellites are Mercator projections. Mercator not only invented this highly useful projection but also used all available sources to fill his maps with information. The result was that the cartographic house headed by him was successful and that his name has become famous.

### Modern times

After Mercator the developments in cartography came faster and faster. Jean B. B. d'Anville (1697–1782) made over 200 maps of many parts of the world. He based his success upon rejecting all unproven information (Figure 11-6). In this way he established a reputation for accuracy which outlasted him by centuries. In fact, his map of China was utilized in 1926 as a base map for the first survey of the Chinese Republic!

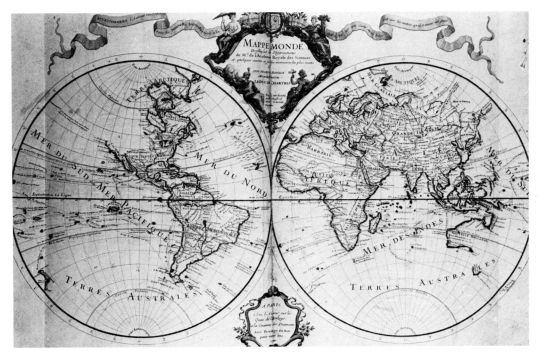

**Figure 11-6 D'Anville Eastern and Western Hemispheres map, 1760**
Jean Baptiste Bourguignon d'Anville produced this map of the hemispheres in 1760. His maps are models of clarity and accuracy, leaving spaces blank for want of information rather than substituting decorative or fanciful details. (Op. cit. Figure 11-3, maps 161–162, page 34)

Thus, map making progressed along two fronts. One problem was to find out and show more and more earth information; the other was to find better and more accurate ways of projecting curved onto planar surfaces. Perhaps most important of the developing breed of cartographers was Johann Henrich Lambert, who in 1722

*specifically stated the objective of preservation of spherical properties as a desideratum for plane maps. He continued by listing some of the possible properties to be preserved including the preservation of local similarities. Although Mercator's projection and the stereographic projection already did this, Lambert treated the subject more generally. An even greater innovation on Lambert's part was the attempt to characterize these properties in analytical form (as non-linear differential equations) and to suggest general solutions, including oblique cases. He thus literally invented an infinite number of specific map projections. Lambert's analytical approach has characterized all important work on map projections since that date.[5]*

Nationalism and continuing warfare in Europe stimulated further advances in cartography. The French Academy of Triangulation (1710–1740) conducted the first topographic survey of any country. The topographic maps prepared by them were used as military aids and soon led other nations into developing similar surveys (Figure 11-7). At about the same time, the young American nation wanted to learn about its own territory and resources. George

[5]J. H. Lambert, *Notes and Comments on the Composition of Terrestrial and Celestial Maps*, translated and introduced by W. R. Tobler, Michigan Geographical Publication 8, Department of Geography, University of Michigan, Ann Arbor, 1972, p. x.

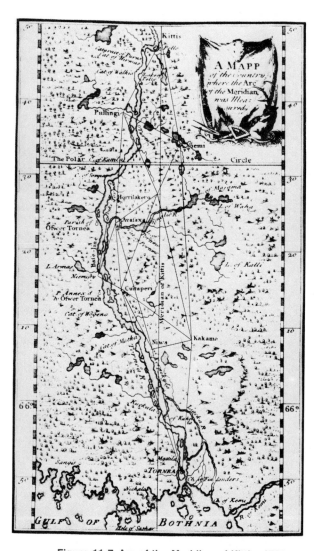

**Figure 11-7 Arc of the Meridian of Kittis, 1759**
From 1736 to 1743, a party of French astronomer-surveyors measured an arc of the meridian from Tornea to Kittis. Together with similar measurements in other parts of the world, this survey established the shape of the earth and led to the adoption of the metric system, based on the earth's circumference. (Peabody Institute Library; op. cit. Figure 11-3, map 170, page 35)

Washington himself was a land surveyor early in his career. At first, the New World was parceled out in a system of *metes and bounds.* That is, natural markers in the landscape—rocks, trees, outstanding cliffs and hills, etc.—were used to designate the corner points on the boundaries of properties and political units. This accounts for the irregularly shaped territories of all sorts east of the Appalachian Mountains. But the inadequacy of this method of dividing up the resources of the continent became apparent, and by the year 1785 the newly formed Congress had established a systematic method of land surveying which resulted in the north-south, east-west orientation of county and state boundaries in the midwestern and western United States. This rectilinear land survey used parallels of latitude as starting lines from which the land was surveyed into townships, each 6 miles square. In every township are 36 sections each containing 1 square mile, or 640 acres (Figure 11-8). Sometimes mistakes were made; sometimes natural barriers prevented consistent coverage; sometimes jogs or offsets in boundaries had to be made to compensate for the narrowing of the globe from the equator to the pole. Nevertheless, this orderly system of dividing up the earth itself as a resource was rapidly adopted as a standard and worldwide technique. At the beginning of the twentieth century a measure of control and standardization was introduced into mapping throughout the world. At that time, the International Geographical Union brought about an agreement among nations to map the world on a uniform basis. Each country was to survey its own area and make the results available to all others. The result was the *Millionth Map,* so called because a single 1:1,000,000 scale was to be used for it. The surface of the earth was divided up into *lunes,* or the equivalent of rectangles drawn on a sphere. Each lune from the equator to 60° is 4° of latitude by 6° of longitude. Poleward of 60° the lunes are 4° by 12°. Although the Millionth Map has yet to be completed—many nations have lagged in its production—it still represents the best available general map for the world. Using the equator (0° latitude) and the prime meridian (0° and 180°

longitude) as baselines, it also maintains standardized symbols and signs for all features shown.

From 1909 until the present, mapping has been developing increasing sophistication. Now satellites photograph the entire earth as a series of parallel and overlapping strips which can then be put together as maps of different kinds. Everyone seems to need and demand more maps. The military, real estate developers, governments, and private individuals from tourists to the League of Women Voters require better and better charts of the places where they live, work, play, and make decisions. Thus, our purpose here has not been to give a description of cartography either in a complete historical sense or with regard to the actual techniques of construction and production. Rather, we have taken these few pages to sketch the way in which all people use maps to express their perceptions of the world in which they live.

**Figure 11-8 United States land survey** The familiar grid pattern of land ownership west of the Appalachian Mountains has its basis in the land survey system adopted by Congress in 1785. (*A*) A base line (east-west) and principal meridian (north-south) are layed out by survey, and their point of intersection becomes the location from which all other locations are referenced. Every six miles, parallel lines of latitude are established as *township* lines in the north-south direction. Similar parallel lines form *ranges* every six miles in the east-west direction. The squares so formed are called *townships*. The location of any township is described relative to the base point by counting township and range by direction: e.g., township 3 north, range 2 west, or abbreviated, T3N, R2W. (*B*) Each township consists of thirty-six square miles, or *sections*, numbered as shown. Townships always contain thirty-six square miles, but lines of meridians converge due to the curvature of the earth. Therefore, corrections are made which offset some adjacent tiers of townships. This results in occasional jogs in section line roads. (*C*) Finally, each section may be subdivided, usually by quarters. This allows one to describe the location and size of a piece of property very precisely: e.g., the north ¹/₂ of the southwest ¹/₄ of the southeast ¹/₄ of Section 5, T3N, R2W, Abbreviation of the description would appear in a title or deed to the land as: N ¹/₂ of SW ¹/₄ of SE ¹/₄, Sec. 5, T3N, R2W.

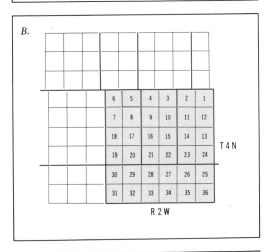

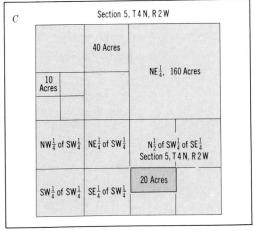

We have shown how preliterate peoples map the features of the earth most important to themselves and their systems of livelihood and survival. In the same manner, we have mentioned the mathematical and intellectual qualities of the Greek cartographers as opposed to the pragmatic Roman road builders. Religious values of medieval maps were supplanted by more and more accurate renditions of the sixteenth and later centuries. Nationalism, military campaigns, and the rush for ownership of the land in a world rapidly filling up with people resulted in new tensions which were in part reduced by partitioning out earth-space itself. This segmentation of the land could be tied down and made firm by mapping, with more and more accuracy. In other words, the sophisticated mapping techniques being developed at the present time are the means to reduce conflicts and conversely to enhance some people's territorial ambitions. Where tension is great, either because of the presence of valuable resources such as minerals or because of overlapping territorial desires, the land has been mapped over and over with increasing accuracy. Where territorial ambitions and resources are slight, as in Antarctica, nations and individuals are willing to settle for abstract and symbolic boundaries such as the pie-shaped segments into which contesting nations divide the Antarctic continent. Examination of the thumbnail sketch of cartographic history which we have just presented reveals a steady sequence from narrow and provincial world views to a single perception of the earth as a place to map cooperatively in its entirety.

### Mental maps

The *mental maps* analyzed by Peter Gould and others illustrate this point. If you, the reader, will stop for a moment and consider the images of the world which you carry in your head and by means of which you navigate from place to place, you will realize that those images incorporate areas of great detail based on intimate knowledge and other places where you could draw nothing meaningful if asked to put it down on paper. The research into mental maps consists of efforts to detect and understand the images of the world that people carry about with them.

One interesting analysis made by Gould was to ask college students at Berkeley (California), Minnesota, Pennsylvania, and Alabama to rank the forty-eight coterminus states in order of their desirability as places in which to live. The resulting values were summed for the four different sets of students and plotted on maps of the United States (Figure 11-9 *A* to *D*). High values show desirable places; low values show undesirable habitats. We will not take the space for a detailed examination of the results, but the reader can take a leisurely look at the maps and hypothesize for himself why the peaks and pits of desirability occur as they do. California is uniformly high (we will return to the implications of this "Golden State image" in Chapter 13). The Great Plains are a low and unappealing plain. Of particular interest is the way in which all Northern and Western students rank the Deep South as a homogeneously undesirable locality, while Alabama students are much more discriminating and see the South as both a varied place and one which is good to live in.

This type of map relies upon a standard surveyed base upon which to plot psychological values gathered by some type of interview or questionnaire. There is another kind of mental map which may possibly show the actual perception of reality held by individuals or groups. In this case, people are asked to draw maps of the territory in which they live. No one can draw a completely accurate map, but the theory is that people will draw those places they know the best with the greatest accuracy and that they will show the things they deem important in preference to those they don't like or even notice. Comparisons can then be made between the relative positions of features on surveyed maps and those same features on the freehand charts. Figure 11-10 shows such a comparison based on maps drawn by migratory Turkish workers temporarily employed in West Germany. The first diagram simply depicts a rectangular grid which, in this case, was placed across the city of Cologne (Köln) in Germany. The second diagram summarizes how the migratory

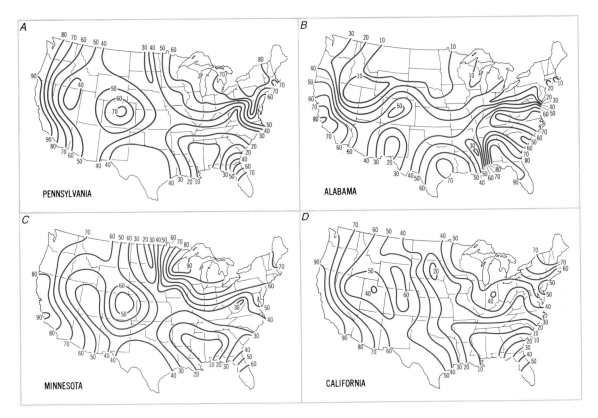

**Figure 11-9 Student mental maps of desirable places to live** Students at
four different universities were asked to rank the states in the United States
according to their desirability as places in which to live. Places with high values
are appealing; places with low values are unappealing. Notice how each group
ranked its own location high in desirability. The maps also show some con-
sensus of opinion about certain desirable and undesirable locations; Colorado,
for example, is generally highly regarded by all groups. (P. R. Gould, "On
Mental Maps," *Michigan Inter-University Community of Mathematical Geographers,* Discussion Paper No. 9, 1966. Reprinted in P. English and R. Mayfield
(eds.), *Man, Space and Environment,* Oxford University Press, New York,
1972)

workers perceived and drew the relative locations of a number of prominent features in the
Cologne cityscape. The nearly rectangular grid
pattern preserved at the center indicates the
greater familiarity of the workers with the center of the city. The distortions at the edges show
how their knowledge of that area diminishes
rapidly with distance from the central business
district.

Both of these maps offer new clues into the
way in which we view our world. If you were to

draw a map of the city in which your campus is
located, it would predictably differ in many
ways from a map of the same area drawn by a
businessman, a homemaker, or an inner-city
child. Much more research needs to be done on
these topics, but the signposts are there. We use
and seek out the places and the things we have
learned to value. We are more and more urban
rather than rural creatures, and more and more
our attention focuses upon cities rather than the
natural resources which support our way of life.

*Resources as
Cultural
Appraisals:
Man's Changing
World View*

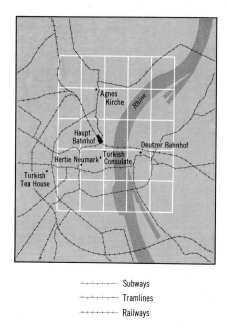

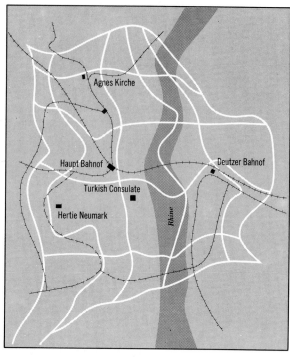

Subways

Tramlines

Railways

**Figure 11-10 Turkish workers' mental map of Cologne, Germany**    A sample of Turkish workers in this German city were each asked to draw a map of the city, showing where they lived, worked, and otherwise traveled. The maps were combined by averaging the locations of prominent landmarks identified by the group and comparing this location with the true position of the landmark as indicated by a large-scale map of the city. A map projection was then calculated that mapped the true position onto the imagined position. The subsequent stretching of an original square grid in the mental map shows the nature of the space in the collected image of the city. (The data collection and analysis were carried out by John Clark, University of Michigan 1972–1973)

If we are to understand the geography of our environments and the problems which beset us in our interactions with the natural as well as the social world, we must use every technique possible to gain useful insights. A knowledge of the way in which we see and map the world around us will help us to understand and, it is hoped, solve man-environment problems, some of which are described in the next chapter.

# 12 | ENVIRONMENTAL IMBALANCE: MAN'S ROLE IN CHANGING THE FACE OF THE EARTH

*For the sword outwears its sheath,*
*and the soul wears out the breast. . .*

(George Gordon, Lord Byron)

Autumn, and the trees shed their leaves; spring and summer with the molting of skin and feather and fur. The carcass in the jungle and the whale awash on the beach. Life is defined by death and cleans itself with a constant sloughing off of old tissue. Even the continents waste into the seas.

In discussing a topic as vast and sensitive as man's role in environmental imbalance we must begin with the idea that the world when healthy is in a constant state of birth, death, decay, and renewal. It is only when the tempo of these processes is changed and when their location is skewed from reasonable to unreasonable distributions that trouble looms. This makes it necessary for us to understand how resources occur in space and time and how policies regarding resource use are tied to human perceptions of their spatial and temporal environments.

## Resources in a Space-time Framework

Once we have agreed to designate a set of tangible materials as resources, we must con-sider the rates at which we utilize the available supply. We must also consider whether the supply is limited or constantly renewable. A basic division of resource types is between renewable and nonrenewable ones. These are also sometimes referred to as *flow* and *fund stocks*. Flow, or renewable, resources are largely biotic. They are usually elements in self-maintaining systems and with proper husbanding can be used over and over. Forest preserves or well-managed fisheries would come under this category. Fund, or nonrenewable resources, exist in a static framework. In other words, the systems that create nonrenewable resources such as mineral deposits operate so slowly that once they are used up, it is unlikely that similar concentrations can be reestablished during a reasonable span of time.

A basic way of looking at renewable and nonrenewable resources is to consider the length of time that must pass before materials can be replaced through the operation of natural processes. Table 12-1 lists some residence times of natural cycles. For example, the average length of time that a molecule of water will remain in the atmosphere before returning to the surface is about 10 days. Terrestrial groundwater requires 150 years to complete its underground circulation. Oxygen incorporated into

**Table 12–1   Residence Times of Some Natural Cycles**

| Earth Materials | Some Typical Residence Times |
|---|---|
| Atmosphere circulation | |
| Water vapor | 10 days (lower atmosphere) |
| Carbon dioxide | 5 to 10 days (with sea) |
| Aerosol particles | |
|     Stratosphere (upper atmosphere) | Several months to several years |
|     Troposphere (lower atmosphere) | One to several weeks |
| | |
| Hydrosphere circulation | |
| Atlantic surface water | 10 years |
| Atlantic deep water | 600 years |
| Pacific surface water | 25 years |
| Pacific deep water | 1,300 years |
| Terrestrial groundwater | 150 years (above 2,500 feet (760 meters) depth) |
| | |
| Biosphere circulation* | |
| Water | 2,000,000 years |
| Oxygen | 2,000 years |
| Carbon dioxide | 300 years |
| | |
| Sea water constituents* | |
| Water | 44,000 years |
| All salts | 22,000,000 years |
| Calcium ion | 1,200,000 years |
| Sulfate ion | 11,000,000 years |
| Sodium ion | 260,000,000 years |
| Chloride ion | Infinite |

*Average time it takes for these materials to recycle with the atmosphere and hydrosphere.
Source: *The Earth and Human Affairs*, compiled by National Research Council, National Academy of Sciences, copyright © 1972 by the National Academy of Sciences. By permission of Harper & Row, Publishers (Canfield Press Division), table 2–1, p. 41.

living tissue will take an average of 2,000 years to pass through the atmosphere, the hydrosphere, and back into the biosphere; while a sodium ion will remain in sea water for 260 million years. What this means is that given enough time almost all nonliving resources are renewable, but that mankind's sojourn on earth has been so brief and our life-spans are so short that for all practical purposes many of such processes are one-way or irreversible.

As we have seen in Chapter 3, our residence on earth is terrifyingly brief when compared with the total length of earth history. It is the lack of temporal fit between cultural processes which consume resources and natural processes which provide them that in many cases creates disharmonies in the man-environment system.

### How men of goodwill disagree

A general pattern emerges when we analyze historical sequences of the human development and use of specific resources. At first there is a period of uncontrolled exploitation in which the "cream of the crop" is stripped off. Surface mineral deposits, large stands of accessible timber, herds of wild game all fall prey at this stage. This exploitation leads to the threatened depletion of the resource. Sometimes people don't care—as in the case of the American bison—and replace one resource with another. But oftentimes, as the "bottom of the barrel" gets closer and closer, policy statements are made and regulatory laws are passed. The first thing that happens in this latter stage is a call for and an inventory of the threatened resource. Inventories usually reveal that rates of resource use are exceeding rates of discovery. This causes a flurry of rhetoric, followed by legislation aimed at restoring the original state of plenty. What we witness here is the positive role of negative feedback.

## Determining the rate of use

Particularly at this point men of goodwill can violently disagree about what should be done. Resource policies can be quite rational and yet completely opposed to each other, depending upon the perceptions and philosophies of the people involved. The prescriptions that are suggested may include *budgeting the rate of use*. This would be particularly applicable for assigning the rates at which fossil waters or petroleum reserves shall be pumped from the ground. It would also relate to the density of wells drilled per unit area. In this way local capacities may be maintained by regulating the number of users. Too rapid pumping can dry out the space around well shafts and leave valuable liquids stranded or trapped below ground. Slower pumping allows full capillary action and smoother flows, thus assuring maximum extraction over longer periods of time. In some cases the suggested rates of resource use are unrealistically high. The number of whales that can be slaughtered each year according to international agreement among whaling nations is far in excess of the number actually caught. In fact, the quotas are meaningless, for if they were filled for more than one or two years, whales would become extinct. On the other hand, rates of use can sometimes be too slow. In the absence of natural predators, herds of deer may increase too rapidly, with subsequent overcrowding and starvation if hunting quotas are set too low. A further result can be the destruction of vegetation in such areas through excessive browsing.

Opposed to notions of slow, long-term use are those of *maximizing the flow of profit over a short run*. Rapid resource exploitation need not be considered completely negative. Sometimes, the quick profits from one area become investment capital in another. That is, the direct exploitation of natural resources can be the means of gaining necessary wealth. In the United States, the gold fortunes of California are a prime example, but there are many others, including those made from:

Fish and lumber in New England

Coal in West Virginia

Lumber, iron, and copper in Michigan

Silver and gold in Colorado

Lumber in the Pacific Northwest

Grasslands and petroleum in the Great Plains

The soil itself throughout the South and Midwest

Development of these resources took capital which came from (1) foreign investment, (2) Eastern financiers in well-developed areas far behind the advancing frontier, and (3) local money, especially from *nouveau riche* entrepreneurs first on the scene. It was not the miners, or lumberman, or fisherman who got rich. It was the saloon keepers, merchants, and early manufacturers.

In every case the accumulation of capital led to a *takeoff stage* marked by a period of rapid expansion and enormously wasteful exploitation of resources. Only the cream of the land was skimmed off. The buffalo were slaughtered; Michigan white pine and California redwoods devastated; the seas emptied of herring and the whale herds reduced to isolated in-

*Environmental Imbalance: Man's Role in Changing the Face of the Earth*

dividuals. It was a time of *once-and-for-all, git-and-git-out* exploitation of nonrenewable resources. High-grade gold, silver, and copper were taken, shallow petroleum fields emptied, hardwood forests cut; and even now the fossil water from beneath the high plains of eastern Colorado is irrigating fields from subterranean reservoirs never to be refilled.

But we must not be completely harsh in judging the pioneers and early businessmen. Hindsight is easy, and current standards may not have always applied. The rapid exploitation of a resource to exhaustion sometimes yields another cycle of growth *providing the profits earned are reinvested in productive local industries.* For example, the hardwood forests of Michigan and Indiana provided profits which helped to create the furniture industry for which Grand Rapids is famous. In much the same way, white pine and copper fortunes were subsequently invested in the Detroit automobile industry. Early investments in machine industries, buggy manufacturing, and marine engines were an intermediate step between the mines and lumber camps and the assembly lines at Willow Run and River Rouge. But whether or not it is worthwhile to sacrifice one thing for another is something that must be decided—once the facts are gathered— by political and social rather than scientific means.

When a threatened resource is able to renew itself, the rate of use and the rate of renewal become extremely important and subject to debate and legislation. The continuing argument over the national forests is an example of this. How fast should our trees be cut? Are reforestation programs realistic in the estimates of how rapidly new trees can take the place of harvested ones? Is clear cutting (that is, removing every tree from a large tract and then replanting) better than selective logging? Certainly clear cutting is technically easier and cheaper; but will the homogeneous stands of seedlings survive as readily as young trees growing in the shade of older ones? The ideal that all rational users of renewable resources aspire to is that of *sustained yield*.

## Alternative uses of a resource

The idea of multiple use is also important. A river should provide drinking water, recreation, fishing, and aesthetically pleasing views. But can it at the same time serve as a source of hydropower, a waste disposal system, and a source of irrigation water? High dams and large reservoirs have multiple purposes. For example, hydropower production is maximized by maintaining the ponded water at the highest possible level at all times. Conversely, flood control use would require that the level of the reservoir be lowered to make room for anticipated flood crests. These two uses are largely incompatible with one another, and the policy relating to them must be negotiated. The need for negotiation and compromise is also true for forests. If they are to provide timber, how can they also be places for campers, hunters, and naturalists?

## Fitting the policy to the resource

Our perceptions of the spatial and temporal qualities of a resource influence the policies, both public and private, which regulate its use. For example, if a resource is perceived to be in short supply, one response is to *increase the search for new sources.* Sometimes the search goes on inside laboratories. New methods of extracting oil from oil shale, new ways of concentrating low-grade ores, new varieties of faster-growing crops are all examples of this. But spatial search is also important. Oil companies once received large tax rebates for use in exploration and drilling new wells. However, policies encouraging search can be effective only where undiscovered stocks still exist. If the total amount of a resource is known, then a policy of protection and allocation is appropriate. We know where the few remaining sequoia trees are growing, and no amount of search will turn up more. The policy decision for protecting them depends upon the value the public places upon them. The national parks which protect unique scenic spots are the result of such considerations. Relative location can be important in these matters. The nearer a natural wonder is to a highly urbanized area, the more difficult it is to preserve it in a pristine state. It may even be difficult to save it from total destruction if the rent-paying ability of other land uses becomes too competitive.

Controls *regulating local capacities* are also important. We mean by this ways in which undesirable concentrations of wastes can be prevented. These may include setting limits or rates on the amount of sewage dumped into a body of water, or the amount of dust and noxious gases that escape from chimneys. It can also mean *zoning laws* which regulate living densities in urban and rural areas.

In all these methods of controlling our use of the environment the twin elements of time and space are important. And in all these cases, sources of conflict constantly occur. Many of the sources of such conflict are specifically geographic in character, and can be described in terms of boundary definition, territoriality, regionalization, and scale differences. The following examples represent only a partial list: *unowned resources*, or those for which ownership is not clearly established; *moving resources* such as spawning fish, migrating wildlife, or running water which cross from one state or nation into another; and *resources* viewed and used simultaneously *at more than one scale*. An example of the last case would be a forest of rare trees. Local townspeople might wish to log off the area, for that would mean jobs and income and prosperity; conservationists scattered across the entire nation and operating with a world view would see the need to preserve the same trees for the good of everyone.

## Five Processes Affecting Environmental Equilibrium

Whenever resource policies are made and enforced, parts of society will benefit while others suffer. Much depends upon how various groups of users define the costs which they are willing to pay and the profits with which they will be satisfied. The rapid depletion of resources for short-range profit seems rather foolish, but our interpretation of these matters may be overly pessimistic. We base our views on an analysis of the situation emphasizing five processes through which environmental equilibrium can be disturbed. These are (1) diffusion, (2) concentration, (3) destruction, (4) overproduction, and (5) change of state. The first two processes are distinctly spatial in character. The last three refer to site-connected activities although they too have strong situational overtones. In many cases once such processes are understood and once society has agreed upon a set of goals, it is possible to modify or reverse their negative aspects. One of the most important ways in which resources become depleted is when the rates at which they are accumulated in usable concentrations are exceeded by the rates at which they are diffused across the earth's surface in increments too small to to reused.

### Nonreversible diffusion: Mineral depletion[1]

Man's use of minerals is representative of the lack of fit between natural and cultural processes. Although enormous tonnages of the most useful elements exist in the earth's crust, they can be used only when brought together into recoverable concentrations. Table 12-2 lists eight of the most common metals. The total percentage estimated to be in the crust is contrasted with the percent that must occur in an ore for it to be considered minable. The *enrichment factor* represents the number of times above average concentration a metal must occur in order to be mined. In the entire United States in 1966, more than 90 percent of the production of 16 major minerals and elements was accounted for by the output of only 186 mines. This means that usable concentrations of minerals are quite rare. At the same time, the demand for all kinds of minerals and elements is rapidly increasing. Table 12-3 shows the domestic primary production of nine major minerals as well as additional supplies provided from recycled scrap. The total amount used exceeds those two sources in all cases except

[1]The use of the term "diffusion" at this point is very general. A more technical discussion of *diffusion theory* follows later in this chapter.

**Table 12–2  Enrichment Factor for Some Common Metals**

| Metal | Percent in Crust | Percent in Ore | Enrichment Factor* |
|---|---|---|---|
| Mercury | 0.000008 | 0.2 | 25,000 |
| Gold | 0.0000002 | 0.0008 | 4,000 |
| Lead | 0.0013 | 5.0 | 3,840 |
| Silver | 0.00007 | 0.01 | 1,450 |
| Nickel | 0.008 | 1.0 | 125 |
| Copper | 0.006 | 0.6 | 100 |
| Iron | 5.2 | 30.0 | 6 |
| Aluminum | 8.2 | 38.0 | 4 |

*The enrichment factor indicates how many times above its average concentration a metal must be in order to be mined.

Source: *The Earth and Human Affairs,* compiled by National Research Council, National Academy of Sciences, copyright © 1972 by the National Academy of Sciences. By permission of Harper & Row, Publishers (Canfield Press Division), table 4–2, p. 80.

uranium, and the difference must be made up by foreign imports. The last column shows the estimated primary demand in the year 2000. The question is, where will those materials come from?

The search for minerals has led deeper and deeper into the earth, and farther and farther from the world's central markets and industrial regions. At the same time, more and more dilute bodies of ore have been tapped in our

**Table 12–3  Total Annual U.S. Mineral Supplies and Uses**

| | Domestic Primary Production | Old Scrap* | Total Amount Used† | Projected Primary Demand for 2000 A.D. |
|---|---|---|---|---|
| Aluminum | 450,000 | 160,000 | 4,947,000 | 23,800,000 |
| Copper | 1,380,000 | 422,000 | 2,122,000 | 4,860,000 |
| Iron | 49,000,000 | 35,400,000 | 110,200,000 | 138,000,000 |
| Lead | 517,000 | 450,000 | 1,207,000 | 1,390,000 |
| Mercury | 597 | 378 | 1,995 | 2,730 |
| Bituminous coal and lignite | 504,000,000 | 0 | 504,050,000 | 900,000,000 |
| Natural gas (dry) | 446,000,000 | 0 | 465,080,000 | 1,030,000,000 |
| Petroleum (including natural gas liquids) | 508,190,000 | 0 | 668,190,000 | 1,490,000,000 |
| Uranium | 9,515 | 0 | 9,515 | 55,800 |

*Preliminary data for 1971, given in metric tons. Source: First Annual Report of the Secretary of the Interior under Mining and Minerals Policy Act of 1970 (P.L. 91–631), March 1972.

†Including government stockpiling, industry stocks, and exports. The difference between domestic supply and demand is met by foreign imports.

Source: *The Earth and Human Affairs,* compiled by National Research Council, National Academy of Sciences, copyright © 1972 by the National Academy of Sciences. By permission of Harper & Row, Publishers (Canfield Press Division), table 4–1, p. 79.

ceaseless quest. Figure 12-1 shows how this search has progressed in a three-dimensional space defined by distance, depth, and dilution. The distances involved grow greater with every passing year. In the case of iron, the ores of Minnesota are now supplemented and replaced in American mills and foundries by those from Labrador, Venezuela, and West Africa. The Kalamazoo copper ores of Arizona are half a mile deep beneath unmineralized rock. If the substance sought is valuable enough, shafts are sunk to even greater depths. Gold mines in South Africa penetrate to depths of more than 12,000 feet. Copper also provides a good example of the use of more and more dilute mineral concentrations. The first copper used by preliterate peoples were nuggets of pure metal pounded into blades, arrowheads, and symbolic tokens. By 1910 copper was mined and extracted from 2 percent ores. By 1970 the average ore used in the United States contained only 0.6 percent copper, and the Berkeley pit in Butte, Montana, operates successfully on 0.2 percent ore.

In some cases, *revolving stocks* of minerals can be established. Gold and copper, for example, are used over and over again. But in many other situations, metals, once used, are lost forever. Silver, an essential ingredient in the emulsion on photographic film, cannot be recovered. One dramatic result of the demand for silver and the nonrecoverable nature of much of its use is the copper and silver *sandwich coins* which have replaced traditional silver currency in the United States. And as one public official has observed: "The island of Jamaica, a huge exporter of bauxite, is gradually drifting—in the form of a unicellular layer of aluminum beer cans—onto the United States and covering us."

### Concentration and dispersion of iron ores

Although iron constitutes slightly more than 5 percent of the earth's crust, relatively few deposits are sufficiently concentrated to justify mining them. One of the world's great deposits of Precambrian iron ore was found in the nineteenth century in the northern portions of Minnesota, Wisconsin, and Michigan. A series of individual deposits or ranges including the Mar-

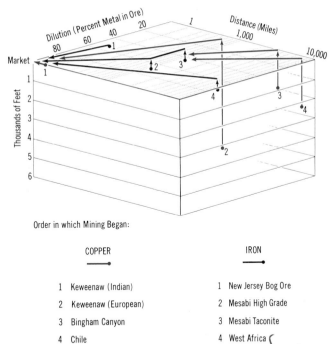

Order in which Mining Began:

COPPER

1   Keweenaw (Indian)
2   Keweenaw (European)
3   Bingham Canyon
4   Chile

IRON

1   New Jersey Bog Ore
2   Mesabi High Grade
3   Mesabi Taconite
4   West Africa

**Figure 12-1 Alternatives for exploiting a hypothetical mineral resource** The possibilities for exploiting any mineral in the earth depend upon the distribution of the element in the environment. The cost of recovery is a function of the depth below the earth's surface, the distance from the market, and the degree of dilution of the material. Technological improvements in search methods, transportation, mining, or processing extend the range of exploitation within limits of natural distribution of the material. Not shown on the diagram are recycled wastes from the market, which become more important as only lower- and lower-grade ores remain. Copper and iron ore examples shown have different histories.

quette, Menominee, Vermillion, Gogebic, and Mesabi have supplied most of the ore produced domestically in the United States for more than 50 years. Among these, the Mesabi range has been by far the most important. In the years between 1848 and 1930 1.5 billion tons of ore were taken from this district. In 1970 more than 70.2 million tons of iron ore were produced; but by 1980 the shipment of high-grade ores will have dropped to about 5 million long tons. Low-grade ores, called *taconite*, are now being

*Environmental Imbalance: Man's Role in Changing the Face of the Earth*

*beneficiated*, that is, brought to economically feasible concentrations, but taconite pellets will amount to no more than 50 million long tons shipped annually from the area. The heyday of those great mines is obviously over. As Table 12-3 shows, the demand for iron will continue to increase and will have to be supplied from other sources.

To appreciate the problem of the lack of fit between iron-concentrating systems, which best match the geologic time scale, and iron-dispersing systems, which fall within the human range of events, we must consider both kinds of processes in the context of the combined man-environment system. The story of the Midwestern iron ore deposits extends back 3.5 billion years and begins when the most primitive forms of life appeared on earth. Prior to that time the atmosphere of the earth consisted of hydrogen, helium, methane, and ammonia. There was very little free nitrogen or oxygen. The waters of the earth's surface, whether they were lakes or oceans, constituted a "dilute organic soup" of molecules probably resulting from the discharge of lightning in that earlier atmosphere. It is unclear how the first organisms actually developed in those seas, but instead of relying of photosynthesis they may have gained energy directly from compounds in the water.

During that early period the atmosphere slowly changed. The present nitrogen- and oxygen-rich mixture which we breathe is a product of processes which altered the original atmosphere, one which would have been poisonous to life as we know it. Hydrogen, which was originally abundant, must have escaped from earth's gravitational field into space in large quantities. At the same time, solar radiation broke down water vapor, ammonia, and methane, thus freeing additional nitrogen and oxygen, which in turn aided the creation of carbon dioxide. Little by little, the proportion of oxygen, carbon dioxide, and nitrogen increased until an atmosphere more like our own developed, along with recognizable plant life.

The oldest preserved life-forms are the scanty fossilized remains of primitive plants. *Stromatolites* are the most common of these. Their pillowy fossils of layered and branched calcium carbonate or silica give us a clue that blue-green algae, or something very similar, lived in those ancient seas. Such plants, however primitive, could perform photosynthesis, giving off oxygen as a by-product. Some of the oldest deposits of stromatolites occur in the Gunflint Chert, among the Precambrian sedimentary rocks on the south shore of Lake Superior. It is the 2-billion-year-old flora of the Gunflint Chert that figure in this tale of concentration and dispersion.

Somehow Precambrian stromatolites existed in the hostile environment of early earth. It has been suggested that colonies of these plants formed oases of oxygen-rich water in those early seas. It is likely that for nearly a billion years the oxygen produced by their photosynthetic processes combined with ions of iron and other dissolved minerals to form relatively insoluble oxides. These oxides precipitated out of solution to form the basis for the iron ore deposits. Evidence of this is that the stromatolites of the Gunflint Chert formation are found in association with banded iron ores. After an incredibly long period of time the concentrations of free ions in the sea would have been used up. When that happened, oxygen could begin enriching the water and atmosphere around those original oases of blue-green algae.

It is possible that the sudden appearance of animal life in the Cambrian period at the beginning of the Paleozoic 600 million years ago was the result of the availability of sufficient free oxygen. It has also been suggested that concentrations of atmospheric oxygen did not become really great until about 200 million years later during the Devonian period, when land life became abundant for the first time. Until then there was not enough ozone ($O_3$) in the atmosphere to effectively shield the earth from intense ultraviolet radiation from space. Only when green plants had provided oxygen in quantities approximating that in today's atmosphere was the ozone shield inpenetrable enough to allow life on land.

During the period from 1 billion to 100 million years ago a vast panorama of life appeared on earth, but it is not our intention to trace its development. We only wish to point out that among the alterations of the earth's surface was an extremely slow leaching process which affected a very small proportion (less than 2 percent) of the iron formations associated with Gunflint Chert. Silica was removed by percolating waters, with an attendant concentration of iron from about 25 percent to approximately 60 percent in the deposits of the iron ranges. Once the iron ore had been reconcentrated, very little happened to the deposits for most of their long history.

Less than a million years ago the continental glaciers of the Pleistocene planed off the area around the iron ranges. Much ore was removed by glacial action and spread out in the moraines and glacial till which covered the center of North America as far south as the Missouri River. That iron, while present in the soil, is too dilute for economic recovery.

Humans now begin to enter the picture. The Paleoindians of America, however, lacked the technology to smelt iron. Therefore, early man had little or no effect on those deposits. But suddenly the pace of activity quickened with the appearance of Europeans in North America. A little less than 500 years have passed since Columbus crossed the Atlantic, and yet, the end of those ores is now in sight. By 1854—less than 1.25 centuries ago—the Marquette iron ore deposit had been discovered and the first ore removed. In 1855 the Soo Locks were opened and ore boats began to ply the Great Lakes. In the same year the Bessemer process significantly increased the iron- and steel-producing potential of the newly established industries growing up along the shores of Lake Michigan and Lake Erie.

The events of the last century began to move at a faster and faster pace. The Menominee Mine opened in 1872. In 1875 coke was introduced as a new and efficient fuel for smelting iron. By 1881 the Soo Locks were enlarged and improved, and in 1892 the greatest mine of all, the Mesabi, was opened. As early as 1909 the Mesabi was providing more than 50 percent of all the iron ore used in the United States, but by 1914 iron ores were already being reconcen-

trated for ease of shipment. By 1930 over half of the ore in the entire region had been mined (1.5 billion tons extracted; 1.4 billion tons remained). And in 1945 the taconite, or low-grade ores, began to be mined. At about the same time, significant imports of iron ore were first made to the United States. By 1970 imported ores equaled nearly 45 million tons, while the Lake Superior district in the same year produced just about 70 million tons. By A.D. 2000 the shipment of conventional ores from that area will be less than 4 percent of the total projected demand for primary ores in the United States.

The lesson should be apparent: 200 years of human activity have used up the materials accumulated by 2 billion years of natural processes. The key to appreciating all this is in the definition of "used up." The world is not less rich today in iron than it was when the first European stood upon the shores of Lake Superior. Only the useful concentrations of iron ore and not the iron itself have been changed. The law of the conservation of matter indicates that, with the exception of atomic reactions, elements are not ultimately altered or destroyed. However, the removal of the iron as ore and its subsequent diffusion across the surface of the earth into a useless film of junk cars, old nails, empty cans, and rust decrease its utility to the point where serious shortages may occur in our lifetime. The use of scrap iron will delay this fate, but an inevitable increment of metal is lost with each production-utilization-abandonment cycle of the system. Even if all the metal lost were eventually to find its way to the sea where it might again be concentrated through geologic or biotic processes, the time required would be of such a magnitude that no one for the next thousand generations could profit from it. Cultural processes of dispersion operating through a hierarchy of industrial production units and retail distribution centers thus represent our first example of environmental degradation.

As we have said, once the natural and cultural processes inherent in such depletion are understood, it may be possible to slow or reverse the direction of change. It is necessary, though, for society to agree upon the price it is willing to pay for reliable resources. Metal is a case in

*Environmental Imbalance: Man's Role in Changing the Face of the Earth*

point. While consumers pay for the tin cans and automobiles they use and eventually throw away, they are unwilling to pay the costs of salvaging waste materials. Cans are almost always thrown into undifferentiated solid waste and lost in land fills. Abandoned automobiles are such a nuisance that cities like New York cannot find enough places to dump them. Moreover, junk dealers are unwilling to pay the costs of hauling scrap cars off the streets to their crushing machines. One way in which much larger quantities of scrap could be recycled, thus slowing the rate at which metals are lost in unusable diffusions, would be to pay a recycling tax or deposit on all manufactured objects. Such a tax could work in two ways. The consumer might get back his money by returning the object to a collecting station, or in the case of automobiles and other large objects, the junk dealer could receive a fee from the state for each piece retrieved. This is a small price to pay for resource security; the alternative is that our children and grandchildren will have to do without many materials unless we act now.

## Concentration

Our previous discussions lead us to the problem of large numbers of people and their use of resources. Where very large populations are concerned, the ultimate resource may be space itself. Numbers alone are not bad; consumption alone is not bad; there is nothing wrong with the production of waste. The problem is the lack of fit between the number of users and the thing used. Whether the resource we are considering is a deposit of iron ore, a school of fish, or adequate land on which to live, we need to balance the carrying capacity of the resource against subtractions from it or additions to it. We have looked at various examples of the destruction or depletion of resources. Now we must look at crowding and pollution, both of which degrade the environment.

**Crowding** An extreme act of crowding took place on June 20, 1756, during the French and British struggle for India, when the Nawab of Bengal allegedly ordered a captured British garrison consisting of 146 prisoners to be placed in a guardroom chamber measuring only 18 by 14 by 14 feet 10 inches. Two small windows offered the only source of air or outlet for heat. When dawn came and the room was opened, all but 23 prisoners had died of heat prostration and foul air in the infamous Black Hole of Calcutta.

This example may seem insignificant, but doom-sayers claim that the world itself is becoming almost as overcrowded. Allowing approximately one square yard of land per person and taking one of the more pessimistic and therefore larger projections of world population growth, estimates show that the last available land will be used up on Friday, November 13, 2104. Obviously, this is an absurd argument; something will happen before such a situation develops.

What is a reasonable limit on the human use of earth-space and what are its implications concerning crowding? What concentrations of wastes can the human ecosystem tolerate and still remain viable? Put still another way, what is the difference between unpleasant and dangerous crowding and what is the difference between distasteful and lethal accumulations of waste? Unfortunately there are no real answers to any of the above questions. A number of experiments with populations of rats, chickens, and other animals show that several kinds of population controls result from excessive crowding. On the one hand, the reduction of per capita food supplies with increasing populations can bring about starvation and/or decreased sexuality. In other animal communities faced with increasing numbers and finite resources there appears to be some sort of population control through endocrine disturbances brought about by crowding. Social disorders have been observed to develop among Norway rats living under excessively crowded laboratory conditions. Dominant males will control feeding and drinking stations at prime times, but even when they absent themselves from guarding the available resources, the less aggressive rats will refuse to partake and quickly perish. Communities like this are referred to as *behavioral sinks*. But the fact clearly remains that human beings with their large brains and tenacious ability to survive, and also their inherited social systems and other pieces of learned behavior we call culture, are not rats in a maze. It is misleading and dangerous to project and predict human behavior under crowded conditions from instances like those described above.

If we look at the densities at which people have lived for long periods of time, we realize that humans are able to withstand considerable inconvenience. In fact, two kinds of clustering take place. On the one hand is the ghettoization of unwilling minority groups, and on the other we have the self-segregation of the upper-middle and upper classes. Of course, the latter seldom live at high densities, and yet millions of people choose to live on the island of Manhattan, where densities in certain very small areas may reach over 300,000 per square mile. (Seldom do more than a few blocks exist with such populations; this figure is a statistical extrapolation). Table 12-4 gives some peak densities for small areas within cities.

**Pollution** There are both advantages and disadvantages associated with clustering at such densities. In addition to the personal stress of communications overload, we must also consider the problems associated with the sloughing off of all manner of excess energy and materials. Noise, heat, waste, and trash all create problems when brought together at nearly unbearable concentrations in cities. One example of this is the production of solid waste in American cities. Garbage, rubbish, and junk produced in America amount to 5 pounds per person per day. The grand total is thus more than 500,000 tons per day, or 185 million tons per year, of municipal waste. Moreover, this is increasing at about 4 percent per year. This is in part because of increasing populations. But at least half of our increased production of rubbish results from increasing per capita consumption.

The disposal of municipal solid waste is the third highest cost that cities face. We spend about 3.5 billion dollars on cleaning our nests at the present time, but the cost will have tripled by 1980. In some places 10 cents of every tax dollar is spent on waste disposal, and of that

Table 12-4    Peak Living Densities in Small Areas

| Area | Density Per Square Mile |
|---|---|
| Calcutta, 1951 | 218,000 |
| London, 1931 | 152,500 |
| London, 1851 | 200,000 |
| N.Y.C., 1900 | 350,000 |
| N.Y.C., 1965 | 318,000 |
| Paris, 1962 | 86,300 |
| N.Y.C.,* 1960 | 7,462† |
| Manhattan, 1960 | 75,150† |

*Standard metropolitan statistical area.
†Average figure for the total area.

money 70 to 80 percent is spent on transportation. This problem remains largely unstudied, and the future is not easy to predict. It is closely linked to the rising amount of synthetic materials which are nonbiodegradable through normal processes of rot or decay. In 1960 wet garbage constituted 5 percent of all wastes, but by 1970 wet garbage had fallen to about half that amount. Paper and single-use paper products have increased their contribution to the national garbage can the most by an impressive 50 to 75 percent. This is due in large part to excessive, poor, and deceitful packaging. More than 35 million tons of discarded packaging must be disposed of each year in the United States. Of this 60 percent is paper, 20 percent glass, 16 percent metal, and 4 percent plastic. In the average city, 1,000 tons of paper and 172 tons of metals are discarded each year by every 10,000 people.

All of this represents a kind of negative moving resource which finds its way onto the common ground. The city and its people are supplied by hierarchies of all kinds which focus all manner of materials and move them into relatively small areas. This represents the basic problem of overconcentration. But the overall problem is greater than might be supposed if we limit our considerations solely to the garbage trucks which munch their way down our streets and alleys every week. Cities are people, millions of people, and people are mouths and stomachs. To feed all those hungry mouths the nation needs livestock of all kinds. In fact in

*Environmental Imbalance: Man's Role in Changing the Face of the Earth*

1970 there were 50 million pigs, 38 million cattle, and 350 million chickens produced in the United States. Feedlots with sometimes more than 50,000 cattle are fairly common in the country today. In fact, there are 256,000 lots of all sizes where cattle are fattened. Since a single animal produces about 80 pounds of urine and excrement each day, the problem of disposing of their wastes equals Hercules' labor of cleaning the Augean stables.

Feedlots are growing progressively larger and larger, thus forcing out smaller and more evenly distributed ones. The spatial implications of this again can be expressed in terms of concentration. For example, six counties in Colorado fed approximately 5 percent of all the cattle fattened in the United States in 1963. Ten years later, in the states composing the basin of the Missouri River there was a per capita equivalent feeder cattle population greater than the human population in terms of the wastes produced. The results of all this range from half a million fish killed in Kansas by runoff from cattle feedlots to

Polluted water draining away from a cattle feed-lot. (Dept. of Agriculture photo)

major nitrogen pollution of municipal water supplies.

If we return to our earlier notions of a hierarchically organized world with supplies from all over the world flowing into central urban locations, the picture becomes even more grim. Frances Moore Lappé reports that 700,000 tons of fishmeal were imported into the United States from Chile and Peru in 1968. Although this material can be used as a human food supplement, it was fed to American animals. However, because of the relative inefficiency of animal metabolisms relatively little found its way into the flesh that was subsequently consumed by humans. Instead, most passed through the animals and on into the streams and lakes of the nation. Just as the trend is to concentrate feedlots and cattle, we may also picture a parallel system of concentration bringing nutrients from the vast waters of the southeast Pacific to a relatively small area half a world away in the central United States. Such concentrations overwhelm the capacities of local systems to carry off wastes and render them harmless through dilution and biodegradation.

It would be possible to continue with example after example of systems which bring together concentrations of materials which cannot be dispersed by natural processes. Typical of these would be the problem of urban sewage disposal around and within the Great Lakes. It seems a small matter when one toilet is flushed somewhere within the watershed of the St. Lawrence River. But when we picture feeder sewer lines linked to trunk lines by the millions, and trunk lines pouring into regional conduits, which in turn empty into rivers feeding the Great Lakes, it comes as no surprise that trouble lies ahead. This combination of human and natural waterways brings nutrients in deadly concentrations to larger bodies of water. Lake Erie is particularly susceptible to this kind of pollution by concentration because of its shallow depth and slow currents. The result is eutrophication in the form of the runaway production of algae. This, in turn, reduces the gaseous content of the waters, suffocates the fish, and generally adds to the unpleasant conditions which threaten to destroy the lake. When we consider that the

Air and water pollution along Lake Erie. Plumes of smoke and dirty water can be seen streaming to the Northeast.

lake has also become the dumping ground for all manner of industrial wastes, it does not seem unreasonable to talk about its imminent death.

Our purpose here is not to discuss lake ecology so much as to point out that the complete tree shape of most hierarchical systems must not be overlooked. In other words, for every system of concentration there should be a system of dispersion of equal magnitude and efficiency. In all the preceding examples the solution lies with considerations of the scale at which things happen. At this point we need to return to our ideas of regional versus local organizations. Most towns and cities operate at local scales when compared with natural systems like rivers and watersheds which exist at regional scales. Little wonder that many small problems result in one enormous one. There is no reason why sewage and other wastes need to destroy the environment through dangerous concentrations. The solution to this problem lies in the creation of regional and national-sized organizations for the disposal of wastes. Certainly, federal regulatory and coordinating agencies such as the Environmental Protection Agency are being created to meet such problems. However, they must resist solving the problem of national waste disposal by dumping everything into the Atlantic and Pacific Oceans. This would be simply failing to face the ultimate facts. The seas cannot cleanse themselves quickly enough of the filth of nations. Only when consumers are willing to pay the costs of adequate sewage treatment and to organize themselves at the same scales as the natural systems upon which they ultimately rely will the environment be truly safe.

### Destruction

Destruction and overproduction are so closely linked that it is useful to talk about these two causes of environmental imbalance in adjoining sections. Sometimes an animal or a plant can simply be overharvested to the point where it can no longer maintain itself. The destruction of the dodo bird and the great auk, which were used to reprovision sailing ships in the eigh-

teenth and nineteenth centuries, are examples of this. Whales and redwood trees face similar fates unless effective protective action is taken soon. In other cases flora and fauna are simply considered as pests and are wiped out as quickly as possible. Everything from bison to rain forests have met this fate. The defoliation of large tracts of jungle in Southeast Asia which were considered a nuisance during the Indochina War will have consequences reaching centuries into the future. Shameful as it is to admit, humans have also actually viewed other humans as pests to be killed and driven from the land. We do not refer to the barbarisms of open warfare but rather to scattered attempts by European settlers in North America, Australia, and Africa to eliminate local hunting and gathering groups by poisoning them or hunting them down.

Extinction sometimes goes unnoticed; but at other times the removal of a species from the ecosystem allows some other plant or animal which was originally held in check by it to experience a population explosion, that is, overproduction. This removal of natural enemies can create monsters out of the mildest creatures through their sheer increase in numbers.

**Insecticides in Sabah** The complexities of extinction and overproduction and the role man plays in those processes are illustrated by the consequences of the use of insecticides in Sabah, one of the Borneo states of Malaysia. Commercial coco production was introduced to this equatorial area in 1956. The approach used was typical of man's heavy-handed approach to nature. Large tracts of jungle were cleared, and commercial timber was removed. The remainder of the forest was either burned, or cut and left to rot, or allowed to stand as shade for the coco seedlings. At the present time, coco plantations exist as islands surrounded by secondary growth with the original forest just beyond.

The disrupted territory at the edge of the forest served as an ideal environment for fugitive and volunteer plant species. At the same time, the insects associated with that vegetation have a similar ability to move into new situations and to survive and multiply. As a result,

*Physical Geography: Environment and Man*

286

the coco plantations soon fell prey to a series of insect pests which rapidly destroyed more than 20 percent of the coco trees. A ring bark borer, *Endoclita hosei*, which appeared first, was followed by leaf-eating caterpillars, aphids, and mealybugs. By 1961, bagworms, *Psychidae*, which are tent caterpillars and the larvae of moths, threatened the coco fields with a particularly serious outbreak.

During the period of increasing insect infestation, vigorous attempts were made to control all the insects. First, inspection and hand spraying was tried. But this proved too costly and had to be abandoned. After that, massive spraying with DDT, BHC, lead arsenate, and dieldrin was carried out. At this point, the peculiar spatial characteristics of the insects involved became very important. Bagworms encase themselves in a tough silk sack covered with twigs and pieces of leaf. When threatened, they retreat completely inside this cover, and while feeding, only a small portion of their bodies extrude. Other pests which also feed on the coco plant bore into its bark and branches and remain covered. In every case, the insects which prey on plants are relatively sedentary, moving about as little as possible and feeding in local concentrations. The natural enemies of these pests are predatory and parasitic and must explore over large territories in order to find their prey. Thus, the coco pests are protected by bags and tunnels and are unlikely to wander into concentrations of insecticides unless those poisons are brought directly to the place where they are feeding. On the other hand, the insects which control them live exposed lives and cover large territories, thereby optimizing the possibility of their contact with insecticides. The results might have been predicted. The natural enemies of the coco pests were wiped out; the pests themselves were little affected by the poisons, and a runaway overpopulation of insects resulted from man's destruction of first the forest and then the controlling predators.

Gordon R. Conway in his account of the battle to save the Sabah coco plantations points out that the real solution was the cheapest and easiest one. Almost all spraying was stopped, and some of the trees from the secondary growth encircling the fields which had served as hosts for certain of the coco pests were cut down. The predators were able to reestablish their numbers and to move into the plantations and to reduce the coco pests to a level which can be dealt with by human laborers on a selective basis. In this case, man through destruction of a native environment triggered unexpected insect population growth which could be controlled only through reestablishing some kind of natural equilibrium.

## Overproduction

At other times, a plant or animal may be introduced into a new land far from its original home. The new arrival may find an open ecologic niche with conditions ideal for its growth and no natural enemies. Again, unchecked population growth can occur with overwhelming results. Perhaps the most famous example of this is the introduction of rabbits into Australia. First brought there from England by early settlers for sport and food, the rabbits quickly began to multiply, with disastrous results. Entire regions were stripped of grass, and valuable water sources were consumed and contaminated. The sheep and cattle industries soon felt the impact of these invaders, and measures were taken to wipe them out. Wholesale hunting proved ineffective. Next rabbit-proof fences were built across the countryside for thousands of miles, but they too proved useless. Biological warfare was also tried with the introduction of myxoma virus. Three years after myxomatosis first decimated the rabbit population the survivors had developed an immunity to the disease and their populations once more increased. Similar stories can be told about the introduction of the mongoose into Jamaica, the European starling into the United States in 1891, and the American gray squirrel into the British Isles in 1876. Figures 12-2 and 12-3 show the diffusion and spread of the latter two of these creatures from their starting points to the far corners of their new homes. At this point, it is useful to discuss the phenomenon of diffusion as it is viewed by geographers.

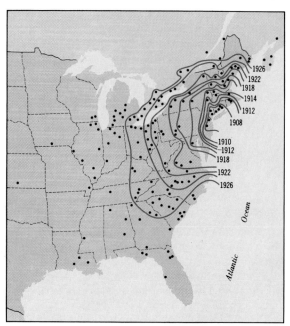

**Figure 12-2 Spread of the range of the European starling (*Sturnus vulgaris*) in the United States and Canada from 1891 to 1926** (From T. R. Detwyler (ed.), *Man's Impact on the Environment*, McGraw-Hill, New York, 1971; after M. T. Cooke, *The Spread of the European Starling in North America*, Circular, U.S. Dept. of Agriculture, 40:1–9, 1928)

## Diffusion

When gold was discovered at Sutter's Mill in California, relatively few people knew of it for the first week or two. This was because, regardless of the discoverer's desire for fame or secrecy, there were not many persons nearby to receive and pass on the message. As word reached San Francisco and finally found its way to the Eastern states, more and more carriers and recipients became available. The news *diffused* to every corner of the land, and the gold rush was on. The rate at which that information spread was a function of the propensity of the people to exchange information and the population distribution of those involved. Once the idea reached densely populated areas, it spread rapidly. We can imagine how this process works if we consider that each person who receives the message immediately becomes a new sender. As more and more senders are created, the news descends from all sides upon those still ignorant of it. It follows that the greater the overlap, the faster it will spread.

**Figure 12-3 Introduction and spread of the American gray squirrel in the British Isles** (Monica Shorten, "Introduced Menace," *Natural History Magazine*, reproduced with permission December, 1964. Copyright © The American Museum of Natural History, 1964, p. 44).

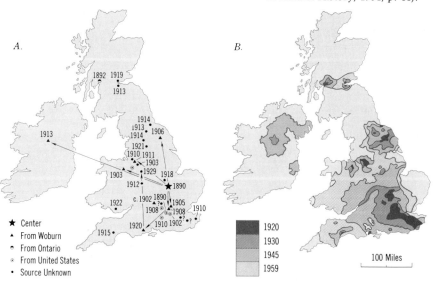

Geographers are concerned primarily with the spatial aspects of diffusion processes. They share this interest in particular with anthropologists who have studied the idea of *culture hearths* and *culture areas*. A culture hearth can be thought of as some location on the earth's surface inhabited by a group of people whose inventiveness creates ideas, traits, customs, and inventions which move or diffuse outward until people in possibly remote areas also utilize them. Similarly, a culture area is a geographically defined portion of the earth's surface the population of which shares a set of ideas, traits, customs, and inventions which distinguish that group from all others. The distinction between the two is the *originality* of things moving outward from the hearth; the inhabitants of a culture area in most cases share a collection of things which have come from all over, though a few may have originated at home.

In this respect, culture hearths are thought of as generating *diffusion waves* which, moving outward from their source, carry "messages" like ripples expanding outward from a rock thrown into a lake. While these ideas were originally discussed in qualitative terms by anthropologists Clark Wissler and A. L. Kroeber, it was the Swedish geographer Torsten Hägerstrand who presented quantitative measures of diffusion. Hägerstrand's treatment of diffusion looks at the process over a number of time intervals. At every succeeding time the "wave" moves farther through space from its source. Models such as this are *distance-dependent* and *spatial* in character.

Another type of diffusion study derives more from the work of sociologists who view diffusion as an interaction process between individuals which can take place between peers without reference to their geographic locations or by moving up and down social or administrative hierarchies. Such models are *hierarchical and nonspatial* except in the sense that movement from one level of a hierarchy to another represents changes in scale. Another aspect of diffusion studies concerns whether the number of people sharing the thing diffused increases with each move or time period or whether the number of such "knowers" remains constant.

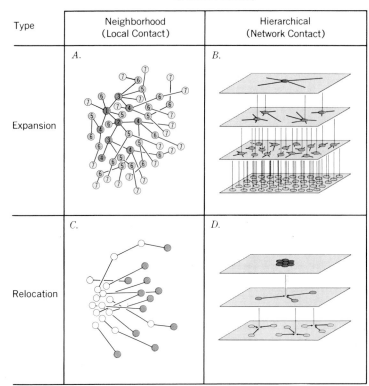

SPATIAL PROCESS

**Figure 12-4 Types of diffusion**  Examples of each type of diffusion include: (*A*) neighborhood spread of a rumor or disease, (*B*) spread of a national fad, (*C*) shift of frontier homesteads, (*D*) migration to a metropolis.

Various combinations of these traits are possible. For example, the movement of styles from one metropolitan center to another and then outward from those points to surrounding cities and thereafter to towns is both hierarchical and spatial. Figure 12-4 presents four basic combinations: in two the "knowers" remain constant in number but move either spatially or hierarchically; in the other two the numbers increase with each time period although the process can be either distance-dependent or hierarchically structured. The diffusion processes where numbers increase have particular significance in terms of human and other ecosystems. The examples given below illustrate how environ-

*Environmental Imbalance: Man's Role in Changing the Face of the Earth*

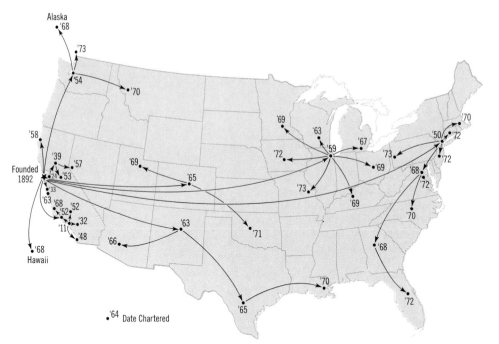

**Figure 12-5 Growth of the Sierra Club—an example of hierarchical diffusion** The hierarchical diffusion linkages are defined wherever a descendent chapter is formed from territory previously assigned to antecedent chapter. An example is the Mackinac Chapter territory in the state of Michigan, which was formerly a portion of the Great Lakes Chapter, including eleven states with chapter headquarters in Chicago.

ments can be affected both positively and negatively by the growth and diffusion across space and time of ecological elements.

**The Sierra Club** Our first example of spatial hierarchical diffusion is the growth of the Sierra Club, and organization dedicated to helping people explore and protect wilderness areas (Figure 12-5). The club was founded by John Muir and a group of friends in San Francisco in the year 1892. For many years it remained only in that vicinity. However, individual membership became increasingly widespread. In 1906 the members living in the Los Angeles area wished to form their own local chapter. Thereafter, there was a short-range diffusion of chapters in California. By 1930 a chapter was opened in New York City. Several years later the Great Lakes chapter located in Chicago was or-

ganized. In 1960 the club members decided that they could become an effective national organization, and further hierarchical growth took place from New York, Chicago, and other early centers. At the present time the Sierra Club is among the major environmental lobby groups in the United States.

**The Brazilian honey bee** Figure 12-6 shows the distance-dependent diffusion of the Brazilian honeybee *Apis mellifera adansonii*. Technically speaking, these bees are not "knowers" but, rather, an expanding life-form. However, the diffusion they follow in hiving off and swarming to new locations is analogous to the spread of disease or rumors or even the spread of knowledge of domesticated plants among early man. This bee was originally imported into Brazil from Africa in 1956. In 1957, twenty-six swarms were accidentally released into São Paulo, Brazil. The rapid spread of these insects is of particular importance because of their aggressive and unpredictable behavior, their danger to man and animals with whom they come in contact, and their ability to replace other bees. These bees have spread rapidly through the mechanism of hiving off and swarming to new homes over relatively short distances. This process is a distance-dependent one typical of contagious expansion diffusion. If this bee were to reach North America, it might well mean the end of the beekeeping industry in the United States, since their viciousness would necessitate their destruction. The resulting loss to American agriculture through inadequate plant pollination has been estimated at as much as 5 billion dollars annually. A committee of the National Research Council has monitored the advance of these bees across South America. The committee points out that control of the bees in the vastness of the Amazon basin is beyond anyone's means. They recommend that every effort be made to stop them at the Isthmus of Panama by placing new types of breeding bees with more desirable genetic characteristics in their path. Some colonies will also have to be killed or given new, less dangerous queens in order to control the menace in the heavily populated areas of South America. In any event, knowledge of diffusion processes provides scientists with better means of becoming ecological watchdogs.

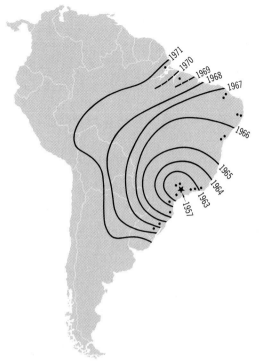

**Figure 12-6 Spatial diffusion of the Brazilian honeybee** This honeybee, which is unusually vicious, poses a threat to the people in the areas it invades. There is a proposal to try to halt its advance at the Isthmus of Panama, thus preventing its spread to Central and North America. (Gerald S. Schatz, "Countering the Brazilian Honey Bee: Aftermath of a Biological Mishap," *News Report, National Academy of Sciences,* vol. 22, no. 8, October, 1972)

**The Southern corn leaf blight epidemic** Lest we think that Sabah or Brazil are too remote or that examples like these are too rare, let us point out another example of man's tinkering, this time in the United States.

In 1970 the United States corn crop was reduced by more than 700 million bushels. This was the result of a wide-ranging epidemic of the Southern corn leaf blight, a disease caused by *Helminthosporium maydis.* This loss used up most of the reserve stocks of corn and forced prices up. Sorghum, barley, oat, and wheat reserves were used to offset the deficit. The export of grains from the United States was

affected, and thus, reverberations of this incident were felt around the world.

Both the causes of the epidemic and the rapid response of agriculturalists which restored corn production to normal by 1972 are very much connected to aspects of both location and environment. The American corn crop depends almost entirely upon hybrid varieties which have the advantages of increased yields and the disadvantage of susceptibility to unexpected disease. In the decade preceding its virulent outbreak, corn leaf blights of all kinds destroyed no more than 2 or 3 percent of the annual crop. Even hybrids were considered to be resistant to

*Environmental Imbalance: Man's Role in Changing the Face of the Earth*

such blights. A few reports of susceptibility to blight were reported in 1961, but only in 1968 did the loss of 200,000 pounds of seed corn indicate any serious problem. By 1969 seed fields and hybrid test fields showed indications of an association between hybrid corn with T cytoplasm and the corn leaf blight. By then, however, the development of a virulent strain was well underway and fields of early seed corn in Florida were attacked. By spring of 1970 cornfields in Florida, Georgia, Alabama, and Mississippi were being devastated. The blight had reached all across the Gulf states and up the Mississippi Valley into the Midwest by July. August found it in Wisconsin, Minnesota, and Canada. The dramatic losses which resulted have already been mentioned.

The destruction that resulted from the corn leaf blight is clearly attributable to man. We have bred new varieties and planted homogeneous stands of crops across entire nations. We have thereby created new environments for the overproduction of disease. Again, destruction and overproduction go hand in hand.

On the positive side of things, the problems of genetic homogeneity and hybrid susceptibility are being more and more widely recognized. Plant breeders and seed growers throughout the world attempt to produce a succession of varieties on a regular basis so that there will always be a ready supply of new genotypes as older ones become vulnerable to diseases. The spatial organization of the system helps ensure its success. At present there is an extensive network linking agriculturalists throughout the world. Nations exchange plant information on a regular basis and report the pathogenic responses of plants as soon as they are identified. Once a pathogen is found, growers send their own varieties of the susceptible crops to the infected area in order to identify resistant types. In this way, they prepare well in advance for the invasion of their own fields by virulent strains of disease. Obviously, it is far better to bring a few seeds to the disease than to introduce the disease to new areas for testing purposes. While this technique represents only a small part of an overall program of research

and control, the importance of its spatial attributes are undeniable.

Meanwhile, new seed corn lacking T cytoplasm was quickly brought into production from experimental stocks. Some farmers shifted out of corn production for the summer of 1971, while the manual detasseling of some fields of corn also checked the virulence of the blight. Altogether, the organized response of agricultural organizations throughout this country and around the world checked what could have been an even more serious disaster. Nevertheless, the problem of man-made homogeneous environments still remains and someday may be responsible for disasters exceeding our capacity to control.

### Change of state

When talking about the dangers of concentration, we have emphasized natural wastes like animal manure. In reasonable quantities manure and other by-products of organic processes can benefit the environment. This is because such compounds are *biodegradable*. This means that they are easily changed from one physical and chemical form to another and that they are quickly rendered harmless by naturally occurring processes. Their basic ingredients are recycled and become available for another round of use.

All biological processes change the state of environmental materials. Life creates complex organic compounds from simple elements. These compounds consist of large molecules, some of which store and transform energy, while others carry genetic codes enabling them to reproduce. All of this results in larger life-forms which, in turn, further order the environments they occupy. But it should be noted that the complete cycle of such systems includes the efficient return of each compound's basic materials into reusable forms. Human activities threaten earth environments by creating artificial compounds which do not easily break down into their basic elements. This is what we mean when we speak of *change of state*.

Man's manufacturing of artificial compounds

may be the most dangerous and challenging of all the processes which bring about environmental imbalance. This is particularly true if such materials are allowed to escape into the general environment. Radioactive wastes from atomic power plants are an example of this. The coming decades will see a large increase in the size and number of atomic plants brought into service (Table 12-3). The problem of disposing of their wastes will grow proportionately. Part of the trouble with such materials is that they are produced in solid, liquid, and gaseous states. This means a variety of measures must be devised in order to control them. At the same time, the radioactive half-life (the time necessary for the intensity of radioactive emissions to drop by one-half) of these meterials ranges from a few minutes to many years (Table 12-5). We have not yet developed containers for such wastes that will resist corrosion and natural forces long enough for the contained materials to neutralize themselves. Consequently, dangerous substances may escape into the environment and later be reconcentrated by some of the spatial processes we have already described.

Any nation which wishes to use atomic reactors must face the technological and administrative challenges waste disposal of this kind presents. Fortunately, the problem is well recognized by official and private organizations, and research funds have been made available with which to develop regulatory procedures and control techniques. Such concepts as 100 percent reliable engineering standards and safeguards against sabotage and the theft of fissionable materials, as well as licensing, monitoring, and inspecting atomic facilities, are all being developed. Nevertheless, mistakes have happened and can still happen. For example, we have mentioned the effect of radioactive fallout on the vegetation and animals of the tundra. Similar events in the future are all too possible.

Other important man-made materials which present environmental problems include the biologically inert plastics or polyesters from which containers, drinking straws, film packs, and thousands of other items are made. Produced in prodigious volume and diffused across

**Table 12–5  The Half-Lives of Some Radioactive Materials**

| Element | Half-Life |
|---------|-----------|
| Nitrogen-16 | 8 seconds |
| Bromine-85 | 3 minutes |
| Lead-214 | 27 minutes |
| Sodium-24 | 14.8 hours |
| Iodine-131 | 8 days |
| Polonium-210 | 138 days |
| Strontium-90 | 28 years |
| Cesium-137 | 33 years |
| Radium-226 | 1,622 years |
| Carbon-14 | 5,600 years |
| Uranium-238 | 4.5 billion years |
| Thorium-232 | 10 billion years |

the countryside, these materials cannot be attacked by decay organisms and do not rot away. This produces a new burden of solid waste that is aesthetically displeasing and may have unforeseen consequences on the environments where such things accumulate. A significant concentration of buoyant plastic materials such as cigarette filter tips and styrofoam cups has already been reported at the center of the great oceanic gyres. No one is certain what effect such concentrations will have on marine biosystems, but inevitably our children or grandchildren will pay the consequences if there are any.

Even more of a problem and on a par with radioactive wastes are synthetic materials which are active in life processes. More often than not, such materials are dangerous to plants and/or animals. Indeed, these active materials include insecticides, herbicides, and other poisons specifically created to kill unwanted life-forms. Unfortunately, such agents often kill many types of life against which their use was never intended. If the biologically active agents are themselves nonbiodegradable, they can be particularly menacing. DDT and other polychloride chemicals are in this class, as are the residues of 2,4,5-T and 2,4-D, which were the major defoliants used in the Indochina war. 2,4-D is also sold as a commercial herbicide throughout the United States, where 57 million

*Environmental Imbalance: Man's Role in Changing the Face of the Earth*

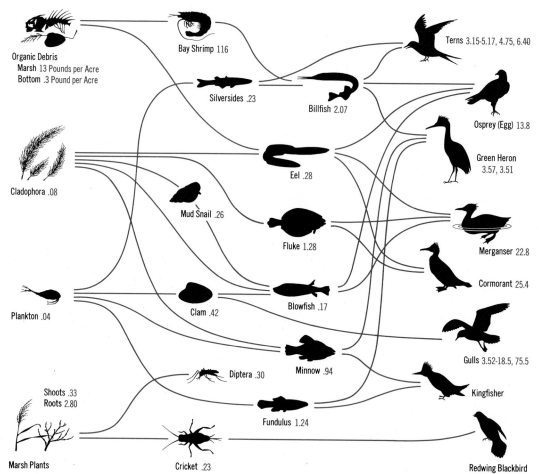

Organic Debris
Marsh 13 Pounds per Acre
Bottom .3 Pound per Acre

Bay Shrimp 116

Terns 3.15-5.17, 4.75, 6.40

Silversides .23

Billfish 2.07

Osprey (Egg) 13.8

Cladophora .08

Eel .28

Green Heron
3.57, 3.51

Mud Snail .26

Fluke 1.28

Merganser 22.8

Plankton .04

Clam .42

Blowfish .17

Cormorant 25.4

Diptera .30

Minnow .94

Gulls 3.52-18.5, 75.5

Shoots .33
Roots 2.80

Kingfisher

Fundulus 1.24

Marsh Plants

Cricket .23

Redwing Blackbird

**Figure 12-7 Concentration of DDT in a Long Island estuary food web** Numbers indicate residues of DDT and its derivatives (in parts per million, wet-weight, whole-body basis) found in each kind of organism. (George M. Woodwell, "Toxic Substances and Ecological Cycles," copyright © 1967 by Scientific American, Inc. All rights reserved.)

**Table 12-6  DDT Residues, Selected Sample (in Parts per Million Net Weight of Whole Organism)**

| Organism | Tissue | DDT Residues PPM |
|---|---|---|
| plankton (Long Island, New York) | | 0.040 |
| shrimp | | 0.16 |
| crickets | | 0.23 |
| flying insects, mostly Diptera | | 0.30 |
| sheepshead minnow | | 0.94 |
| trout (New Zealand) | whole body | 0.7 |
| black duck | | 1.07 |
| plankton (California) | | 5.3 |
| bass (California) | | 4 to 138 |
| grebe (California) | fat | up to 1600 |
| green heron | | 3.57 |
| penguin (Antarctica) | fat | 0.015 |
| herring gull (New York) | | 7.53 |
| osprey (New York) | abandoned egg | 13.8 |
| osprey (Connecticut) | egg | 6.5 |
| bald eagle (Missouri) | egg | 1.1–5.6 |
| dolphin (Florida) | blubber | 220. |
| | | |
| *Man* | | |
| (England) | | 2.2 |
| (France) | | 5.2 |
| (U.S. average) | | 11.0 |
| (Israel) | | 19.2 |

From G. M. Woodwell, "Toxic Substances and Ecological Cycles," *Scientific American*, vol. 216, no. 3, March 1967, pp. 24–31; and Charles F. Wurster, Jr., "Chlorinated Hydrocarbon Insecticides and the World Ecosystem," *Biological Conservation*, vol. 1, no. 2, 1969, pp. 123–129.

pounds were used in 1968. In 1969 more than a million pounds were used on turf alone, and in the decade from 1960 to 1970 more than half a billion pounds were spread on vegetation in the United States. Residues of these materials can be found in the tissues of most large herbivorous animals found in the vicinity of their use. It is uncertain just what effect their presence will have in the long run, but researchers have already induced malformed births in laboratory animals by exposing them to these substances. The deadly role of DDT is even better known. Its presence in the bodies of birds results in their laying thin and soft-shelled eggs which cannot hatch. A number of species are threatened by this, including the American bald eagle.

DDT and similar agents are called "hard" pesticides because they are not easily broken down into harmless forms by biological action. As a consequence, they become widely dispersed in the environment and then slowly reconcentrate by moving up food chains until they reach lethal dosages in animals such as eagles and other predators high on the chain (Figure 12-7 and Table 12-6). Of course, man is at the top of the food chain and can suffer contamination from eating the flesh of large fish and animals.

These dangers have been recognized, and

*Environmental Imbalance: Man's Role in Changing the Face of the Earth*

Steel mills pouring out smoke. This represents the "tragedy of the commons," where large organizations take advantage of a common good—in this case the atmosphere. (Environmental Protection Agency photo)

A much too common scene along American roadways. Here, individuals abuse the common good by dumping trash. This "tragedy of the commons" involves individuals as well as corporations. (U.S. Dept. of Agriculture photo)

some action has been taken to prohibit the uncontrolled use of such chemicals. Yet the ultimate control of "hard" compounds can be only partially achieved through a technical program. Society must also decide what it values most and act accordingly. Part of the increase in agricultural production in recent years is the result of the lavish use of pesticides and herbicides. The reduction of their use would help make the environment safer, but crop yields might well decline. This is not always the case, as the example of Sabah shows, but it is more than likely that less food would still be produced. Society, therefore, needs to decide what it wants most, more food or a chemically safe environment.

## The Tragedy of the Commons

Extinction and destruction are also the fate of commonly owned resources. This statement sounds harsh and extreme, but there is hardly an exception to the rule thus stated. Garrett Hardin has called this dilemma the *tragedy of the commons*. The tragedy can be illustrated with the following example.

What is it worth to a stockman to add one more steer to a pasture which he shares with other herders, particularly if they are strangers? He must match the benefits against the costs. The benefit is that he has the products of one additional animal. The cost is that the pressure of one more animal unit is directed against the grass resource. The benefit is entirely his; the cost is shared out among all the strangers. Any rational stockman concludes that the only sensible thing to do is to add another animal to the commons. In a society composed of rational men, each man is locked into a system that compels him to increase his herd without limit although the world within which he operates is limited. This is the destiny toward which all men rush. Each pursues his own interests in a society that believes in the freedom of the commons. "Freedom in a commons brings ruin to all."[2]

[2]G. Hardin, "The Tragedy of the Commons," *Science*, vol. 162, Dec. 13, 1968, p. 1245.

Different societies have devised different methods of trying to deal with the tragedy of the commons, but success has been evasive. Usually ruin is avoided by removing a resource from common ownership. If this action or something comparable is not undertaken, the resource is destroyed. This was the case with the passenger pigeon, which became extinct because each person killed as many as he wanted. It was nearly so with the American bison. Such a fate now threatens most commercial saltwater fish and sea mammals which are currently being driven to extinction.

The spatial attributes of this problem are interesting. The notion of private property plus explicit rules of inheritance removes resources from common ownership. One of the best ways to establish possession of a resource is to tell where it is spatially located and to claim it. This is the time-honored method of *staking a claim*. The identification of spatial jurisdiction both reduces conflict over ownership and allows for reasonable rates of use. Private ownership is an accepted institution in many if not most modern societies. But private ownership works well only if the resource is stationary. The dilemma persists if the resource can move. By crossing property lines a moving resource becomes a common good again. This was the argument between Robin Hood and the Sheriff of Nottingham over the deer in Sherwood forest. Robin Hood was poaching whenever he shot a deer in the King's forest. Now if he had been a freeman with a small parcel of his own land at the edge of the forest, and if a deer from the forest had wandered onto his property, where he then killed it, the act would have been legal by the *rule of capture*. This rule holds that if you can catch a wild thing on your property, it becomes your possession.

It is wise to remember that the law is a set of rules of conduct. It maintains order by following a set of precedents. When a new situation arises and a decision must be made, the blind following of precedent may not yield reasonable results. This is what happened in the petroleum industry when the rule of capture was applied.

## Black gold rushes

In the 1920s and 1930s in the United States the destructive exploitation of petroleum fields was a common occurrence. Let us examine such situations more closely. In those decades the discovery of a new oil field was followed by runaway drilling of wells. This not only destroyed any other use of the surface but also made it impossible to recover the maximum amount of petroleum from the deposit. Petroleum and gas are usually under considerable pressure, the uncontrolled release of which can result in a "gusher." If the pressure can be controlled, it can be used to bring petroleum to the surface. Drilling too many wells releases the energy of the trapped gases, and the recoverable amount of oil declines. Figure 12-8 shows the circumstances that lead to this type of exploitation. Someone discovers an oil deposit under his property. His neighbors notice his success and figure that the oil is under their property as well as his. They dig wells which cause the oil to flow more slowly in the first well since the pressure has been reduced. The first man realizes that his rate of recovery is declining, and he digs more wells on his property to get a larger share of the common fund. The neighbors retaliate with still more wells, and the result is ruin for all. Less oil is recovered in total; capital and labor are wasted, and the market is glutted, causing a drop in the price of oil. The owners will try to sue each other, but the law, looking for a precedent, quotes the "rule of capture" and holds petroleum to be like a wild thing that has wandered onto a man's property where, once captured, it becomes his possession.

## Unit pool operations

The law has been slow to change although some progress has been made in the United States. The above sort of exploitation of oil deposits has now nearly disappeared. It is instructive to consider how more rational exploitation was achieved. One method of rational development, the unit pool operation, was

STAGE 1 – Discovery!

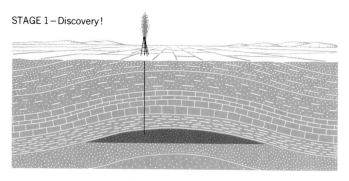

STAGE 2 – Rush to "Capture" Oil

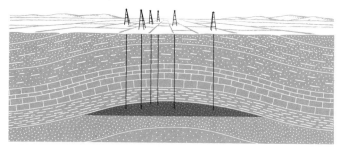

STAGE 3 – Wasteful Exploitation

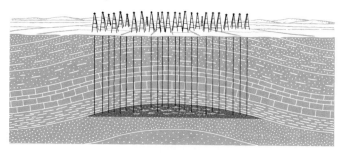

**Figure 12-8 Runaway exploitation of a moving resource** Conflicts between private and group interests occur in attempting to exploit a resource which can move across property lines. This often results in a destructive rush to "capture" the resource. Other moving resources, such as fish, whales, water and air, tend to be subject to destructive exploitation because they too tend to move across property lines or into jurisdictions where ownership is in doubt.

*Physical Geography: Environment and Man*

298

started in 1924. The extent of the oil pool or reservoir underlying a region is determined, and all the owners at the surface agree on a single plan of "producing" the field as a whole. Each owner is allotted a share determined by as fair and just a method as can be devised. You may note several problems with this procedure. There is the technological problem of deciding where the real boundaries of the pool are located. The procedure is also at variance in several ways, with commonly held notions of free enterprise, particularly those notions held by the free-wheeling individuals who were engaged in oil exploitation in the first quarter of the twentieth century. On the one hand, some infringement of antitrust laws seemed to be implied, and on the other, a degree of coercion seems necessary. To escape the tragedy of the commons by mutual agreement requires that every single party accept the allocation made by the joint operation. A single dissent would ruin the agreement. Such voluntary agreement cannot be expected. Laws had to be passed to coerce people into accepting the joint operation, and then the rules had to be administered by some authority.

The oil industry tried other methods such as accepting laws regarding the spacing of oil wells and the allocation to each well of its "allowable" proportion of total production. These plans are usually under the jurisdiction of a regulating agent, usually the state. The total output from each political unit is then regulated to maintain stable crude oil prices. Since petroleum is an international commodity, the procedure would not appear to have much chance. However, the industry is characterized by the presence of a few very large international corporations, and their oligopolistic control of volume and prices results in a fairly orderly market place. Nevertheless, the petroleum industry's solution is not very suitable as a model for other types of exploitation. In the first place, the "stability" they achieve is for their own benefit and not for the benefit of the general public. Secondly, the procedure does not work for an industry with many independent firms such as deep sea fisheries.

Another method of avoiding the dilemma of

the commons is to acquire mineral rights over such a large territory that the private jurisdiction of a single company very likely exceeds the physical limits of the oil deposit. Where private rulers control their own territories, as in Saudi Arabia and the Persian Gulf sheikdoms, the head of state traditionally owns all the land and the oil companies have found it convenient to deal with the single owner. This is obviously not a question of the commons. In every case territoriality is an important issue. In fact, spatial considerations are vital in determining the ownership of most resources. Thus there is scarcely an example of resource depletion which does not have spatial implications. This also means that in order to make the right decisions regarding their use, resource systems must be viewed in their totality. In the next and last chapter we attempt to look at several kinds of geographic systems, including urban ones, and to suggest ways in which the processes of environmental decline can be reversed.

# 13 | GEOGRAPHY: LOCATION, CULTURE, AND ENVIRONMENT

Our intention in writing this book has been to present the interaction of human and natural systems from a geographic point of view. In doing so, we have discussed many events that contribute to environmental decline and destruction. It is not our intention, however, to become doomsayers, croaking out "Nevermore," like Poe's raven. Undeniably, earth's environments are in serious trouble, but it is our belief that the spatial and natural processes inherent in geographic systems are neutral in character. That is, they are inherently neither good nor bad, but are only what people make of them. We also believe that most processes of environmental decline are reversible. This means the fate of the earth's environments, and, therefore, of ourselves, is up to each and all of us. Our well-being lies in the realm of perception, behavior, and education. There is still time, if we act now, to leave a good earth to our grandchildren. Geographic knowledge of how the world functions should help us appreciate this and suggest to us critical points where individual efforts may be most effective.

## Human Environments as Geographic Systems

The following pages present five problem environments in terms of the ideas found through- out this book. These examples are drawn from the past and present and from different cultures. We hope that the geographic insights their analysis provides will suggest guidelines for positive action in other situations. The first case describes how two different cultures obtain and use water from alluvial fans in very different ways. Adjustments to the environmental system in which the actions of each group take place clearly derive from their particular cultures.

### Alternative systems of arid-land water procurement

Water for farming in arid lands is scarce, but reliable supplies can be accumulated in several ways by natural systems. One such source of water is located within alluvial fans found at the entrances to valleys or canyons along the flanks of mountains. Even though streams in these locations may flow only intermittently, lenses of fresh water are often present under the loose materials of the alluvial fans (Figure 13-1). Water enters systems such as these when warm winds are forced to rise over the mountains. The rising air mass subsequently cools and yields orographic rainfall. The rain runs off and accumulates in the valleys, whence it flows downstream. If there is sufficient runoff, some will be disgorged into intermontane basins to form shallow, ephemeral lakes which quickly evaporate, leaving behind saline silt playas. Water that ponds on these playas is of no use for agriculture or drinking because it is salty. Another portion of the rainfall infiltrates the surfaces of the streambeds and moves underground into the alluvial fans. There the overlying alluvium protects it from quickly evaporating in the hot sun. Water thus accumulated can remain for long periods before slowly disappearing from the silts, sands, and gravels where it is stored. Evaporation is retarded under such conditions and the water remains fresh while slowly flowing through the fans to the lower playas. A saltwater table is usually found below the fresh water, but because fresh water weighs less, the lens of fresh water rides upon the heavier saline solution. People in arid lands all over the world have recognized the nature of this resource system and take various actions to utilize the stored water.

The one place man can enter this system is on the alluvial fans. In the western United States one can still find old windmills located on alluvial fans, usually a little to one side of the valley exit to avoid damage from flash floods (Figure 13-2A). Nowadays windmills may be replaced by electric or gasoline-driven pumps. Under these circumstances at least two philosophies of resource use can be practiced: quick pumping for short-term benefits and budgeted water removal for long-term returns. In the case of alluvial fan water deposits, there is a steady renewal of the resource but at a slow rate.

Quick pumping might be practiced to provide water for stock or crops over a critical drought period which the farmer anticipates will be a short one. Once the lens of water is gone, it may take many months to completely restore it.

**Figure 13-1 Arid lands agriculture** In arid lands, a lens of fresh water sometimes exists under alluvial fans along the flanks of mountains. These circumstances present an opportunity for men to put the water to their own use. The natural system is one of concentrating mountain water runoff followed by rapid dissipation through evaporation from the salty playas.

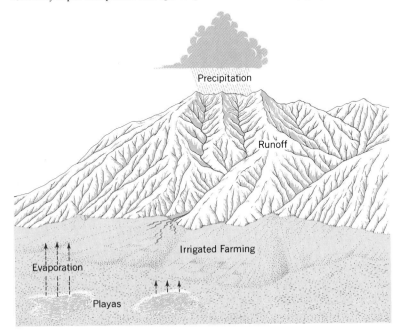

Precipitation

Runoff

Irrigated Farming

Evaporation

Playas

*Geography: Location, Culture, and Environment*

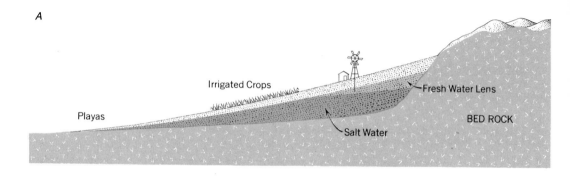

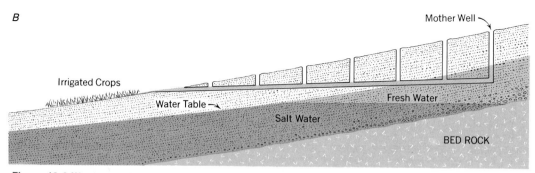

**Figure 13-2 Western U.S. alluvial fan *A*.** In the Western U.S. arid regions, the fresh water under alluvial fans is exploited by employing wind power or, more commonly now, electric or diesel pumps. ***B*. Qanats.** In Persia and many other places in the Middle East and North Africa, qanats, or "horizontal wells," are used to tap the alluvial fan water. Gravity is the source of power. The practice is very ancient and requires considerable engineering skill.

For long-term returns, it is likely that the water will be used for stock rather than for crop irrigation. In that case the number of cattle which can be raised on the surrounding steppe or desert range may be controlled by the volume of water which the alluvial fan can produce. That, in turn, depends upon the size of the catchment basin, the height of the mountains, the relative continental location of the entire system, the permeability and porosity of the ground materials, and the capacity and rate of the pumps involved. At the same time, the cattle can move only a few miles from the watering place before they must return for a drink. Thus, the range utilization which accom-

panies this water use is constrained by special environmental circumstances. If the rancher were to try to increase his herd size by increased pumping, he could only temporarily add to the number of animals dependent upon the system. The temptation presented by pumps utilizing fossil fuels for energy is that it is easy to temporarily increase the flow of water and more difficult to foresee the longer-range consequences.

In Iran and other parts of the Middle East and North Africa, similar environmental conditions are found, but the resource is utilized in a significantly different way. In those areas "horizontal wells" called *qanats* are dug in order to tap the freshwater lens (Figure 13-2*B*). A *qanat* is a series of vertical well shafts connected by a slightly sloping tunnel, or *drift*, which leads downhill and out into the valley or basin in front of the fan. The bottoms of the shafts are connected by a tunnel with a very slight gradient. The water reaches the surface by means of gravity flow where the tunnel intersects the slope of the topographic surface. The trick is to keep the gradient of the tunnel shallow enough to prevent erosion and steep enough to encourage a regular flow. The size of the village supplied by such a system (Figure 13-3) is regulated by the amount of water thus made available. The alluvial fan serves as a reservoir which evens out the intermittent and irregular flow resulting from unpredictable desert rains.

The qanat system of irrigation was apparently an ancient Persian invention, and has been utilized there since at least 714 B.C. At present, there are between 20,000 and 40,000 qanats in Iran with an aggregate length of approximately 100,000 miles. Some 20 to 25 percent of all the crops grown each year in that country depend upon qanat water. The advantages are obvious and include low expenditures of fossil fuels and capital equipment to maintain the system. The disadvantage is that these tunnels easily collapse and are difficult to restore. Labor inputs are high although this is in a nation where underemployment is a problem. More important, the work is extremely dangerous for the craftsmen who specialize in digging the tunnels by hand.

The two systems which we have just described have come in direct contact during the earthquakes of the last decade in Iran. At that time, miles of tunnels collapsed, leaving many settlements without water. Those populations might well have died or had to migrate had not relief teams arrived with equipment for drilling deep tube wells. The new wells were equipped with gasoline pumps and the flow of water actually increased. However, the prosperity which followed may well be ephemeral for the reasons cited in the American example. The rate of water removal is apparently far in excess of the natural system's ability to restore water to the alluvial reservoir. Increased cropland has

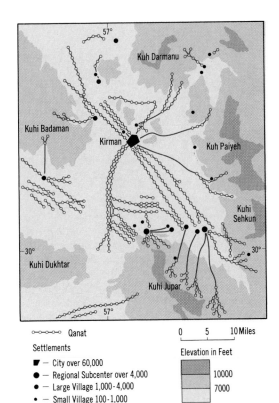

**Figure 13-3 Qanats of Kirman Basin** Qanats often extend for miles. Notice that the source, or "mother well," is usually at the mouth of a draw or canyon. (Paul Ward English, *City and Village in Iran: Settlement and Economy in the Kirman Basin,* Madison: The University of Wisconsin Press, © 1967 by the Regents of the University of Wisconsin, p. 32)

facilitated parallel increases in local populations, and we are left wondering what the years ahead will bring.

The lesson in this case is clear. Man must assess and respect the rate at which natural systems operate if he attempts to merge his activities with them. If he can speed up the overall system without creating shortages somewhere along its length, well and good. But the rates of concentration and dispersal within man-environment systems must be kept in equilibrium throughout all the subsystems, particularly the culturally induced ones, if long-term benefits are to follow.

*Geography: Location, Culture, and Environment*

## The "Great American Dust Bowl"

The use of soils and the history of land occupance associated with them are understood most easily in terms of the natural and human systems we have described throughout this text. Soils are defined by the use to which humans put them. They are defined as well by their location relative to the rest of the world productive system. The special properties of the natural environments under which they form are also important. Let us illustrate this by talking briefly about the countryside in Oklahoma and northern Texas, once known as the "Dust Bowl." That is a semiarid region with short-grass steppeland best suited for grazing moderate numbers of cattle. Chestnut and brown soils predominate; to the east are the belts of chernozems and prairie soils. Farthest away are the Eastern podzolics. Now *by coincidence* (and we emphasize this point), the major markets of the nation are also in the Northeast and are also far from the chestnut-brown soils of the dry, high plains. All this is shown in Figure 13-4.

In the first quarter of the twentieth century events halfway around the world were helping to create the infamous dust bowl of the Depression era. World War I had ended with victory for the Allies but a total disaster within Russia. The war had devastated much of western European Russia and had decimated the generation of young men who made up the rural work force. Revolution followed close on the heels of war, and the famous granaries of the Ukrainian chernozem soils lay idle for long periods. Not only did Russia cease to be an exporter of grain during those years, but famine stalked the land.

Even America was involved in trying to restore the Russian economy, and Herbert Hoover, later the thirty-first President of the United States, directed famine relief there for five years. Meanwhile, better times returned to much of the rest of the world. The Roaring Twenties with their boom economy and runaway stock market put new money in people's pockets. In fact, the new prosperity reached much of what we now call the developing nations of the world. There were increased demands for food, and there was money to pay for it. Wheat became a more valuable commodity than it had ever been before. Part of its value was its relatively short supply, which came in part from the loss of production in the emerging Soviet Union.

In the preboom economy of the United States, the low rent-earning ability of wheat had made it uncompetitive with other crops. In accordance with the theories of Thünen its low rent per acre and good transportability meant that it was produced on the outer margins of the rings of agriculture surrounding the northeastern American market (Figure 13-5). This zone of production at that time coincided with the chernozems and other soils of the better-watered plains of Kansas and Nebraska. But when wheat prices soared in response to increased demands on the world market, the competitive position of wheat altered. However, it was still unable to replace corn, cotton, or other crops of the inner rings, for their market prices were also rising. The end result was that the Thünen-like zones of production extended farther into the American heartland, and that the unleached, nutrient-rich but hitherto un-

Figure 13-4 Soil types and vegetation and use sequence from humid East to arid West in the United States

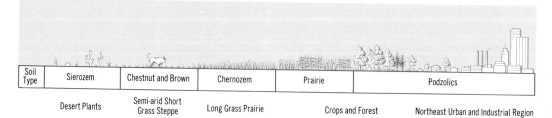

| Soil Type | Sierozem | Chestnut and Brown | Chernozem | Prairie | Podzolics |
|-----------|----------|--------------------|-----------|---------|-----------|
| | Desert Plants | Semi-arid Short Grass Steppe | Long Grass Prairie | Crops and Forest | Northeast Urban and Industrial Region |

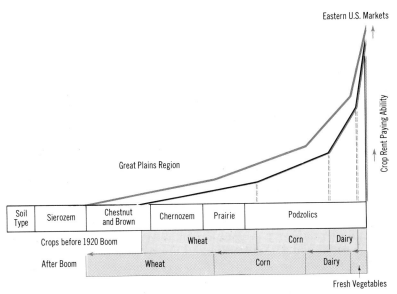

**Figure 13-5 Thünen-type model of the expansion of agriculture into chestnut and brown soil regions of the Great Plains, 1920–1930** By coincidence, the chestnut brown soils of the Great Plains were distant from the large Eastern markets in the United States, and before the 1920s the area was beyond the agricultural margin. The rise in world and national food demand and food prices during the 1920 boom created a market for crops from this region. In terms of the Thünen agricultural model, the wheat ring expanded out to this formerly marginal land. However, climatic and market fluctuations led to a disastrous decline in the farming system and, in the following decade, to the creation of the Dust Bowl condition during the Depression years.

farmed chestnut and brown soils now assumed an added allure.

Many veterans of the war had returned home with their savings and mustering-out pay and with a strong desire to be independent farmers. The opportunities offered by the semiarid lands of the Great Plains were quickly noted, and many a farmer went west to buy a small farm with his savings. This condition was also enhanced by the fact that the 1920s were favored with a sequence of relatively wet years in the area of the chestnut-brown soils. And so it happened that the natural grass cover of these areas was plowed under and wheat was planted everywhere. It is an established rule that the drier the area, the less reliable will be its rainfall. Thus, when rain fell for two or three years running in the semiarid west, it was no guarantee that those favorable conditions would last. But the small farmers didn't know that and were caught by surprise when the first years of the thirties ushered in one of the worst droughts on record.

Not only did the weather turn unfavorable in the decade after the chestnut-brown soils came into wheat production, but the great bubble of prosperity burst in October 1929 with a disastrous decline in the stock market. The Great Depression began, and the price of wheat fell very soon thereafter. Once more the Thünen rings adjusted to a new set of conditions. Production shrank back toward the market, and it was no longer profitable to raise wheat so far away. But that scarcely mattered, since the developing Dust Bowl conditions of the 1930s made farming next to impossible in Oklahoma.

*Geography: Location, Culture, and Environment*

And so the small farmers lost their land as their mortgages came due. The terrible and tragic trek of the Oklahoma migrants began, and the scenes of social and economic disintegration made vivid in Steinbeck's novel *The Grapes of Wrath* took place.

But the story of the chestnut-brown soils does not end there. After more than a decade the weather turned around again and farming once more became possible in the semiarid west. This time, however, the successful farmer was an entrepreneur with considerable capital and cost accounting skills at his command. Small farms had been consolidated into giant wheat ranches, and the businessmen-farmers who controlled them anticipated losses perhaps two years out of three and took this into account in their budgeting. Wheat price subsidies and the soil bank which paid farmers not to plant certain crops became a kind of insurance, while modern farming methods also increased prosperity in the area. At the same time, new deep wells have been drilled throughout this region, and some of the former wheat land is now under irrigation and yielding rich harvests of sugar beets and alfalfa. Prosperity is once more the key word.

We will end this account with a thought for the future. The water brought up from the deep wells of the semiarid Great Plains is a nonrenewable resource. Although it exists in large quantities, the aquifers in which it is stored were filled thousands of years ago at the end of the Pleistocene when the continental glaciers melted. Subsequently, less moisture is available and no significant recharging of those deep reservoirs is currently taking place. When the water is exhausted in the foreseeable future, what will happen next to this region? Will water be brought there from thousands of miles away? Will the land revert to modern wheat farming with one good crop for two or three bad ones? Or will some unsuspected use be found for the land? These are among the questions that a knowledge of physical and human geography helps clarify.

### The Irish famine

The Irish famine and typhoid epidemic of 1845–1847 and an occurrence of blue babies in Decatur, Illinois, in the 1960s make an interesting comparison. Unlike as the two events might at first seem, they have much in common. Both were closely tied to environmental conditions; both depended upon the interpretation of scientific investigations; both hinged upon human value systems and attitudes. The first was an outright disaster and tragedy. The latter case may show mankind's ability to find rational solutions to difficult problems. Our strategy will be to describe the famine and then to describe the situation in Decatur with frequent references to the Irish case.

The "Irish" potato was first domesticated in South America. The particular variety which became so important on European tables apparently came from islands off the coast of Peru. After returning explorers introduced the potato to Europe, its utility was eventually recognized and its use soon spread. Within the next 200 years it became a staple food for many societies. The potato was particularly welcome in places like Ireland where cool, moist conditions, high humidity, and acidic soils made many other crops impractical. Also, its high yields per acre meant that a poor man's large family could be sustained on a small piece of ground. The potato was not what could be described as a reliable crop, however, and failures of varying intensity had plagued the Irish fields from as early as 1728. Total disaster struck finally in 1845, when the entire Irish potato crop was destroyed by a blight caused by the fungus *Phytophthora infestans*. The crop of 1846 fared no better, and by the winter of 1847 the crowded poor were so weakened and brought low that they fell easy victim to a typhoid epidemic that killed many who might have survived near starvation.

The spread of the blight from America to Europe, with outbreaks in north Germany, the Isle of Wight, parts of the English mainland, and later in Ireland, is a good example of spatially dependent *contagious diffusion*. The fungus attacked both stored potatoes and those still in the ground, turning them into a black, soggy, and totally inedible mass. The Reverend M. J. Berkeley, typical of the "gentleman" scientists of eighteenth and nineteenth century England, recognized the fungus as the source of the blight and reported on it in the *Journal of the Horticultural Society* of London in 1846. His ideas contradicted the generally accepted notion that the fungus was the effect and not the cause of the blight and were rejected by the scientific community with no effort to prove them right or wrong. If the truth had been accepted, the crops might have been saved by spraying them with Bordeaux mixture and other copper compounds.

At the same time, the Irish peasantry were ill prepared to withstand the pressures placed upon them by crop failures and· a hostile or at best indifferent occupation by the English, who had dominated Ireland since Elizabethan times. The Irish population was reduced in almost all cases to an ignorant and agricultural servitude, particularly after Oliver Cromwell had crushed their uprising against him. The penal laws, 1695 to 1829, had prevented Irish Catholics from purchasing land and from participating in the military, in commerce, or in formal systems of education. At the same time, large estates were carved from Irish lands and given to absentee English landlords as rewards for service to the crown. Rents levied against the tenants were high and had to be paid with produce such as grains, since money was scarce. A farmer who failed to pay his rent was almost certain to be evicted and had little choice except to flee to the cities if he and his family were to have any chance of survival. One of the few kinds of aid available to the huddled urban poor was *outdoor relief*, working on public projects in the worst kind of winter weather. To top it all off, the English Parliament thoroughly believed in *laissez faire* economics and thought that an uncontrolled economy would force prices low enough for the poor to enter the market economy, something which did not occur. One incredible consequence of all this was that during the famine, Ireland continued to export grain in payment of rent.

All of this paved the way for the typhoid that followed. The sequence has been described by George Carey and Julie Schwartzberg.

*1. The potato failure led to famine.*
*2. Famine led to indigence and the selling of property for food.*

*3. The sociopolitical structure drove people to urban centers.*
*4. Outdoor winter work relief in the towns led to further weakness and illness.*
*5. The ragged, weakened people in the urban centers huddled together for warmth.*
*6. As they huddled together, lice spread through their rags, carrying typhus.*[1]

The result of all this was wholesale migration to North America and England of those most able to survive. The population of Ireland was significantly reduced, and it assumed an age structure with the elderly predominating and late marriages the rule.

In this case special factors in the natural and human environments encouraged reliance on monocropping for subsistence purposes. At the same time, the social and political organization of the local Irish and occupying English prevented quick, rational responses to the situation. The overall effect can be pictured as a wave of blight moving eastward out of North America which in turn triggered a diffusion of Irish culture to the United States and other places. The attitudes and values associated with the various participating groups are of particular importance in understanding the situation.

*In the Irish case, the ruling class identified with the outlook and perspectives of the English squire. Since they were, by choice, urban dwellers, frequently living abroad, the west country Irish peasant was outside of their realm of perception. From this stemmed illusions such as the naive assumption that most Irish families were within the cash economy, and that fiscal policies might alleviate the disaster.*

*The inflexible adherence to preconceived ideology was not limited to the governing classes, however. Scientists committed to a specific theory of fungus propagation rejected a hypothesis which would have proven correct regarding the cause of the blight, eschewing the possibility of empirical testing in favor of a test by dogma.*

*On the other hand, the world view of the Irish*

[1]George W. Carey and Julie Schwartzberg, *Teaching Population Geography*, Teachers College, New York, 1969, pp. 27–28.

*peasant had been rigidly constricted and circumscribed by generations of policy directed towards limiting his access to media of communication and circumscribing his arena of choice, so that although he was caught in the maelstrom of the disaster, he lacked the facilities for fully communicating his plight to the government whose views were formed behind the filtered lenses of their faulty perceptions.*[2]

### Nitrate pollution in an Illinois community

Barry Commoner in his book *The Closing Circle* discusses an ecologic problem facing the people of Decatur, Illinois, and its agricultural hinterland. It is useful to compare his example with that of the Irish famine because of similarities underlying the two systems. On the other hand, the two events stand in vivid contrast to one another because of the very different human responses which their situations evoked.

The community of Decatur provides goods and services to the farms which surround it. In many ways the prosperity of those farms underlies the prosperity of the town. Modern farming techniques necessitate the use of large amounts of nitrogen fertilizer which dramatically increase corn yields. At the same time, unused nitrogen washes from the fields into the streams, lakes, and wells upon which the farm and urban populations depend for drinking water. There is strong evidence that when the nitrate level in water rises above the recommended limit of 45 parts per million, infants, particularly girl babies, may suffer and even die from oxygen starvation; that is, they may become "blue babies." This happens when dissolved nitrates are converted by intestinal bacterial action into nitrite. Nitrite is taken up by the hemoglobin in the blood, and the methemoglobin which results prevents oxygen from being carried through the infant's system in sufficient quantities. The citizens of Decatur are, therefore, faced with a real dilemma: must they give up their agriculturally based prosperity in order to have healthy children? The

following account outlines their situation in greater detail and reveals the importance of maintaining good communication channels in human ecosystems.

The soils of the Midwest are well suited for the mixed grain-livestock farming that typifies the area. In previous decades green manure from harvested crops, as well as the manure of animals, was returned each year to the fields. This restored vital amounts of organic nitrogen to the soil and kept crop yields reasonably high. In recent years, however, the situation has changed. It has become too expensive and inefficient for the farmers to return waste products to the land. As a result, natural fertility has declined and the soil must be restored by large inputs of man-made fertilizers. As more and more nitrogen is applied, crop yields increase dramatically (Table 13-1). It should be noted though that the first 100,000 tons applied to Illinois fields produced a 40 percent increase in yield, while the next 300,000 tons produced an additional increase of only 36 percent. In other words, a point is reached where each added increment of fertilizer produces less of an increase than the one before. Nevertheless, the farmers must use heavy applications of nitrogen fertilizers to make a profit. As you will recall from the discussion of Thünen's theory of land use, this is a high-rent area for agriculture. In fact, corn must return about 80 bushels per acre for the farmer to break even. His profits are made by the additional bushels of corn he can force from the land by adding more and more fertilizer. If he can produce 95 bushels per acre, he makes 15 bushels profit. But the extra fertilizer he adds to get those additional bushels of profit is used less and less efficiently. The nitrogen which is not taken up by the plants simply washes off the land.

Now let us pause for a moment and compare this and the previous example. In both nineteenth century Ireland and twentieth century Illinois relative location and site conditions played important parts in the development of the respective patterns of land use. In both cases, climate and soil required either special crops or special technology for optimum yields. From the point of view of a possibilist, many

[2]Ibid.

choices were open to the people occupying the land, but in more pragmatic terms potatoes and corn were good answers to the question of which crops to plant. Not the least of their choice depended upon the need for high yields. In the first case, potatoes supported large populations of rural poor on small tracts of land. In the latter, high production costs necessitate large yields for reasonable profits.

Once more considering Decatur and its problems, the effects of excess nitrate are not immediately apparent, and for a long time there was doubt in some people's minds that the high level of nitrate in Illinois rivers was directly the result of the increased use of fertilizers. Barry Commoner had scarcely presented a paper on this topic at the annual meeting of the American Association for the Advancement of Science when his conclusions were challenged by representatives of the fertilizer industry as well as some agricultural scientists who had helped perfect the use of nitrogen fertilizers on Midwestern farms. This is an example of what we have called "when men of goodwill disagree." The issue at question was not the agricultural benefits resulting from fertilizer use but rather the importance of the long-range consequences possibly associated with that use. Despite the fact that some of those involved maintained a "My mind is made up. Don't disturb me with the facts" attitude, there was sufficient freedom and objectivity of thought to launch a thorough investigation of the matter, and various groups within the scientific community established the clear-cut relationship between intensive fertilization and the increased nitrogen content of Illinois streams. This is a far cry from the closed minds and indifference with which Reverend Berkeley's conclusions were greeted by English savants a century and a half earlier.

Commoner's next step was to communicate directly to the people concerned. In 1970 he and others reported their findings to health department officials, farmers, and agronomists at a special seminar held in a local high school. This "grass roots" method of action was an effective one. Though not everyone could immediately accept the implications of those reports, the basic goodwill and cooperation of the entire

**Table 13-1    The Relationship of Crop Yields to the Application of Fertilizer Nitrogen**

| Time Period | Amount Applied to Illinois Fields | Yields per Acre, Bushels |
|---|---|---|
| 1945–1948 | Minimal amounts | 50 |
| 1958 | 100,000 tons | 70 |
| 1965 | 400,000 tons | 95 |

Source: Barry Commoner, *The Closing Circle*, Bantam Books, New York, 1972, p. 81.

community in the face of environmental threat soon became apparent. As a result of that meeting, efforts were intensified to find alternative ways to maintain crop yields at high levels without the overuse of nitrogen. At the same time, once the farmers were made aware of the possible side effects of their fertilizers, they indicated their willingness to substitute those alternative farm methods wherever possible.

The issue is still far from resolved, but the situation around Decatur stands in sharp contrast to the narrow-minded and predetermined attitudes of the nineteenth century politicians responsible for the ineffectual response to the great potato famine. Here, then, is another insight into effective environmental management: every effort must be made to maintain and increase the level and efficiency of communication flow and information handling. In other words, the need for education and research concerning critical issues never ends. Well-informed minds will not do away with differences in opinion about what is best, but they vastly improve the chances of working out compromises acceptable to all.

### Los Angeles as a human ecosystem

In order to understand the Los Angeles region in terms of the ideas introduced throughout this book, let us first diagram the general system which it represents. The interlocking flow of natural and cultural processes and elements which constitute the system known as Los Angeles is shown in Figure 13-6. On the far left are listed major characteristics of the atmos-

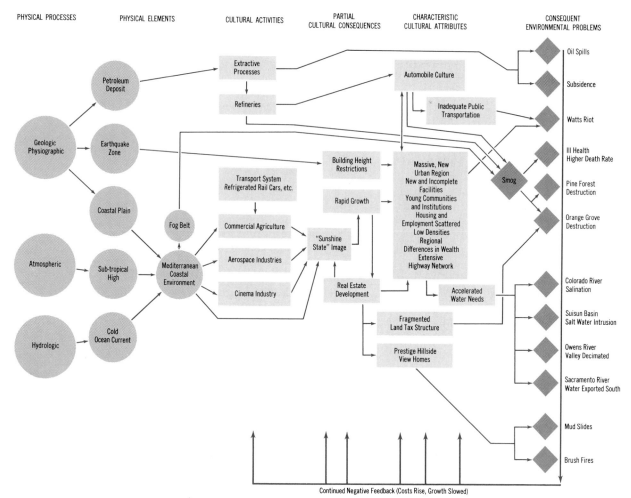

**Figure 13-6 Los Angeles basin physical/cultural system** A systems diagram of the Los Angeles Basin, including both natural and social elements. The many inputs combine in complex ways, some of which create unexpected and unwanted consequences. These act as negative feedbacks to further influence the system. The diagram is not meant to be complete and does not imply that any equilibrium state is in sight.

phere, hydrosphere, and lithosphere in Southern California. There are an infinite number of things that might be shown, but we have chosen a few which we think are critical. First is the land itself, a coastal plain backed by, among others, the San Bernardino and San Gabriel Mountains of the Transverse Range. Behind the mountains to the east the Mojave Desert mer-

ges with the higher deserts of Nevada and Arizona.

Geologically speaking, this is an unstable area with numerous fault zones either penetrating it or located on its boundaries. These include the San Gabriel fault complex, the Newport-Inglewood system, and the Santa Ynez, which crosses both Ventura and Santa Barbara

Counties. Tremors are a frequent occurrence and major earthquakes have occurred in almost every decade. However, despite the presence of fault zones, the sedimentary rocks of the Los Angeles lowlands contained valuable petroleum deposits.

The general location of Southern California at 34° north latitude places it within the influence of the subtropical high-pressure system. A Mediterranean climate prevails, with mild and often rainy winters and long, dry summers. At the same time, the hydrology of the area reflects its global position as well as more local influences. The California Current that flows offshore from north to south is relatively cool. When the land warms, local offshore breezes move inland, bringing fog which is suitable for certain kinds of agriculture. But the warming air increases its ability to hold moisture and the land beneath is dried out as the cool sea breezes warm and rise over the land.

The result of all this and more is a Mediterranean coastal environment well suited for specialized agriculture. Here then are some of the major physical elements and processes which have been discussed elsewhere in this book. Let us begin with them and thereafter by adding human activities follow through some of the sequences of events that affect Southern California.

The felicitous environment of Southern California was particularly good for agriculture, but little except local production took place until the railroads with refrigerator car technology pushed through to the city. The combination of improved transportation and suitable year-round growing conditions allowed the development of commercial agriculture which at first emphasized citrus fruits. The orange seems like a small sun itself, and it was only a matter of a few years until someone first called California the *Sunshine State*. The movie industry, which had its early beginnings in New Jersey and elsewhere in the East, was attracted by the warm and cloudless weather which allowed outdoor filming every month of the year. People who watched the silver screen knew where the "movies" and then the "talkies" were made, and those sunny skies got good publicity. The

aircraft industry also chose Southern California because of the weather, although its arrival was considerably later. Clear skies allowed testing throughout the year, and warm temperatures made it unnecessary to build well-insulated or heated hangers. Aircraft are complex machines requiring skilled workers and engineers, and one of the selling points employment agencies gave for moving out to California became the good climate, which compensated in large part for the state's remote location on the far side of the continent. Real estate developers and the chamber of commerce went to great lengths to promote the "Sunshine State" image. This perception of California as a desirable place to live and work prevails to the present time. This is shown by the high ranking given it on Peter Gould's mental maps described in Chapter 11.

Early rapid population growth began in the 1880s when the Santa Fe Railroad reached Southern California. The original little Spanish settlement, which in 1800 had only 300 residents, had swelled to 50,000 by 1890. (The first oil field began production in 1892 and added to the image of limitless riches.) The second rapid increase in production came soon after the Union Pacific was completed to the city in 1905. At that time, small local industries provided for the population's needs, but the range of those goods was insufficient for them to compete in Eastern markets. Meanwhile, the increased agriculture and the increased population required more water than the coastal plain could provide. The Los Angeles Aqueduct was built to the east side of the Sierra Nevada Mountains in 1913, and the extra water which this brought provided the means for Los Angeles to grow by annexing thirsty communities on its borders. Here was one of the first resource conflicts between the city and the land.

The first search for additional water led the city to the Owens River, 223 miles north at the foot of Mt. Whitney. But the water taken from there was not enough. Los Angeles by the 1920s had purchased water rights in the northern Owens Valley. Crawley Lake Reservoir and the Mono Extension Canal followed by 1935. The result was the near depopulating of the Owens Valley and the death of Owens Lake as its

*Geography: Location, Culture, and Environment*

Plaza of Los Angeles, about 1890. This is how the center of Los Angeles
looked in 1890. Compare this photograph with the more recent
photographs shown below. (Wide World Photos)

The Center of Los Angeles in 1957 and 1973. In 1957, the skyline of Los
Angeles was rather flat and was dominated by the twenty-seven-story City
Hall (white tower, top center). Building height restrictions, inspired by fear
of earthquake damage, had resulted in no building over thirteen stories
high except for the City Hall. The height restrictions were later lifted,
and the downtown financial district now has several buildings over fifty
stories high. Los Angeles is an example of a growing and changing city
landscape which reflects complex interaction between environmental conditions, technological capacity, instutional control, and aspirations of its
citizens. (Wide World Photos)

waters were diverted to the city. Two-thirds of the water used in the modern metropolitan region still comes from that area, but the demand continues to grow and more water is now brought from central California. The resources of one area are being moved to another and millions of people are benefiting from such a transfer, but the end is not in sight. Water is also being brought from the Colorado River via the Colorado River Aqueduct. The removal of those fresh waters from the river plus the addition of mineralized drainage water from irrigation projects has resulted in an increased salt content in the river flow crossing the border into Mexico. In the international agreement regulating use of the river the amount of water that the United States must pass on to Mexico was agreed upon but nothing was said about its salinity. Once again, the finite nature of a resource has led to ill feeling between two groups of well-meaning, but self-serving, people. Even the Sacramento River northeast of San Francisco has been used to meet these increasing water demands. Now, thirsty Southern Californians are eyeing the water resources of the extreme northern part of their state. The manipulation of California's waters has centered, in large part, upon the California Water Plan, which was authorized by the state legislature in 1947. The plan utilizes the Sacramento River, with much of its water being diverted to the south. The Oroville Dam, a major part of the Feather River Project on a tributary of the Sacramento, also provides water for urban populations to the south. But the real prize for which Californians yearn is the abundant waters of the Pacific Northwest. However, the prospective use of their waters outside of their own region is viewed with no enthusiasm by the people of Washington, Oregon, and British Columbia, and the federal government has banned the majority of feasibility studies of such projects. It would seem that the answers to water shortages in the Los Angeles area must come from within its own region, or if not, at least from within the American Southwest.

In this way, we can trace a path through the general system from hydrosphere to climate, on through the influence of man and his percep-

tions of the area as a desirable place to live and of water as a freely expendable resource. We come at last to the negative results brought to water source areas beyond the city. But as water demand continues to grow, word of water shortages diffuses outward to the rest of the nation. This has in a small way formed a deviation dampening mechanism or negative feedback which may discourage further migration to the state. Whether or not it and others will come in time to take the pressure of excess population off the Los Angeles area is debatable, but before translating the question of crowding and the subsequent runaway use of resources into a more general case, let us look at some related problems of pollution and environmental degradation in Southern California.

This time let us begin with the geology and quickly move on to the petroleum deposits and their exploitation. We have already mentioned the discovery of oil in 1892. In the dozen years between 1917 and 1929, fifteen new oil fields were discovered in the Los Angeles basin. The coastal stretch became important with the Huntington Beach field in 1920 and the Signal Hill field in 1921. Refineries and tank farms were quick to follow, and the state became an exporter of petroleum products to Eastern markets via the Panama Canal. In 1936 an important new field was opened at Wilmington near the harbors of Los Angeles and Long Beach. Although the field's production varied in the next three decades, a total of 913 million barrels of oil, 484 million barrels of water, and 832 million MCF of gas (MCF = 1,000,000 cubic feet) have been removed from about 10 square miles of territory. The removal of all that material had the same effect as a leak in a water bed. Subsidence of the surface of the land above the deposits was detectable by 1941. The rate of subsidence has been carefully measured and reached its peak in 1951 with 2.37 feet per year, the same year in which maximum pumping of 140,000 barrels of oil per day from the field was attained. The nearby Signal Hill field also contributed to the sinking of the earth. Altogether the land sank 27 feet at the center of this depression (Figure 13-7).

The solution to this problem was twofold. At

*Geography: Location, Culture, and Environment*

313

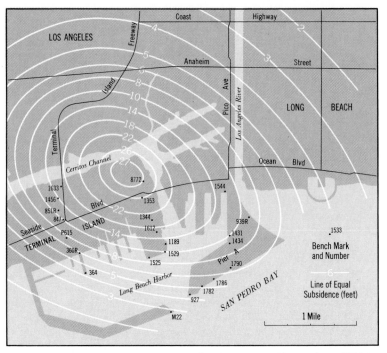

**Figure 13-7 Long Beach harbor area, California, subsidence from 1928 to 1962** (J. F. Poland, and G. H. Davis, "Land Subsidence Due to Withdrawal of Fluids," *Reviews in Engineering Geology,* The Geological Society of America, vol. 2, 1969)

first, levees, bulkheads, and retaining walls were constructed. But dire predictions of further subsidence forced the city and the producers to take a more rational look at the problem. The final solution was to repressurize the field by pumping specially treated water back into the ground. As a result of this action there has been up to 15 percent rebound of the initial subsidence, which has in general been checked. It is interesting to note from a geographic point of view that the runaway exploitation—particularly of Signal Hill—created the most serious obstacle to repressurizing. The deposit was owned and tapped by 117 producers, all of whom had to agree to such drastic measures. Coordinating them and forming them into a single operating unit was much more difficult than the engineering problems involved.

Here we have followed one path through our system showing Los Angeles. Let us retreat partway along it and look at another problem.

The search for oil has taken drilling rigs out to sea. Southern California has its share of offshore wells, particularly in the Santa Barbara Channel opposite the city of the same name. One such well "blew" on January 28, 1969, when workmen pulled the drilling bit from the hole. While wells can usually be quickly capped under these circumstances, the well casing in this instance scarcely penetrated beneath the ocean floor. The escaping oil burst out through shattered rock of an adjacent fault and flooded to the surface. For eleven days, while crews worked around the clock trying to stanch the flow, an estimated 21,000 barrels per day escaped. The slick that spread out on the surface eventually covered 800 square miles. Despite frantic efforts of thousands of conservationists, beach bums, businessmen, dowagers, and the general public, the loss of marine birdlife was extreme. Sea mammals and fish also perished from being coated with the sticky mess. The well was eventually closed off; the oil was finally cleaned up; the dead wildlife was buried or drifted out to sea; but the questions raised by this and other disasters such as the wreck of the tanker *Torrey Canyon* off the coast of Cornwall, England, cannot be overlooked. Safety measures adequate to meet the problems of dangerous geology or wreck-proof ships are too expensive for board chairmen to justify to their stockholders. The result is that too often the costs are pushed off on the general public, who must clean up the mess or go without beaches and birdlife. In terms of what we have said before, the ocean belongs to no one. It is the *commons* all over again, and there is nobody to protect it.

Again we move back along our systems path, this time to the refineries which produce gasoline for the thousands and thousands of automobiles in the city. These same refineries also emit fumes into the polluted skies over Los Angeles. Refinery fumes, hydrocarbon exhaust gases from automobiles, and the famous California fog combine to make an eye-smarting, acrid, debilitating smog. Though few people die from its effects, orange groves have suffered a decline in production and quality of fruit, while vast tracts of pine forest in the mountains beyond the city are dying from the same cause. Most of the blame can now be traced to automobile exhausts, which contribute about 90 percent by weight of the total air pollutants. Pollution control devices on automobiles help to some extent, but the number of cars is increasing so rapidly that major efforts at control result in only small gains. Table 13-2 shows some of the tonnages of pollutants entering the air in Los Angeles County each day (photo, page 3).

Again the commons and its tragedy come to the fore. Which among us is willing or able to give up his car for the good of the public? If this is difficult for the average American, it would be well nigh impossible for the citizens of spread-out, freeway-laced Southern California where no adequate public transport exists.

## Effecting Change in Geographic Systems

What can be done about pollution of the air, the waters, and the land? Even the land itself

**Table 13-2  Emission of Air Pollutants in Los Angeles County**

| Year | Source | Amount, Tons per Day |
|------|--------|----------------------|
| 1950 | Transportation | 6,888 |
|  | Other | 3,240 |
|  | Total | 10,128 |
| 1965 | Transportation | 13,054 |
|  | Other | 1,792 |
|  | Total | 14,846 |
| 1970 (projected) | Transportation | 10,083 |
|  | Other | 1,770 |
|  | Total | 11,853 |

Source: "Summary of Total Air Pollution Data for Los Angeles County: A Report of the Engineering Division," Air Pollution Control District, County of Los Angeles, January 1965, pp. i–ii and 1–41.

has become a dwindling resource as urban sprawl increases. More generally, how can rational change for the better be implemented in geographic systems? Although the nature and sources of trouble vary from one location to another, the basic issues remain the same. The problem may be air pollution in one place and solid waste in another, but in every case, underlying similarities catch our attention. Air pollution is a suitable topic with which to begin our discussion, but let us leave California and consider such matters in a broader context. After all, Los Angeles is everywhere.

An analysis of the spatial and temporal aspects of air pollution suggests some remedial strategies. First, we must recognize that serious localized air pollution episodes occur when the normal diffusion action of the atmosphere fails to dissipate pollutants because of certain meteorological conditions. Such dangerous concentrations are localized in time and space. They take place sporadically and with varying frequency, but not entirely in an unpredictable manner. Given sufficient knowledge of the weather, these events can be forecast (Figure 13-8). As our meteorological knowledge improves and as our capacity to monitor the atmosphere increases, we can expect pollution forecasts to become more accurate. Data gathering and sur-

*Geography: Location, Culture, and Environment*

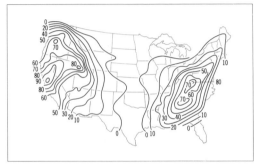

**Figure 13-8 Frequency of high air pollution potential** The number of days that widespread high air pollution potential was forecast throughout the United States are shown in this map from the Environmental Sciences Services Administration. The program began in the East on Aug. 1, 1960, and in the West on Oct. 1, 1963. Between these respective dates and Apr. 3, 1970, there were thirty-nine episodes in the West and seventy-five in the East. The numbers indicate the days a particular area was affected by a high air pollution potential forecast, for example, the area between the line marked "0" and the line marked "10 days." This map does not show the many additional days of bad air pollution weather of less than 75,000 square miles extent, or of less than thirty-six hours duration, or both. (© 1971 Committee for Environmental Information. Reproduced from "Episode 104" in *Air Pollution* by Virginia Brodine by permission of Harcourt Brace Jovanovich, Inc.)

veillance of the total environment thus become important parts of any remedial program. But facts alone do not produce solutions; those must be based on policy agreements among the parties involved.

Our concern here is with the interplay between physical processes and cultural ones. If the physical system temporarily assumes a particular configuration like an inversion layer over a city, it may be possible for the cultural system to respond by changing its configuration during the period of crisis. If over a long period of time the natural system simply cannot meet the demands placed upon it, it may be possible to modify the long-range behavior of the society involved. On the other hand, technological improvements such as emission control devices on automobiles may become economically feasible and remove unmanageable burdens from the natural system.

In the case of air pollution, if early warning devices give sufficient advance notice, measures can be taken such as shutting down incinerators, factories, and power plants which emit large quantities of pollutants into the atmosphere. Traffic can also be curbed, and the populace told not to burn trash or leaves. Control systems similar to this already exist in Los Angeles, the New York–New Jersey region, and elsewhere, although as yet they are not totally successful.

Strategies of this kind represent the least-cost approach to adequately controlling short-duration events. Costs include the disruption of normal affairs and the price of the early warning system. Legal and institutional arrangements must be agreed upon and implemented in such a way that when action is necessary, affected segments of society will accept the disruptions and inconveniences and act responsibly. This includes specific agreements with factories and businesses that they will shut down or cut back their operations. The Los Angeles petroleum refineries have already accepted such measures. But the public must also be involved in the remedy. They must be educated to the dangers of the situation and kept informed. For example, the "murk" index of pollutants contained within the atmosphere is a good daily measure often reported on radio and television. The question is, once the public is informed, how can individuals be persuaded to do their bit, to accept their share of inconvenience? Effecting change in mass attitudes and behavior is perhaps the most difficult thing to do. As Pogo says, "We have met the enemy and he is us."

One tragedy of the commons is the difficulty of charging individuals for their use of a moving resource. The air we breathe has traditionally been considered a "common good." It costs nothing, is found everywhere, and moves unimpeded across jurisdictional boundaries. However, as Roger Revelle points out:

*With an inversion at 1,000 feet over a crowded city, the weight of the column of air available at any given time near the ground is about 100 tons per person. If the air is to remain breathable (carbon dioxide concentration less than 1 percent), the amount required each day for combus-*

*tion of gasoline, fuel oil, diesel oil, coal, and natural gas at rates of use prevailing in 1965 is close to 6 tons per person. Thus, if the inversion process persists, the city air needs to be completely changed at least once every 17 days. Periods of near stagnation lasting four or five days occur several times a year over large parts of the United States. It is clear that, even with present population densities and rates of use of fossil fuels, our cities are coming uncomfortably close to using up all their available air.*

*Under these circumstances, we can no longer think of urban air as a "free Good." Instead, it must be thought of as a natural resource—that is, as part of the natural environment for which the demand is liable to outrun the supply, and to which a cost can be attached.*[3]

## Technological, Legal, and Behavioral Solutions to Environmental Problems

The rights of private ownership as well as typical economic attitudes do not apply when the common good is threatened. For example, as the air becomes polluted, public policy based on informed debate is needed to establish the acceptable quality that the air must have. Further debate and arbitration must determine how that quality shall be reached and maintained. This brings us face to face with three different types of policy suggestions. These can be called *technological, legal, and behavioral solutions to environmental problems.*

For example, Revelle considers and rejects a technological solution which has already become the mode of operation in many cities. "The number of automobiles operating in the city at any one time might be lowered by greatly enlarging the area covered by freeways and parking lots, which would speed up traffic and reduce the time spent in cruising the streets, looking for a place to park. . . ."

Typical of legal solutions is the Federal Environmental Protection Agency's suggestion that gasoline rationing and extra gasoline taxes may be needed to reduce the severity of air pollution in Los Angeles.

[3]Roger Revelle, "Pollution and Cities," *The Metropolitan Enigma*, James Q. Wilson (ed.), Chamber of Commerce of the United States, Washington, D.C., 1967, pp. 86–87.

Washington, D.C., commuter traffic on Highway I-95 in northern Virginia.
So long as people insist on commuting by automobile the diurnal ebb and flow of commuter traffic from suburb to city and back again makes automobile congestion inevitable. Notice the difference in traffic density in the two directions. This picture, however, is of a rather special situation. An attempt is being made to induce people to use commuter buses by giving buses exclusive use of two lanes which can be changed to correspond with the directions of heaviest flow. A commuter bus is shown moving freely on these lanes reserved for it, while the inbound Washington auto and truck traffic is bumper to bumper. (Wide World Photos)

Returning from market along a mountain trail in Colombia, South America. The mode of travel and the quality of the route available greatly affect how far and how often one goes to market.

A third type of behavioral solution was initiated in 1969 to deal with traffic congestion in the nation's capital. The Northern Virginia Transportation Committee and the Metropolitan Washington, D.C., Council of Governments with help from the United States Department of Transportation set aside two traffic lanes on the Shirley Highway leading from Virginia into Washington, D.C., for exclusive use by express buses. This was an attempt to modify the spatial behavior of commuters by rewarding them with quick, comfortable trips if they left their cars at home. By 1972 more people were riding the buses during rush hours than were using private automobiles. Quantitatively, bus passengers increased from less than 2,000 per day in 1969 to more than 18,400 per day in 1972. Such experiments follow B. F. Skinner's ideas of behavior modification through rewards (i.e., faster commuting service) rather than threats (i.e., dire predictions of environmental disaster) or punishment (i.e., rationing and taxes). Another example of behavior modification through rewards further illustrates this point. The San Francisco Toll Bridge Crossing Authority now charges different toll rates for automobiles depending upon the number of people each car carries. The more passengers, the less the toll. As many as 1,400 fewer automobiles per day are now entering the city as the result of this project. A decrease in air pollution inevitably will follow. The additional benefits of lowered congestion and less noise are also obvious.

Let us see how these strategies might apply equally well to the problem of urban water shortages. We have already said that there is sufficient water in the world for everyone, but the introduction of cultural subsystems into the hydrologic cycle can upset its equilibrium. As a result, many cities face increasing water shortages. Again, three types of remedies can be applied. Technologically, bigger and longer aqueducts can be built. But we have already pointed out that interregional water transfers are becoming increasingly difficult to arrange because the people in source areas are unwilling to part with scarce resources. From a geographic point of view transportation costs also tend to make such plans less feasible with every mile of

distance added. In like manner, spatial systems of concentration, even of water, may tend to overload the carrying capacities of local environments.

Legally and economically, water would become more abundant if its true costs were charged. According to Theodore Schad, executive director of the National Water Commission created in 1968 by President Johnson, "Water is so important that the country can't afford in the future to give it away or make it available at less than cost."[4] We agree with Schad's statement when we consider the activities of the Army Corps of Engineers and similar government agencies over the last 50 years. Numerous dams, canals, levees, drainage projects, and other waterworks have been provided at little or no direct cost to their users but rather have been charged against the pocketbook of the general taxpayers. This had led to contradictory situations. For example, in 1972 the government paid farmers to keep upward of 50 million acres of land out of production; at the same time it paid as much as 90 percent of the cost involved in reclaiming 9 million acres of land per year. At city scales, water users are very often charged a flat rate or one based on some arbitrary measure. For example, though New York City meters water consumption of its commerical and industrial consumers, this amounts to only about 25 percent of the total consumption, an average 1.2 billion gallons per day. Other water users are assessed water charges by the front foot of property rather than by the actual amount used. But to charge the true costs of water by legal fiat would go in the face of the voting public.

A third solution is suggested by a program of the Washington, D.C., Suburban Sanitary Commission (WSSC). A predicted and serious water shortage faces suburban Washington, especially every summer. To counteract it the WSSC in April 1972 began an experiment including improved technology and consumer education. A test area consisting of 2,400 households in Cabin John, Maryland, was chosen and

[4]Quoted by Constance Holden, "Water Commission: No More Free Rides for Water Users," *Science*, vol. 180, no. 4082, Apr. 13, 1973, p. 165.

its inhabitants exposed to a thorough campaign of community education. Radio, television, and newpaper spots were used to sensitize the public to the problem. Neighborhood meetings were held to answer questions, slide shows were given, and informative publications were distributed to consumers. Water users were made aware of the relationships that existed between their individual uses of water, the impending shortages, and the resulting overload on sewage treatment facilities.

At the same time, flow reduction devices were installed on faucets, showerheads, and water closets in the 2,400 houses. Preliminary results of a 17-month trial period show that per capita water consumption in the test area has been reduced by 30 to 50 percent. At the same time, carefully developed relationships with the consumers have ensured that few complaints occurred. The result is an encouraging pilot model for similar plans being initiated elsewhere around the nation. It is this combination of several strategies which brings the best results.

## Poco a Poco

The message of woe in our discussion of Southern California was that little by little, *poco a poco*, things can go wrong. Geographically speaking, the tiny contributions of low-order elements in hierarchically ordered space can swell into a flood of trouble. This is undeniably true. Although the organization of space by human cultures is of itself neither good nor bad, the runaway movement of materials through geographic systems can cause imbalances of grave import. But systems depend upon the flow of ideas as well as materials, and ideas can change the character of any system in which man is involved. It is often said that we already have the technology necessary to solve the earth's environmental ills but that we cannot agree on how to use it. The truth is that technology without continuing education and subsequent changes in behavior cannot save the world. The examples in the above pages, however, indicate that much can be done to improve things through a combination of technology, legal and political action, and behavior modification. *Poco a poco* an informed public and a responsible government can reverse environmental decline if the right approach is chosen. We have tried in this book to make suggestions about such an approach from a geographic point of view. The rest of the job is up to you.

## CHAPTER 1

### Cited References

Estes, R. D., "Predators and Scavengers: Stealth, Pursuit and Opportunism on Ngorongoro Crater in Africa," *Natural History*, vol. 76, Part I, February 1967, and Part II, March 1967.

Nystuen, John D., "Identification of Some Fundamental Spatial Concepts," *Papers of the Michigan Academy of Science, Arts and Letters*, vol. 48, 1963, pp. 373–384.

### Selected Readings

Abler, Ronald, John S. Adams, and Peter Gould, *Spatial Organization*, Prentice-Hall, Inc., Englewood Cliffs, N.J., 1971. A comprehensive and readable presentation of current ideas in geographic analysis for advanced students. Many interesting examples.

Bunge, William, *Theoretical Geography*, Lund Studies in Geography, ser. C., no. 1, The Royal University of Lund, Sweden, Dept. of Geography, 1966. An imaginative and influential book defining the scope and purpose of theoretical geography during a period of rapid change in geographic methodology.

Manners, Ian R., and Marvin W. Mikesell (eds.), *Perspectives on Environment*, Publication No. 13, Association of American Geographers, Washington, D.C., 1974. The essays collected in this volume set the scene for the present text and outline how human and natural geographic systems form a single entity; a key reference.

Miller, G. Tyler, Jr., *Energy and Environment: Four Energy Crises*, Wadsworth Publishing Co., Belmont, Calif., 1975. Paperback. This book provides a brief and attractively packaged discussion of the sources of energy available to humankind; current; good diagrams.

Odum, Eugene P., *Ecology*, Holt, Rinehart and Winston, Inc., New York, 1963. This paperback is a useful and concise introduction to the terminology and techniques of ecological study.

Odum, Howard T., *Environment, Power, and Society*, Wiley-Interscience, New York, 1971. A stimulating book relating man's use of power to all phases of life and society. Packed with ideas but not light reading.

Russwurm, Lorne H., and Edward Sommerville, *Man's Natural Environment: A Systems Approach*, Duxbury Press, North Scituate, Mass., 1974. Paperback. A shorter (23 articles) reader on the subject of ecological systems; suitable for outside reading with this text.

Watson, J. W., "Geography–A Discipline in Distance," *The Scottish Geographical Magazine*, vol. 71, no. 1, April 1955, pp. 1–3. A call for explicit theory in geography.

CHAPTER 2

## Cited References

Nordbeck, Stig, "The Law of Allometric Growth," *Michigan Interuniversity Community of Mathematical Geographers*, Discussion Paper No. 7, 1965, available from University Microfilms, Ann Arbor, Michigan.

Tobler, Waldo, "Satellite Confirmation of Settlement Size Coefficients, *Area I*, vol. 3, 1969, pp. 31–34.

Watts, David, *Principles of Biogeography*, McGraw-Hill Book Co., New York, 1971.

## Selected Readings

Dubos, Rene J., "Humanizing the Earth," *Science,* vol. 179, no. 4075, Feb. 23, 1973, pp. 769–772. An interesting essay based on the United Conference on the Human Environment, held in Stockholm in June 1972.

Hall, Edward T., *The Hidden Dimension*, Doubleday and Co., Garden City, New York, 1966. A discussion of proxemics and of attitudes toward the use of personal space.

Schoener, Thomas W., "Resource Partitioning in Ecological Communities," *Science*, vol. 185, no. 4145, July 1974, pp. 27–39. "Research on how similar species divide resources helps reveal the natural regulation of species diversity."

Sommer, Robert, *Personal Space–The Behavioral Basis of Design*, Prentice-Hall, Inc., Englewood Cliffs, N.J., 1969. Report of psychological studies of behavior and perception of immediate personal space.

Steinhart, Carol, and John Steinhart, *Energy: Sources, Use, and Role in Human Affairs*, Duxbury Press, North Scituate, Mass., 1974. A detailed account of the sources and amounts of energy utilized in a wide variety of human activities.

Stevens, Peter S., *Patterns in Nature*, An AtlantMonthly Press Book, Little, Brown and Co., 1974. The book presents many forms of spatial organization in nature; beautifully illustrated by figures and photographs.

Thomas, Lewis, *The Lives of a Cell: Notes of a Biology Watcher*, The Viking Press, New York, 1974. A charming and stimulating series of essays worth reading for both their literary and scientific merit.

CHAPTER 3

## Cited References

Bunge, William, *Field Notes*, Discussion Paper No. 1, The Detroit Geographical Expedition, Detroit, Mich., July 1969.

Chorley, Richard J., and Barbara Kennedy, *Physical Geography: A Systems Approach*, Prentice-Hall, Englewood Cliffs, N.J., 1972.

Fletcher, Colin, *The Man Who Walked Through Time*, New York, Knopf, 1968. (Also in paperback, Random House, 1972.)

Johnson, Ross B., "The Great Sand Dunes of Southern Colorado," *Geological Survey Research, 1967*, United States Geological Survey Professional Paper 575-C, pp. C177–C183.

Maruyama, Magoroh, "The Second Cybernetics: Deviation-Amplifying Mutual Causal Processes," *American Scientist*, vol. 51, 1963, pp. 164–179.

### Selected Readings

Berry, Brian J. L., "The Geography of the United States in the Year 2000," *Ekistics*, vol. 29 no. 174, May 1970, pp. 339–351. A geographic view of the future of the United States, taking into account society's vastly increasing capacity for communication.

Buckley, Walter (ed.), *Modern Systems Research for the Behavioral Scientist*, Aldine Publishing Co., Chicago, 1968. A collection of some of the best, most understandable articles on general systems theory. A good place to begin studying this subject.

Clark, Colin, *Population Growth and Land Use*, Macmillan and Co., Ltd., London, 1967. A reference book on world population patterns and urban density values.

Curry, Leslie, "Chance and Landscape," *Northern Geographical Essays in Honor of G. H. J. Daysh*, ed. by J. W. House, Oriel Press, Newcastle-upon-Tyne, 1966, pp. 40–55. This essay relates ideas of landscape format on, systems theory, and probability theory in a highly readable manner.

Doxiadis, Constantinos A., *Ekistics*, Oxford University Press, New York, 1968. Imaginative development of spatial concepts with good diagrams and maps of modern city structure.

Ehrlich, P. R., and A. H. Ehrlich, *Population, Resources, Environment: Issues in Human Ecology*, Freeman Co., San Francisco, 1970. Analysis of population dynamics and human pressures on world resources.

McArthur, Norman, "The Demography of Primitive Populations," *Science*, vol. 167, no. 3921, pp. 1097–1101. A note of warning is sounded in this article concerning attempts to estimate populations without adequate data.

Zelinsky, Wilbur, *A Prologue to Population Geography*, Prentice-Hall, Inc., Englewood Cliffs, N.J., 1966. A precise and relatively brief discussion of the historical, cultural, economic, and political factors that account for different types of populations and their distributions.

## CHAPTER 4

### Cited References

Hobbs, P. V., and E. Robinson, "Atmospheric Effects of Pollutants," *Science*, vol. 183, no. 4128, Mar. 8, 1974, pp. 909–915.

Oort, Abraham, H., "The Energy Cycle of the Earth," *The Biosphere*, A Scientific American Book, W.H. Freeman and Co., San Francisco, 1970, pp. 14–23.

Riehl, Herbert, *Introduction to the Atmosphere*, McGraw-Hill Book Co., New York, 1965.

Trewartha, Glenn T., Arthur H. Robinson, and E. H. Hammond, *Physical Elements of Geography*, 5th ed., McGraw-Hill Book Co., New York, 1967

Van Riper, Joseph, *Man's Physical World*, 2d ed., McGraw-Hill Book Co., New York, 1971.

### Selected Readings

Brodine, Virginia, *Air Pollution*, Harcourt Brace Jovanovich, Inc., New York, 1973. An overall review of air pollution problems which clarifies the subject without being too technical.

*References*

Cicerone, Ralph J., R. S. Stolarsk, and S. Walters, "Stratospheric Ozone Destruction by Man-Made Chlorofluoromethanes," *Science*, vol. 185, no. 4157, Sept. 27, 1974, pp. 1165–1166. This is the original report on the danger of using freon-loaded spray cans. Technical, but a milestone in sounding the danger to the ozone layer.

Hare, F. K., *The Restless Atmosphere*, Harper & Row Publishers, Inc., New York, 1963.

Kuenen, P. H., *Realms of Water*, John Wiley & Sons, Inc., New York, 1955. A smooth blend of technical and literary styles, worth reading despite its age.

National Tuberculosis and Respiratory Disease Association, *Air Pollution Primer*, New York, 1969. A simple but good layman's guide to terminology and problems relating to air pollution.

Scientific American, *The Ocean*, A Scientific American Book, W.H. Freeman and Co., San Francisco, 1969. Ten essays covering a wide range of subjects on the physical and biological processes of the world's oceans.

———, "Energy and Power," special issue, vol. 224, no. 3, September 1971. Eleven essays on energy, ranging from sources throughout the universe to those utilized by a hunting society.

Strahler, A. N., *Introduction to Physical Geography*, John Wiley & Sons, Inc., New York, 1965. A well-illustrated standard physical geography text.

———, *The Earth Sciences*, 2d ed., Harper & Row Publishers, New York, 1971. A fuller treatment of the same physical geography material.

The Doors, "Horse Latitudes," *Strange Days*, Elektra Records No. EKS 74104, New York, 1967.

## CHAPTER 5

### Cited References

Chorley, R. J. (ed.), *Water, Earth and Man*, Methuen & Co., Ltd., London, 1969.

Foster, E. E., *Rainfall and Runoff*, The Macmillan Co., New York, 1949.

Mason, B. J., *Clouds, Rain and Rainmaking*, Cambridge University, Cambridge, 1962.

Thornthwaite, C. E., and J. R. Mather, *The Water Balance, Publications in Climatology*, vol. 8, no. 1, Drexel Institute of Technology, Laboratory of Climatology, 1955.

Tuan, Yi-Fu, *The Hydrologic Cycle and the Wisdom of God—A Theme in Geoteleology*, University of Toronto, Department of Geography Research Publication No. 1, 1968.

Wolman, Abel, "Water Resources, A Report to the Committee on Natural Resources of the National Academy of Science: National Research Council," *Publication 1000-B*, National Academy of Science, National Research Council, Washington D.C., 1962.

### Selected Readings

Blumenstock, David I., *The Ocean of Air*, Rutgers University Press, New Brunswick, N.J., 1959. A very readable book relating the atmosphere and the working of its climates and weathers to the concerns of man.

Gates, David M., *Man and His Environment: Climate*, Harper & Row, Publishers,

Inc., New York, 1972. A brief nontechnical paperback which provides a good description of climate and its relationship to man.

Horvath, Ronald J., "Machine Space," *The Geographical Review*, vol. 64, no. 2, 1974, pp. 167–188. This describes how cities gradually become more and more hard-surfaced in order to accommodate increasing numbers of automobiles.

Penman, H. C., "The Water Cycle," *The Biosphere*, A Scientific American Book, W.H. Freeman and Co., San Francisco, 1970, pp. 38–45. A clear, precise, well-illustrated discussion of this complex system.

Vonnegut, Kurt, *Cat's Cradle*, Delacorte Press, New York, 1971. Vonnegut at his best, with the geography of the imagination much in evidence.

## CHAPTER 6

### Cited References

Ahlmann, H. Wilson, *Glacier Variations and Climatic Fluctuations*, The American Geographical Society, New York, 1953.

Carter, Douglas B., Theodore H. Schmuddle, and David M. Sharpe, "The Interface as a Working Environment: A Purpose for Physical Geography," *Commission on College Geography Technical Paper No. 7*, Association of American Geographers, Washington, D.C., 1972.

Chang, Jen-hu, *Climate and Agriculture*, Aldine Publishing Co., Chicago, Ill., 1968.

Dole, Stephen H., *Habitable Planets for Man*, Blaisdell Publishing Co., a Division of Ginn and Co., Waltham, Mass., 1964.

Landsberg, Helmut, *Physical Climatology*, Gray Printing Co., Inc., DuBois, Pa., 1958.

Sellers, William D., *Physical Climatology*, The University of Chicago Press, Chicago, Ill., 1965.

Thornthwaite, C. W., and J. R. Mather, "Instructions and Tables for Computing Potential Evapotranspiration and the Water Balance," *Publications in Climatology*, vol 10, no. 3, Drexel Institute of Technology, Laboratory of Climatology, Centerton, N.J., 1957.

———, "An Approach toward a Rational Classification of Climate," *Geographical Review*, vol. 38, no. 1, 1948, pp. 55–94.

### Selected Readings

Murchie, Guy, *Song of the Sky*, The Riverside Press, Cambridge, Mass., 1954. A lovely and personal description of everything that goes on in the atmosphere.

Shapley, Harlow (ed.), *Climatic Change: Evidence, Causes, and Effects*, Harvard University Press, Cambridge, Mass., 1953. Twenty-two essays on a wide variety of subjects relating life processes to those of climate and climatic change.

Sloane, Eric, *Look at the Sky, and Tell the Weather*, World Publishing Co., New York, 1970. Sloane at his anecdotal best, giving a combination of technical and folksy information about the weather.

*References*

Stewart, George R., *Storm*, Random House, New York, 1941. A novel, the central figure of which is a great storm moving slowly across the continent.

CHAPTER 7

## Cited References

Callendar, E., and R. Rossmann, "Sea Levels during the Past 35,000 Years," *Science*, vol. 162, Dec. 6, 1968, pp. 1121–1123.

Dickenson, William R., "Global Tectonics," *Science*, vol. 168, June 5, 1970, pp. 1250–1259.

———, "Plate Tectonics in Geologic History," *Science*, vol. 174, Oct. 8, 1971, pp. 107–113.

Heezen, Bruce C., "200,000,000 Years Under the Sea: The Voyage of the U.S.N.S. 'Kane,'" *Saturday Review*, Sept. 7, 1968, pp. 63–88.

Hunt, Charles B., *Physiography of the United States*, W.H. Freeman and Co., San Francisco, 1967.

Knopoff, L., "The Upper Mantle of the Earth," *Science*, vol. 163, Mar. 21, 1969, pp. 1277–1287.

Miller, D. H., *The Energy and Mass Budget at the Surface of the Earth*, Association of American Geographers, Washington, D.C., 1968.

Scientific American, *Continents Adrift*, A Scientific American Book, W.H. Freeman and Co., San Francisco, 1972.

Wolman, M. G., "A Cycle of Sedimentation and Erosion in Urban River Channels," *Geografiska Annaler*, vol. 49A, 1967, pp. 385–395.

## Selected Readings

Engel, A. E. J., "Time and the Earth," *American Scientist*, Winter 1969, pp. 458–483. A thorough review of the earth's history in terms of the varying estimates of time required for certain events to have taken place.

Golomb, Berl, and Herbert M. Eder, "Landforms Made by Man," *Landscape*, vol. 14, no. 1 (Autumn 1964), pp. 4–7. A speculative review of man's impact on erosion and other landform processes, as well as a call for further study of the subject.

Lessing, Doris, "Report on the Threatened City," *The Temptation of Jack Orkney and Other Stories*, Alfred A. Knopf, Inc., New York, 1972. Frightening science fiction account of an unsuccessful attempt to warn San Franciscans that an earthquake is about to destroy the city.

Leopold, Luna B., M. G. Wolman, and J. P. Miller, *Fluvial Processes in Geomorphology*, W.H. Freeman and Co., San Francisco, 1964. A standard reference work necessary for every serious student of stream processes.

Morisawa, Marie, *Streams—Their Dynamics and Morphology*, McGraw-Hill Book Co., New York, 1968. A clear and concise work useful for its reasonable length yet complete treatment of the subject.

McAlester, A. Lee, *The History of Life*, Prentice-Hall, Inc., Englewood Cliffs,

N.J., Foundations of Earth Science Series, 1968. A concise, readable account of life on earth from its beginnings. Evocative comments on the role of blue-green algae and the ozone shield in shaping life environments.

## CHAPTER 8

### Cited References

Branson, Branley Allan, "Stripping the Appalachians," *Natural History*, vol. LXXXIII, no. 9, November 1974, pp. 52–61.

Chorley, Richard J., R. P. Beckinsale, and A. J. Dunn, *The History of the Study of Landforms, or the Development of Geomorphology*, Vol. I, *Geomorphology Before Davis*, Vol. II, *The Life and Work of William Morris Davis*, Methuen, London, 1973.

Dalrymple, E. A. Silver, and Everett D. Jackson, "Origin of the Hawaiian Islands," *American Scientist*, vol. 61, no. 3, pp. 294–308.

Drury, William H., and Ian C. T. Nisbet, "Inter-relations Between Developmental Models in Geomorphology, Plant Ecology, and Animal Ecology," *General Systems Yearbook—1971*, pp. 57–68.

Inman, Douglas L., and Birchard M. Brush, "The Coastal Challenge," *Science*, vol. 181, no. 4094, July 1973, pp. 20–32.

LeGrand, H. E., "Hydrological and Ecological Problems of Karst Regions," *Science*, vol. 179, no. 4076, March 1973, pp. 859–864.

Schuberth, Christopher J., "Barrier Beaches of Eastern America," *Natural History*, vol. LXXIX, no. 6, June-July 1970, pp. 46–55.

Stoddart, D. R., "Variations on a Coral Theme," *Geographical Magazine*, vol. XLIII, no. 9, pp. 610–615.

### Selected Readings

Butzer, Karl W., *Environment and Archeology: An Introduction to Pleistocene Geography*, Aldine Publishing Co., Chicago, 1964. A worldwide survey of environmental conditions during the time of early human development. While this book deals specifically with problems of interest to archaeologists, it also provides a detailed and scholarly picture of Pleistocene events and conditions which have helped shape so many of today's landscapes.

Carter, Luther J., "Strip Mining: A Practical Test for President Ford," *Science*, vol. 186, no. 4170, p. 1190. This one-page summary report has long since been superseded by new developments. However, it is typical of the reporting on important environmental issues found in *Science*, and is a reminder of the usefulness of all such entries.

Clark, Thomas H., and Colin W. Stearn, *Geological Evolution of North America*, The Ronald Press Co., New York, 2d ed., 1968. A sound historical geology outlining the geological and paleontological events of North America.

Cotton, C. A., *Landscape—as Developed by the Processes of Normal Erosion*, New Zealand, Christchurch, 2d ed. 1948, Whitcombe and Tombs, Ltd.

Davis, William Morris, assisted by William H. Snyder, *Physical Geography*, Ginn and Co., Boston, 1899. This book, and the one preceding it by Cotton, are among the classics of writing on geomorphology. While superseded in many ways, they are still well worth reading.

Eardley, A. J., *Structural Geology of North America*, Harper & Brothers Publishers, Inc., New York, 1951. A large book rich with diagrams and maps.

Fenneman, Nevin M., *Physiography of the Eastern United States* and *Physiography of the Western United States*, McGraw-Hill Book Co., New York, 1931 and 1938. These companion volumes contain classic descriptions of the North American continent.

Hunt, Charles B., *Natural Regions of the United States and Canada*, W.H. Freeman and Co., San Francisco, 1974. A revision of an earlier work on the physiographic regions of North America, this book now includes discussions of vegetation and climate. A well-illustrated account at both continental and regional scales.

Lobeck, Armin K., *Things Maps Don't Tell Us*, The MacMillan Co., New York, 1962. A delightful and scholarly account of how to interpret landforms from the maps showing them. This is a combination of geography and geology at their best.

McGinnies, William G., B. J. Goldman, and Patricia Paylore, *Deserts of the World*, The University of Arizona Press, 1968. This volume draws together much of the known information on arid-land processes and environments. Its summaries of landform processes, climate, flora, fauna, etc. are exceptionally useful.

Tank, Ronald W., *Focus on Environmental Geology*, Oxford University Press, London, 1973. "A collection of case histories and readings from original sources," this well-organized book has a truly human-environmental approach not usually found in works on geology.

Van Riper, Joseph E., *Man's Physical World*, 2d ed., New York, McGraw-Hill Book Co., 1971. The revised second edition of this book is well oriented to the newer systems approach to physical geography. The text is clearly written and better organized than many current physical geography books.

Westing, Arther H., and E. W. Pfeiffer, "The Cratering of Indochina," *Scientific American*, vol. 226, no. 5, May 1972, pp. 20–29. A discussion of how war can alter and damage the landscape, perhaps beyond recovery.


CHAPTER 9


**Cited References**

Anderson, Edgar, "Hybridization of the Habitat," *Evolution*, vol. 2, no. 1, 1948, pp. 1–9. Also Bobbs-Merrill Reprint Series G-5.

Buckman, Harry O., and Nyle C. Brady, *The Nature of Properties of Soils*, 7th ed., The Macmillan Co., New York, 1969.

Budowski, Gerardo, "Tropical Savanna, A Sequence of Forest Felling and Repeated Burnings," *Tursialba*, vol. 6, 1956, pp. 23–33. Also Bobbs-Merrill Reprint Series G-29.

Chang, Jen-hu, *Climate and Agriculture, An Ecological Survey*, Aldine Publishing Co., Chicago, Ill., 1968.

Edwards, Clive A., "Soil Pollutants and Soil Animals," *Scientific American*, vol. 220, no. 4, April 1969, pp. 88–99.

McNaughton, S. J., and L. L. Wolf, "Dominance and the Niche in Ecological Systems," *Science*, vol. 167, Jan. 9, 1970, pp. 131–137.

Strahler, Arthur N., "Climate, Soils, and Vegetation," Part III of *Physical Geography*, 3d ed., John Wiley & Sons, Inc., New York, 1969.

Watts, David, *Principles of Biogeography*, McGraw-Hill Book Co., New York, 1971.

## Selected Readings

Albrecht, William A., "Soil Fertility and Biotic Geography," *Geographical Review*, vol. 47, 1957, pp. 86–105. Also Bobbs-Merrill Reprint Series G-4. A discussion at a regional scale of the relationship of soil fertility to crop nutrition for animals and men.

Anderson, Edgar, *Plants, Man and Life*, University of California Press, Berkeley, 1967. Discusses man's symbiotic relationship with weeds and crop plants.

Brown, Lester R., *Seeds of Change*, Frederick A., Praeger, Inc., New York, 1970. A discussion of the "green revolution" brought about by new types of crops, with some cautionary advice.

Darling, F. Fraser, and J. P. Milton (eds.), *Future Environments of North America*, The Natural History Press, Garden City, New York, 1966. Projections into the future of the impact of man on North American environments.

Hunt, Charles, B., *Geology of Soils*, W.H. Freeman and Co., San Francisco, 1972. A comprehensive account of soils and soil-forming processes.

Hutchinson, G. Evelyn, "Eutrophication," *American Scientist*, vol. 61, no. 3, May-June 1973, pp. 269–279. A thorough review of the study of eutrophication from C. A. Weber (1907) to the present. Good references.

Kellog, Charles E., "Soil," *Scientific American*, vol. 185, July 1950, pp. 30–39. Outlines the relationship between soils and vegetation.

Odum, Eugene P., *Ecology*, Holt, Rinehart and Winston, Inc., New York, 1963. A brief, well-written technical account of biosystems, from microscopic to regional scales.

## CHAPTER 10

## Cited References

Burton, I., and R. W. Kates, "The Floodplain and the Seashore: A Comparative Analysis of Hazard-zone Occupance," *Geographical Review*, vol. 54, 1964, pp. 366–385.

Curry, Leslie, "Climate and Economic Life: A New Approach," *Geographical Review*, vol. 42, 1952, pp. 367–383.

Flannery, Kent V., "Archeological Systems Theory and Early Mesoamerica," *Anthropological Archeology in the Americas*, Washington, D.C., 1968.

Grigg, David, *The Harsh Lands*, Macmillan and Co., Ltd., London, 1970.

Hirst, Eric, "Food Related Energy Requirements," *Science*, vol. 184, no. 4133, April 1974, pp. 134–138.

Isaac, Erich, *Geography of Domestication*, Prentice-Hall, Inc., Englewood Cliffs, N.J., 1970.

Kolars, John, "Locational Aspects of Cultural Ecology: The Case of the Goat in Non-western Agriculture," *Geographical Review*, vol. 56, 1966, pp. 577–584. Also Bobbs-Merrill Reprint Series G-241.

Leeds, Anthony, and Andrew P. Vayda, *Man, Culture, and Animals—The Role of Animals in Human Ecological Adjustments*, American Association for the Advancement of Science, Pub. 78, Washington, D.C., 1965.

MacNeish, Richard S., "Ancient Mesoamerican Civilization," *Science*, vol. 143, no. 3606, Feb. 7, 1964, pp. 531–537.

———, Paul C. Mangelsdorf, and Walton Galinat, "Domestication of Corn," *Science*, vol. 143, 1964, pp. 538–545.

Moore, Omar Khayyam, "Divination—A New Perspective," in Andrew P. Vayda (ed.), *Environment and Cultural Behavior*, The Natural History Press, Garden City, N.Y., 1969, pp. 121–129.

Pimentel, David, et al., "Food Production and the Energy Crisis," *Science*, vol. 182, no. 4111, pp. 443–449.

Struever, Stuart (ed.), *Prehistoric Agriculture*, American Museum Sourcebooks in Anthropology, Natural History Press, Garden City, N.Y., 1971.

White, G. F., *Choice of Adjustment to Floods*, University of Chicago Department of Geography Research Paper No. 93, Chicago, Ill., 1964.

Ward, Robert M., "Decisions by Florida Citrus Growers and Adjustments to Freeze Hazards," in G. F. White (ed.), *Natural Hazards: Local, National, Global*, New York, Oxford University Press, 1974, pp. 137–146.

## Selected Readings

Adams, Robert McC., *The Evolution of Urban Society*, Aldine Publishing Co., Chicago, Ill., 1966. Early Mesopotamia and prehispanic Mexico: a comparative study of one of the great transformations in the history of man.

Barth, Frederick, "Ecologic Relationships of Ethnic Groups in Swat, North Pakistan," *American Anthropologist*, vol. 58, 1956, pp. 1079–1089. Also Bobbs-Merrill Reprint Series A-9. An analysis of the adjustment of ethnic groups to ecological niches located on steep slopes. Emphasizes the importance of interaction between the ethnic groups with different resource bases.

Braidwood, Robert, and Gordon R. Willey, *Courses toward Urban Life*, Aldine Publishing Co., Chicago, Ill., 1962. Archaeological considerations of some cultural alternatives.

Brown, Lester R., and Gail W. Finsterbusch, *Man and His Environment: Food*, Harper Row, & Publishers, Inc., New York, 1972. Concerned with the development, current state, and immediate future of world food production.

Burton, I., and R. W. Kates (eds.), *Readings in Resources Management and Conservation*, University of Chicago Press, Chicago, Ill., 1965. Selected readings on the origins and development of public conservation policy.

Carol, Hans, "Stages of Technology and Their Impact upon the Physical Environment: A Basic Problem in Cultural Geography," *Canadian Geographer*, vol. 8, no. 1, 1964, pp. 1–8. A theory of the interaction between technology and environment.

Chisholm, Michael, *Rural Settlement and Land Use*, Aldine, Chicago, Ill., 1970.

*References*

(First published in 1962 by Hutchinson University Library, London.) A modern interpretation of Thünen's agricultural theory by a well-known geographer. Many empirical examples at different scales are presented.

Dunn, E. S., *The Location of Agricultural Production*, University of Florida, Gainesville, Fla., 1954. A modern interpretation of agricultural location theory based on Thünen's theory.

Gould, Peter R., "Man Against His Environment: A Game Theoretic Framework," *Annals of the Association of American Geographers*, vol. 53, no. 3, September 1963.

Hall, Peter (ed.), *Von Thünen's Isolated State*, Pergamon Press, London, 1966. Translation of the major analytical portions of Thünen's original work.

Higbee, Edward, *Farms and Farmers in an Urban Age*, The Twentieth Century Fund, New York, 1963. Lively discussion of modern American farming and consequences of government agricultural policy.

Jefferson, M., "The Civilizing Rails," *Economic Geography*, vol 4, 1928, pp. 217–231. An old, imaginative article on the role of transportation in the development of the modern worldwide society.

Porter, Philip, "Environmental Potentials and Economic Opportunities—A Background for Cultural Adaptation," *American Anthropologist*, vol. 67, 1965, pp. 409–420. Also Bobbs-Merrill Reprint Series G-186. Applies water balance concepts to the evaluation of agricultural uses in traditional African societies and analyzes the survival strategies employed by these societies. An excellent study.

Stewart, George R., *American Ways of Life*, Doubleday, Garden City, N.Y., 1954. Delightful book describing origins of typical American traits as consequences of cultural diffusion and adaptation to the North American natural environment.

## CHAPTER 11

### Cited References

Barrows, Harlan H., "Geography as Human Ecology," *Annals of the Association of American Geographers*, vol. 13, 1923, pp. 1–14.

Clarkson, James D., "Ecology and Spatial Analysis," *Annals of the Association of American Geographers*, vol. 60, 1970, pp. 700–716.

Glacken, Clarence J., *Traces on the Rhodian Shore*, University of California Press, Berkeley, 1967.

Gould, Peter, and Rodney White, *Mental Maps*, Penguin Books, Pelican No. A1688, New York, 1974.

Huntington, Ellsworth, *Mainsprings of Civilization*, John Wiley & Sons, Inc., New York, 1945. A chapter of this book, "Regions and Seasons of Mental Activity," pp. 343–367, is reprinted in Fred E. Dohrs et al. (eds.), *Outside Readings in Geography*, Thomas Y. Crowell Co., New York, 1955, pp. 146–156.

Moorehead, Alan, *Cooper's Creek*, Harper & Row Publishers, Inc., New York, 1964.

Murphey, Rhoads, "Man and Nature in China," *Modern Asian Studies*, vol. 1, no. 4, 1967, pp. 313–333.

Platt, Robert S., "Environmentalism Versus Geography," *The American Journal of Sociology*, vol. 53, no. 5, March 1948, pp. 351–358.

Robinson, Arthur H., *Elements of Cartography*, John Wiley & Sons, Inc., New York, 1953.

Saarinen, Thomas Frederick, *Perception of the Drought Hazard on the Great Plains*, University of Chicago Department of Geography Research Paper No. 106, Chicago, Ill., 1966.

―――, "Attitudes Toward Weather Modification: A Study of Great Plains Farmers," *Human Dimensions of Weather Modification*, Research Paper No. 105, ed. by W. R. Derrick Sewell, Department of Geography, University of Chicago, Chicago, Ill., 1966.

―――, *Perception of Environment*, Association of Geographers, Washington, D.C., 1969.

Schaefer, Fred K., "Exceptionalism in Geography: A Methodological Examination," *Annals of the Association of American Geographers*, vol. 43, 1953, pp. 226–249.

Semple, Ellen, C., *Influences of Geographic Environment*, Henry Holt and Co., Inc., New York, 1911.

Spate, O. H. K., "Toynbee and Huntington: A Study in Determinism," *The Geographical Journal*, vol. 118, part 4, December 1952, pp. 406–428.

―――, "How Determined Is Possibilism?" *Geographical Studies*, vol. 4, 1957, pp. 3–7.

Spoehr, Alexander, "Cultural Differences in the Interpretation of Natural Resources," *Man's Role in Changing the face of the Earth*, ed. by William L. Thomas, Jr., University of Chicago Press, Chicago, Ill., 1956.

Sprout, Harold, and Margaret Sprout, *The Ecological Perspective on Human Affairs—With Special Reference to International Politics*, Princeton, The Princeton University Press, 1965.

Tuan, Yi-Fu, "Attitudes Toward Environment: Themes and Approaches," *Environmental Perception and Behavior*, Research Paper No. 109, ed. by David Lowenthal, Department of Geography, University of Chicago, Chicago, Ill., 1967.

―――, "Discrepancies between Environmental Attitude and Behavior: Examples from Europe and China," *Canadian Geographer*, vol. 12, no. 3, 1968, pp. 176–191.

Vidal de la Blache, Paul, *Principes de géographie humaine*, Paris, 1922. Trans. as *Principles of Human Geography*, Henry Holt and Co., Inc., New York, 1926.

Ward, Robert M., *Cold and Wind Hazard Perception by Orange and Tomato Growers*, Michigan Geographical Publication No. 9, Department of Geography, University of Michigan, Ann Arbor, Mich., 1973.

## Selected Readings

Brown, Lloyd A., *The Story of Maps*, New York, Bonanza Books, 1949. One of the standard references to the history of cartography. Well written and entertaining.

Downs, Roger M., and David Stea (eds.), *Image and Environment*, Chicago, Aldine Publishing Co., New York, 1973. A collection of studies on mental mapping by a geographer and a psychologist.

*References*

**331**

Ittelson, W. H., *Environment and Cognition*, Seminar Press, New York, 1973. "This volume is perhaps the first to bridge the gap between contemporary psychology and environmental perception and cognition." Contributions from many fields, including geography.

Kish, George, *History of Cartography*, Harper & Row, Publishers, Inc., New York, 1973. This collection of 220 color 35mm slides accompanied by an instructor's guide is an outstanding teaching aid as well as an indispensible reference.

Marsh, G. P., *Man and Nature*, Scribner and Sons, New York, 1864; ed. by David Lowenthal and reprinted by Belknap Press, Cambridge, Mass., 1964. This is the pioneering work on man's modification of the earth and the concept of conservation.

Monkhouse, F. J., and H. R. Wilkinson, *Maps and Diagrams*, London, Methuen and Co., Ltd., 1973. An exhaustive work with emphasis on cartography as an illustrative medium.

Platt, John R., *Perception and Change: Projections for Survival*, The University of Michigan Press, Ann Arbor, Mich., 1970. Deals with the need to control the effects of modern technology in order to create peace and a livable environment.

Rothwell, Stuart C., *A Geography of Earth Form*, Wm. C. Brown Co., Publishers, Dubuque, Iowa, 1968. This brief paperback (113 pages) covers the specific subjects of earth shape, locations on the earth's surface, map projections and elements, and the earth in the solar system. A concise beginners' guide.

Thrower, Norman J. W., *Maps and Man*, Prentice-Hall, Inc., Englewood Cliffs, N.J., 1972. "An examination of cartography in relation to culture and civilization."

Tuan, Yi-Fu, "Man and Nature," *Commission of College Geography Resource Paper No. 10*, Association of American Geographers, Washington, D.C., 1971. A good introduction to perception of nature studies.

Vayda, Andrew P., (ed.), *Environment and Cultural Behavior*, Natural History Press, Garden City, N.Y., 1969. "Ecological studies in cultural anthropology."

Winch, Kenneth, (ed.), *International Maps and Atlases in Print*, R. R. Bowker Co., Ann Arbor, Mich. 1974. More than 8000 entries describing available maps and atlases.

## CHAPTER 12

### Cited References

Conway, Gordon, R., "A Consequence of Insecticides," *Natural History* supplement, February 1969, pp. 46–51.

Glaser, Peter E., "Power from the Sun: Its Future," *Science*, vol. 162, 1968, pp. 857–861.

Jackson, J. F., *Landscapes*, ed. by E. H. Zube, University of Massachusetts Press, Amherst, Mass., 1970.

Jones, Lawrence W., "Liquid Hydrogen as a Fuel for the Future," *Science*, vol. 174, October 1971, pp. 367–378.

McAlester, A. Lee, *The History of Life*, Englewood Cliffs, N.J., Prentice-Hall, Inc., 1968.

Singer, S. Fred, "Global Effects of Environmental Pollution," *Science*, vol. 162, December 1968, p. 1308.

Smith, J. E., *"Torrey Canyon" Pollution and Marine Life*, University Press, Cambridge, England, 1968.

Woodwell, George M., Paul P. Craig, and Horton A. Johnson, "DDT in the Biosphere: Where Does It Go?" *Science*, vol. 174, December 1971, pp. 1101–1107.

## Selected Readings

Commoner, Barry, *The Closing Circle*, Alfred A. Knopf, Inc., New York, 1971. Well-written analysis of ecological problems with prescriptions for technological, political, and social counteractions.

Deshler, Walter, "Livestock Trypanosomiasis and Human Settlement in Northeastern Uganda," *Geographical Review*, vol. 50, 1969, pp. 541–554. An example of the interrelatedness of man's actions and biotic processes in an African setting.

Detwyler, Thomas R., *Man's Impact on Environment*, McGraw-Hill Book Co., New York, 1971. A survey and critical commentary on environmental impact studies. Many detailed and informative examples are given.

Ehrlich, Paul R., and Anne H. Ehrlich, *Population, Resources, Environment— Issues in Human Ecology*, W. H. Freeman and Co., San Francisco, 1970. Relationship of overpopulation and population dynamics to demands on food, resources, and the environment. A world view of the social, political, and technological action needed for a brighter prospect for mankind.

Gould, Peter. *Spatial Diffusion*, Association of American Geographers Commission on College Geography, Resource Paper No. 4, Washington, D.C., 1969. This is a good review of diffusion studies. Most of the resource papers in this series provide a useful view of some aspect of geography. They are a good source for a student of geography.

Hägerstrand, Torsten, "Migration and Area," survey of a sample of Swedish migration fields and hypothetical consideration of their genesis. *Migration in Sweden, A Symposium*, Lund Studies in Geography No. 13, The Royal University of Lund, Sweden, 1957. An original and excellent study of migration. It is also a very good example of a scientific study of social phenomena.

———, *Innovation Diffusion as a Spatial Process*, trans. and postscript by Allan Pred, University of Chicago Press, Chicago, Ill., 1967. Another seminal work by this famous Swedish geographer. All subsequent geographical studies of diffusion have drawn heavily upon it.

Hardin, Garrett, "The Tragedy of the Commons," *Science*, vol. 162, December 1968, pp. 1243–1248. An exposition of the difficult institutional and ethical problem associated with the exploitation of "common" resources.

Montague, Katherine, and Peter Montague, *Mercury*, Sierra Club, San Francisco, 1971. An account of the deadly accumulation of mercury in the marine food chain.

Thomas, W. L. (ed.), *Man's Role in Changing the Face of the Earth*, University of Chicago Press, Chicago, Ill., 1956. This large collection of papers on all aspects of man's impact on the environment is the result of a symposium on the topic. The quality of the papers varies greatly, but many are very good.

*References*

Wagner, Philip L., and Marvin W. Mikesell, *Readings in Cultural Geography*, University of Chicago Press, Chicago, Ill., 1962. A useful set of readings on culture areas, on origins and dispersals, and on cultural elements relating to ecology and the landscape.

Wagner, Richard H., *Environment and Man*, W. W. Norton and Co., Inc., New York, 1971. A well-written, informative book on man's increasing impact on his environment. The author is a botanist.

Whiteside, Thomas, *The Withering Rain: America's Herbicidal Folly*, E. P. Dutton & Co., Inc. New York, 1971. A description of defoliation in Vietnam and use of herbicides in American agriculture.

Wood, Nancy, *Clearcut*, Sierra Club, San Francisco, 1971. A criticism of the most common method of lumbering in the United States.

## CHAPTER 13

### Cited References

Jacoby, Louis R., *Perception of Air, Noise and Water Pollution in Detroit*, Michigan Geographical Publication No. 7, Department of Geography, University of Michigan, Ann Arbor, Mich., 1972.

Landsberg, Hans H., *Natural Resources for U.S. Growth*, The Johns Hopkins Press, Baltimore, Md., 1964.

Nicholson, Max, *The Environmental Revolution*, McGraw-Hill Book Co., New York, 1970.

Steinhart, Carol E., and John S. Steinhart, *Blowout*, Duxbury (Wadsworth) Press, Belmont, Calif., 1972.

### Selected Readings

Albaum, Melvin, *Geography and Contemporary Issues: Studies of Relevant Problems*, John Wiley & Sons, Inc., New York, 1973. A collection of geographic studies of poverty, racism, pollution, and crowding.

Bernarde, Melvin A., *Our Precarious Habitat*, W. W. Norton and Co., Inc., New York, 1970. A review of a wide range of problems created by man's use of his environment.

Detwyler, Thomas R., and Melvin G. Marcus, *Urbanization and Environment*, Duxbury Press, Belmont, Calif., 1972. A collection of essays on the interaction between urbanization and the physical environment.

Fabricant, Neil, and Robert M. Hallman, *Toward a Rational Power Policy: Energy, Politics, and Pollution*, George Braziller, New York, 1971. A factual approach written in terms of the practical problems facing metropolitan planners.

Farber, Seymour, "Quality of Living—Stress and Creativity," in F. Fraser Darling and John P. Milton, *Future Environments of North America*, Natural History Press, New York, 1966.

Holdren, John P., and Paul R. Ehrlich (eds.), *Global Ecology*, Harcourt Brace Jovanovich, Inc., New York, 1971. A set of readings on a multiplicity of ecological crises, with emphasis on global problems.

Hoyle, Fred, "Welcome to Slippage City," *Element 79,* A Signet Book, The New American Library, New York, 1967, pp. 50–65. Fantasy fiction, but the account of Los Angeles couldn't be more true.

Ward, Barbara, *Spaceship Earth*, Columbia University Press, New York, 1966. A popular work calling for a balanced world ecology and a revision of the notion of material progress.

Watts, May Theilgaard, *Reading the Landscape*, New York, The Macmillan Co., 1970. A literate and stimulating discussion of scenes in the American landscape. Everyone who loves the land should read this book.

Wilson, James Q. (ed.), *The Metropolitan Enigma: Inquiries into the Nature and Dimensions of America's "Urban Crisis,"* Chamber of Commerce of the United States, Washington, D.C., 1967. An informative and comprehensive series of essays. "Pollution and Cities," by Roger Revelle, pp. 78–121, is of particular interest.

# INDEX